Master Electrician's Review

FIFTH EDITION

Master Electrician's Review

FIFTH EDITION

Richard E. Loyd

THOMSON

DELMAR LEARNING

Australia   Canada   Mexico   Singapore   Spain   United Kingdom   United States

THOMSON
™
DELMAR LEARNING

Master Electrician's Review, 5th Edition
Richard E. Loyd

Vice President, Technology and Trades SBU:
Alar Elken

Editorial Director:
Sandy Clark

Acquisitions Editor:
Patrick Kane

Development Editor:
Jennifer A. Thompson

Marketing Director:
David Garza

Channel Manager:
William Lawrensen

Marketing Coordinator:
Casey Bruno

Production Director:
Mary Ellen Black

Production Manager:
Larry Main

Production Editor:
Ruth Fisher

Library of Congress Cataloging-in-Publication Data:

Loyd, Richard E.
    Master electrician's review / Richard E. Loyd. — 5th ed.
        p. cm.
    Includes index.
    ISBN 1-4018-7989-6
        1. Electric engineering—Examinations, questions, etc. 2. Electricians—Licenses—United States. I. Title.
    TK169.L692 2005
    621.319'24'021873—dc22
                            2004059883

I dedicate this book to my wife and business partner, Nancy. She is my inspiration and number one fan. Her assistance is invaluable. I would also like to dedicate it to all my wonderful circuit rider friends. Their expertise in the *National Electrical Code®* and the electrical industry continue to provide me with the latest information on our great industry.

Table of Contents

CHAPTER FOURTEEN

Foreword

ELECTRIC PER SE

Electricity! In ancient times it was believed to be an act of the gods. Not even Greco-Roman civilizations could understand this thing we know today as electricity. It was not until about A.D. 1600 that any scientific theory was recorded, when after 17 years of research William Gilbert wrote a book on the subject titled *De Magnete*. It was almost another 150 years before major gains were made in understanding electricity so that people might some day be able to control it. At that time, Benjamin Franklin, whom we often refer to as the grandfather of electricity or electrical science, and his close friend, Joseph Priestley, a great historian, began to gather the works of the many worldwide scientists who had been working independently. (One has to wonder where this wondrous industry would be today if the great minds of yesteryear had had the benefit of our great communication networks.) Benjamin Franklin traveled to Europe to gather this information, and with that trip our industry began to surge forward for the first time. Franklin's famous kite experiment allegedly took place about 1752. Coupling the results of that with other scientific data gathered, Franklin decided he could sell installations of lightning protection to every building owner in the city of Philadelphia. He was very successful in doing just that!

We credit Franklin for the terms "conductor" and "nonconductor," which replaced "electric per se" and "non-electric."

With the accumulation of the knowledge gained from these many scientists and inventors of the past, many other great minds emerged and lent their names to even more discoveries. Their names are familiar as other terms related to electricity. Alessandro Volta, James Watt, André Marie Ampère, George Ohm, and Heinrich Hertz are names of just a few of the great minds that enabled our industry to come together.

The next major breakthrough came only a little more than 100 years ago. Thomas Edison, the holder of hundreds of patents promoting direct current (dc) voltage, and Nikola Tesla, the inventor of the 3-phase motor and the promoter of alternating current (ac)

Electricity is here! *(Reprinted with permission by Underwriters Laboratories, Inc.®)*

The 1893 Great Columbian Exposition at the Chicago World's Fair. The Palace of Electricity astonished crowds; it also astonished electricians by repeatedly setting fire to itself. *(Reprinted with permission by Underwriters Laboratories, Inc.®)*

voltage, began an inimical competition in the late 1800s. Edison was methodically making inroads with direct current all along the Eastern Seaboard, and Tesla had exhausted his finances for his experimental project with alternating current in Colorado. Then the city of Chicago requested bids to electrically light the great Columbian Exposition of 1893. Lighting the world's fair in Chicago, which was celebrating 300 years since Columbus discovered America, required building generators and nearly a quarter million lights. Edison teamed up with General Electric and submitted a bid based on the Edison lightbulb powered by direct current. George Westinghouse, already successful in the electrified railroad, the elevator, and about 400 other inventions, joined with Nikola Tesla and submitted the lowest bid—about one third of the Edison-General Electric bid. Westinghouse and Tesla were awarded the contract. They based their bid on the Westinghouse Stopper Light, powered by alternating current. The installation was made by Guarantee Electric of Saint Louis, Missouri. This firm is still one of the oldest and largest electrical contractors in business today. It is said that thousands came out to see this great lighting display. But almost as spectacular as the lighting were the arcing, sparking, and fires started by the display, with accidents occurring daily! It is said that Edison was embellishing the dangers of alternating current by electrocuting dogs while promoting direct current as being much safer. But the

public knew better, because plenty of fires and accidents were also occurring with direct current.

About this time three very important events took place almost simultaneously. The Chicago Board of Fire Underwriters hired an electrician from Boston as an electrical inspector for the exposition, William Henry Merrill. Mr. Merrill saw the need for safety inspections and started Underwriters Laboratories in 1894. In 1895 the first nationally recommended code was published by the National Board of Fire Underwriters (now the American Insurance Association). The 1895 *National Electrical Code® (NEC®)** was drafted through the combined efforts of the architectural, electrical, insurance, and other allied interests based on a document that resulted from actions taken in 1892 by Underwriters National Electrical Association, which met and consolidated various codes that were already in existence. The first *National Electrical Code® (NEC®)* was presented to the National Conference on Electrical Rules, which was composed of delegates from various national associations who voted to unanimously recommend it to their respective associations for adoption or approval. A group from the city of Buffalo, New York, visited the world's fair and, based on that visit, hired George Westinghouse and Nikola Tesla to build ac hydro-generators to be powered by Niagara Falls. Alternating current (ac) had won out over direct current (dc) as this country's major power source. The rest is history!

Preface

The primary purpose of this book is to provide electrical students with a concise, easily understandable study guide to the 2005 edition of the *National Electrical Code®* and the application of electrical calculations. This text is written recognizing that the electrical field has many facets, and the user may have diverse interests and varying levels of experience. However, the interests of the relatively experienced electrician preparing to enter the electrical industry as a legally competent licensed electrical contractor, the master electrician, the journeyman electrician, the electrician-in-training, and especially the advanced electrical student apprentice were all carefully considered in the text's preparation.

It provides a ready source of the basic information on the *NEC®*. This text may be used for reference or as study material for students and electricians who are preparing for an examination for licensing. It is especially designed as an aid to those studying for the nationally recognized examinations, such as Experior and the Southern Building Code Congress International (SBCCI). It will also help anyone preparing for any electrical examination, or it will provide a quick, easily understood study guide for those needing to update themselves on the *National Electrical Code®* and basic electrical mathematical formulas and calculations. The text is brief and concise for easy application for classroom or home study.

METRICS (SI) AND THE *NEC®*

The United States is the last major country in the world not using the metric system as the primary system of measurement. We have been very comfortable using English (U.S. Customary) values. This is changing. Manufacturers are now showing both inch-pound and metric dimensions in their catalogs. Plans and specifications for governmental new construction and renovation projects begun after January 1, 1994, have been done using the metric system. You may not feel comfortable with metrics, but metrics are here to stay. You might just as well get familiar with the metric system.

The *NEC®* and other *NFPA* Standards are becoming international standards. All measurements in the 2005 *NEC®* are shown with metrics first, followed by

---

STUDY RECOMMENDATIONS

- Read Chapter One carefully.
- Take the Sample Examination honestly. Allow yourself three hours and grade yourself honestly. This examination will point out to you the areas where you are weak—for example, speed, grounding, or services.
- Study Chapter Two through Chapter Thirteen and take the Question Review following each of those chapters.
- Take the 50-question Practice Examination in Chapter Fourteen honestly; allow two hours for the examination.
- Evaluate the areas in which you are still weak.
- Take the Practice Examinations in Chapter Fourteen until you feel you have overcome your weak areas.
- Take the Final Examination in Chapter Fourteen; allow three hours, and if you have studied, you will find that you are ready. Good luck!

Each lesson is designed purposely to require the student to apply the entire *NEC®* text, and not specific chapters or articles. It has been found that when studying for a timed, open-book examination, the student must gain proficiency in the Table of Contents, the Index, and the ability to move quickly from cover to cover to find the correct answer to each question in a timely fashion.

---

the inch-pound value in parentheses—for example, 600 mm (24 in.). Ease in understanding is of utmost importance in *Master Electrician's Review*. Therefore, inch-pound values are shown first, followed by metric values in parentheses—for example, 24 inches (600 mm). A *soft metric conversion* is when the dimensions of a product already designed and manufactured to the inch-pound system are converted to metric dimensions. The product does not change in size. A *hard metric measurement* is when a product has been designed to SI metric dimensions. No conversion from inch-pound measurement units is involved. A *hard conversion* is when an existing product is redesigned into a new size.

In the 2005 edition of the *NEC®*, existing inch-pound dimensions did not change. Metric conversions were made, then rounded off. Where rounding off would create a safety hazard, the metric conversions are mathematically identical. For example, if a dimension is required to be 6 feet, it is shown in the *NEC®* as 1.8 m (6 ft). Note that the 6 ft remains the same, and the metric value of 1.83 m has been rounded off to 1.8 m. *Master Electrician's Review* reflects these rounded off changes, except that the inch-pound measurement is shown first—that is, 6 feet (1.8 m).

TRADE SIZES

A unique situation exists. Strange as it may seem, what electricians have been referring to for years has not been correct! Raceway sizes have always been an approximation. For example, there has never been a ½ inch raceway! Measurements taken from the *NEC®* for a few types of raceways show the following:

Trade Size	Inside Diameter
½ Electrical Metallic Tubing	0.622 inch
½ Electrical Nonmetallic Tubing	0.560 inch
½ Flexible Metal Conduit	0.635 inch
½ Rigid Metal Conduit	0.632 inch
½ Intermediate Metal Conduit	0.660 inch

You can readily see that the cross-sectional areas, critical when determining conductor fill, are different. It makes sense to refer to conduit, raceway, and tubing sizes as "**trade sizes**." *NEC®* 90.9(C)(1) states that *where the actual measured size of a product is not the same as the nominal size, trade size designators shall be used rather than dimensions. Trade practices shall be followed in all cases.* Master Electrician's Review* uses the term "trade size" when referring to conduits, raceways, and tubing. For example, instead of referring to a ½ inch EMT, it is referred to as trade size ½ EMT.

The *NEC®* also uses the term **metric designator**. A ½ inch EMT is shown as metric designator 16 (½). A 1 inch EMT is shown as metric designator 27 (1). The numbers 16 and 27 are the metric designator values. The (½) and (1) are the trade sizes. The metric

designator is the rigid steel conduit's inside diameter, in rounded off millimeters (mm). Here are some of the more common sizes of conduit, raceways, and tubing. A complete table is found in the *NEC®*, Table 300.1(C). Because of possible confusion, this text uses only the term "trade size" when referring to conduit and raceway sizes.

METRIC DESIGNATOR AND TRADE SIZE	
Metric Designator	Trade Size
12	⅜
16	½
21	¾
27	1
35	1¼
41	1½
53	2
63	2½
78	3

Conduit knockouts in boxes do not measure up to what we call them. Here are some examples.

Trade Size Knockout	Actual Measurement
½	⅞ inch
¾	1³⁄₃₂ inches
1	1⅜ inches

Outlet boxes and device boxes use their nominal measurement as their trade size. For example, a 4 inch × 4 inch × 1½ inch does not have an internal cubic-inch volume of 4 × 4 × 1½ = 24 cubic inches. Table 370.16(A) shows this size box as having a volume of 21 cubic inches. This table shows trade sizes in two columns in both millimeters and inches.

In this text, a square outlet box is referred to as 4 × 4 × 1½ inch square, 4 inch × 4 inch × 1½ inch square, or trade size 4 × 4 × 1½. Similarly, a single-gang device box might be referred to as a 3 × 2 × 3 inch box, a 3 inch × 2 inch × 3 inch deep box, or a trade size 3 × 2 × 3 box.

Trade sizes for construction material will not change. A 2 × 4 is really a name, not an actual dimension. A 2 × 4 will still be referred to as a 2 × 4. This is its trade size.

In this text, measurements directly related to the *NEC*® are given in both inch-pound and metric units. In some instances, only the inch-pound units are shown. This is particularly true for the examples of raceway and box fill calculations, load calculations for square foot areas, and on the plans (drawings).

Because the *NEC*® rounded off most metric conversion values, a computation using metrics results in a different answer when compared to the same computation done using inch-pounds. For example, load calculations for a residence are based on 3 volt-amperes per square foot or 33 volt-amperes per square meter.

- For a 40 × 50 foot dwelling: 3 VA × 40 ft × 50 ft = 6000 volt-amperes.

- In metrics, using the rounded off values in the *NEC*®: 33 VA × 12 m × 15 m = 5940 volt-amperes.

The difference is small, but nevertheless it is different.

To show calculations in both units throughout this text would be very difficult to understand, and would take up too much space. Calculations in either metrics or inch-pounds are in compliance with the *NEC*®, 90.9(D). In 90.9(C)(3) we find that metric units are not required if the industry practice is to use inch-pound units.

It is interesting to note that the examples in Chapter 9 of the *NEC*® use inch-pound units, not metrics.

ACKNOWLEDGMENTS

The author would like to give special thanks to:
Allied Tube and Conduit
American Iron and Steel Institute (AISI)
Calculated Industries, Inc.
Carlon/Lamson and Sessions Company
Cooper B-Line Inc.
Cooper Bussman
Cooper Industries
Crouse-Hinds EMC
Pyrotenax USA Inc.
Square D Company
Taymac Corporation
Underwriters Laboratories, Inc.
Unity Manufacturing

Applicable tables and section references are reprinted with permission from NFPA 70-2005, the *National Electrical Code*®, copyright © 2004–, National Fire Protection Association, Quincy, MA 02269. This reprinted material is not the complete and official position of the NFPA on the referenced subject, which is represented only by the standard in its entirety.

The author and Thomson Delmar Learning would like to acknowledge the time and effort of the reviewers. Our thanks go to:

Michael Forister
Charles Trout

About the Author

Richard E. Loyd is a nationally known author and consultant specializing in the *National Electrical Code®* and the model building codes. He is president of his own firm, R & N Associates, located in Sun Lakes, Arizona. He and his wife, Nancy, travel throughout the country presenting seminars and speaking at industry-related conventions. He also serves as a code expert at 35 to 40 meetings per year at the International Association of Electrical Inspectors (IAEI) meetings throughout the United States. Mr. Loyd represented the National Electrical Manufacturers Association (NEMA) Section 5RN Steel Rigid Conduit and Tubing as an *NEC®* consultant from 1986 through April 1997. Mr. Loyd presently represents the Steel Tube Institute (STI) of North America as an *NEC®* and Model Codes consultant. He represents the American Iron and Steel Institute (AISI) on ANSI/NFPA 70 in the *National Electrical Code®* as a member of Code Committee Eight (8), which is the panel responsible for raceways, and Code Committee Five (5), which is the panel responsible for grounding. Mr. Loyd is actively involved doing forensic inspections and investigations on a consulting basis and serves as a special expert in matters related to the application of codes as they relate to installations and safety. He is a member of the Arizona Chapter of the Electrical Inspectors Association. He is chairman of the Electrical Section of the National Fire Protection Association (NFPA).

He is an active member in the International Association of Electrical Inspectors (IAEI); NFPA; the Institute of Electrical and Electronics Engineers (IEEE), where he is presently a member of the Power Systems Grounding Committee (Green Book); the International Conference of Building Officials (ICBO); the Southern Building Code Congress International (SBCCI); and the Building Officials and Code Administrators International (BOCA). Mr. Loyd is currently licensed as a master contractor/electrician in Arkansas

(license #1725) and Idaho (license #2077) and is an NBEE certified master electrician.

Mr. Loyd served as the chief electrical inspector and administrator for the State of Idaho and for the State of Arkansas. He has served as chairman of NFPA 79 "Electrical Standard for Industrial Machinery"; as a member of Underwriters Laboratories Advisory Electrical Council; as chairman for Educational Testing Service (ETS), a multistate electrical licensing advisory board; and he served as associate editor for Intertec Publications (*EC&M* Magazine) for 10 years. He is a master electrician and former electrical contractor. He has served as chairman of the National Board of Electrical Examiners (NBEE). He is an accredited instructor for licensing certification courses in Idaho, Oregon, and Wyoming and has taught basic electricity and *National Electrical Code®* classes for Boise State University (BSAU) in Boise, Idaho.

Chapter One

OBJECTIVES

After completing this chapter, you should know:

- The methodology of the development of standard multiple-choice examinations

- The most effective methods to successfully take this type of examination

- How to properly prepare for the examination

- Your strengths and weaknesses based on a test of your initial electrical academic skills depending on the results of the examination at the end of this chapter

EXAMINATIONS AND NATIONAL TESTING ORGANIZATIONS

Examinations are an old phenomenon. They have been administered in many disciplines to evaluate competency in the field of specialty. Until recently they were prepared by others within the chosen field, based on the knowledge of each person preparing the examination. Today many examinations are professionally developed by state agencies specializing in examination preparation and by national testing organizations.

In recent years many local jurisdictions, such as city, county, and state, have dropped their electrical examinations in favor of using the national testing agencies that prepare and administer electrical examinations for those local entities. However, there are still many local jurisdictions developing and administering electrical examinations much in the same manner as has been done since the beginning of electrical examinations. These examinations are often prepared and administered by local electrical inspectors and are often developed with question material related to areas in which the inspector often finds violations or to areas in the *National Electrical Code*® (*NEC*®) that the inspector feels are most difficult to understand. Rarely do local examinations fully evaluate an entry-level craftsman regarding his or her competency in this new career field. National testing organizations, such as

Experior or Southern Building Code Congress International (SBCCI), have carefully taken the advantages of localized testing methodology and incorporated the latest state-of-the-art technology to develop a test that is both fair to the taker and a good evaluation of the taker's knowledge in the specific career field. These testing organizations use those experienced in the trades in pools and in sit-down committees to develop a task analysis to determine what the examination will evaluate and then develop questions based on those tasks. These organizations strive to develop question criteria that are clear and not confusing, questions that have only one answer, and questions that an entry-level candidate entering the journeyman electrical field or the master contractor field should know as minimum standards.

These examinations are generally multiple-choice written examinations. The study arrangement in this book is designed to prepare the user to successfully pass these national examinations. However, it also should provide those preparing for local tests on these same subjects.

DEVELOPING AN ITEM (QUESTION) BANK

The development of examinations that effectively measure the minimum competency of an entry-level candidate is a process that uses many concepts. National testing firms—such as Experior (National Headquarters, 1260 Energy Lane, St. Paul, MN 55108), for construction trades multistate licensing, and Southern Building Code Congress International (SBCCI) (900 Montclair Road, Birmingham, AL 35213-1206)—have developed examinations that have become standards in the industry today. The first step in developing an examination is a task or job analysis. The task analysis must evaluate the ability, knowledge, and skill needed to perform the tasks related to the job in a way that will not endanger public health, safety, and welfare. The analysis must be relevant to the actual electrical practice in the field. The first step is then to establish the criteria, the subject areas of the trade that are most important and need to be tested. Two acceptable methods for determining these various job-related tasks would be (1) to survey licensed persons from the construction trades who perform these tasks in the field on a day-to-day basis, and (2) to assemble a group of competent experts in these fields to list the content outline and determine the degree of importance of each task to be evaluated by the examination and create a blueprint that shows which percent of the exam is designated for each subject task. This blueprint ties the test to the job performance, and the test outline shows the importance of each subject. When done properly, the task analysis will substantiate the validity of the exam.

The next task is to hold workshops or group meetings to develop an item (question) bank for a group of questions. This task of developing the item bank is normally performed by a group of interested experts working in workshops to develop the item bank based on the *NEC®* and other electrical related texts or reference books. As the item bank is developed, it is important that each item be clear, precise, and have only one correct answer. On completion of the item bank, there should be a pool of 500 to 800 questions. These questions are entered into the computer, each based on the task analysis for each job task and weighted according to the difficulty level of each question. The electrical board whose members come from contract states subscribing to these national tests then determines the passing scores, the item selection, and test form assembly. When this has been completed, the examination is ready for administration. Following the test administration to the candidates, a post-test examination analysis is then conducted in order to determine the effectiveness of the examination. In the post-test analysis, which is conducted following every exam, the effectiveness of each item is examined. Inadequate items are returned to the workshop to be rewritten, corrected, or eliminated from future exams. Each item is evaluated from developed criteria from previous exams, and the difficulty level is verified. Post-examination analysis provides critical and important information for future workshops and the validity of the exams. This analysis provides information that cannot be gained in any other manner. Information continues to be developed throughout the life of the item bank. This information is fed into the computer and continually proves and improves the validity of future examinations. The candidates themselves provide the most important information in the continuing development and improvement of the examination.

PREPARING FOR AN EXAMINATION

The first step in preparing for an examination is to obtain the examination information from the authority having jurisdiction (AHJ) to which you are about to make application for an examination. National testing firms furnish each AHJ a bulletin containing generic material that applies nationwide, with specific unique material related only to the jurisdiction in which you are about to take an examination. This bulletin contains information that must be read carefully about the eligibility and procedure for registering for the test. This includes your verification of experience and education requirements and usually requires an application fee. It also contains the deadlines that are important for submitting the application and fees and the examination dates. For instance, you may be required to have your application and fees submitted up to 60 days before the examination date. The fees vary from state to state. They generally require the application fee to be made to the jurisdiction and the exam administration fee to be made directly to the national testing firm. However, this procedure varies, and the instructions must be read carefully. Once your application has been approved, your name will be placed on the roster for the administration of the exam at the testing center that you have selected or has been assigned to you. This will generally be the testing center closest to your home address unless you have indicated that you would like to be tested in another location. A schedule of examination dates and locations for the year is included in the candidate bulletin or is available from the local jurisdiction. You will be mailed an admission letter. Do not lose the letter. It will be required when you arrive at the test center or test location to take the exam. If you fail to bring this letter, you will not be permitted to take the exam. Also, it is very important that you read this bulletin to determine what materials are required and what materials are allowed, usually the *NEC®* or handbook in the edition in which the test is developed.

Warning: Verify which *NEC®* edition is being used for the examination (2002 or 2005). Check allowable reference books before studying for the examination. Some agencies permit only the softcover *NEC®*, others permit the *NFPA® Handbook* and other reference books. Verify the pencil type and the type of calculator allowed. Do not wait until test day. Be prepared.

Other electrical reference books may be permitted. You are usually required to bring your own pencils and a silent, handheld, nonprogrammable, battery-powered calculator. Some jurisdictions also require a hands-on practical test. If it is required in the area in which you are making application, the outline of that work should also appear in the bulletin. Read the outline carefully. The practical test can mean the difference between passing and failing the examination. One such jurisdiction required that the candidate bring conduit, a conduit bender, hacksaw, reaming tool, tape measure, and continuity or ohmmeter. Also contained in the candidate bulletin will be the requirements for candidates with physical disabilities. Documentation of the disability must be submitted at the time the application is made. Special considerations are generally permitted to accommodate religious or disability needs as permitted by the Americans with Disabilities Act (ADA).

Finally, the candidate bulletin contains a content outline; the number of questions in each content area; the study reference material that the test has been developed from, such as the *National Electrical Code* book, *Alternating Current Fundamentals* by Duff and Herman (Thomson Delmar Learning), *Direct Current Fundamentals* by Loper and Tedsen (Thomson Delmar Learning), *The American Electrician's Handbook*; and a sample of questions. It is suggested that you cover the answers and take this sample test in much the same manner as you would take the real examination. This will give you an idea of what the test will be like and will familiarize you with the format of the test if you are not already familiar with multiple-choice examinations. This is an important part of preparing for the electrical examination. Take the candidate bulletin seriously.

The next step in preparing for the examination is to study this book. The review questions at the end of each chapter are taken from the *NEC®* and not necessarily only from the content of each chapter. The *NEC®* is a reference document. As a reference document, it is necessary that you develop a strong proficiency in using the Table of Contents and the Index to be able to quickly locate the articles and sections of the *Code* you need. For each review question that you answer, you should not only research and find the correct answer, but you should also reference the *NEC®* section in which the answer was found. As you go through all the review material, take the sample tests throughout the document. This will enable you to pass the test the first time you take it. Anything less may prove disappointing.

> Each lesson is designed purposely to require the student to apply the entire *NEC®* text, and not just specific chapters or articles. It has been found that when studying for a timed, open-book examination, the student must gain proficiency in the Table of Contents, the Index, and the ability to move quickly from cover to cover in order to find the correct answer to each question in a timely fashion.

You should be aware that if you fail the examination, some jurisdictions require a waiting period of six months to one year before you can retake it.

The next section of this book contains a sample examination similar to the examination that you will take in the jurisdiction that you are studying for and similar to the examination given by many national testing organizations. This test should be taken in a quiet room. You should time yourself for three hours. The examination should be taken all at one time. To do otherwise will not give you a true evaluation of your present knowledge and of the areas in which you are weakest. Learning which areas are your weakest will enable you to concentrate your efforts and study time to gain the competency needed to pass your examination.

You may have purchased this book for various reasons. Some individuals study merely to improve their understanding in their chosen field, but most do so because they are required to do so to pass their performance evaluation on their job or to pass an evaluation in a state or local examination. The specific reason may be unimportant. The important thing is that you are about to undertake the study of electricity that, if done properly, will improve your competency in this field. The better you prepare, the better you will do on any evaluation or examination. Preparedness has more than one effect; for example, your IQ can fluctuate 30 to 40 points between any given dates.

The best way to improve individual performance is to reduce the anxiety level. To improve your study habits, consistent study time is important. You should allow yourself the same amount of time each day or each week to study this course. Consistent study time improves your retention.

This book and other Thomson Delmar Learning books related to this subject are designed to improve an individual's understanding of how to use the *NEC®* as a reference document and how to use the basic electrical formulas as applicable. They duplicate as nearly as possible the test format used by many national testing organizations and local jurisdictions. Whether you are preparing for an examination or just improving your skills, good study habits will give you the maximum efficient use of your time.

Basic Minimum Preparedness Tools

By the time test day arrives, you should have the achieved the following:

1. The *NEC®* is a reference book. Most examinations recognize this and are "open-book exams." However, it is still necessary that you are comfortable using it as a tool.

   - Be comfortable using the *NEC®* Table of Contents. Understand each chapter heading and where and how to find things quickly.

   - Be comfortable using the *NEC®* Index using key words.

   - Have a thorough understanding of *NEC®* Article 90—Introduction. This article gives you the Code purpose, what is covered and not covered, Code arrangement, enforcement, rules, and planning.

   - Have a thorough understanding of *NEC®* Article 100—Definitions. This article contains special definitions that are specific to this document, different from the definition you find in a standard dictionary, such as "bathroom," "accessible," and "location."

   - Have a thorough understanding of *NEC®* Article 110—Requirements for Electrical Installations. This article covers general requirements applicable to all installations.

   - Have a thorough understanding of *NEC®* Article 300—Wiring Methods. This article covers general wiring method requirements applicable to all types of wiring methods.

2. Basic electrical mathematical formulae are a must if you are to be successful.

   - voltage drop

   - Ohm's law

   - common alternating current (ac) formulae

If you have achieved this basic necessary knowledge and you have completed this book, you should have no problem passing any standard electrical exam.

Get a good night's sleep the day before the test.

ALTERNATING CURRENT		
To Find	Single Phase	Three Phase
Amperage when "HP" is known	$\dfrac{HP \times 746}{E \times \%EFF \times PF}$	$\dfrac{HP \times 746}{E \times \%EFF \times PF \times 1.73}$
Amperes when "KW" is known	$\dfrac{KW \times 1000}{E \times PF}$	$\dfrac{KW \times 1000}{E \times PF \times 1.73}$
Amperes when "KVA" is known	$\dfrac{KVA \times 1000}{E}$	$\dfrac{KVA \times 1000}{E \times 1.73}$
Kilowatts	$\dfrac{E \times I \times PF}{1000}$	$\dfrac{E \times I \times PF \times 1.73}{1000}$
Kilovolt-Amperes "KVA"	$\dfrac{E \times I}{1000}$	$\dfrac{E \times I \times 1.73}{1000}$
Horsepower	$\dfrac{E \times \%EFF \times PF}{746}$	$\dfrac{E \times I \times \%EFF \times PF \times 1.73}{746}$

$$\text{Percent Efficiency} = \frac{\text{Output}}{\text{Input}}$$

$$PF = \frac{\text{Power (Watts)}}{\text{Apparent Power}} = \frac{KW}{KVA}$$

Test Day

You should have the following to take with you to the examination. Check your Candidate Information Bulletin to confirm allowable materials.

- Admission mailer (letter).

- Photo I.D., such as a driver's license.

- *NEC®* book (current adopted edition) and any other approved reference material that you feel will benefit you in taking the test. *Note:* Your reference material will be checked for handwritten notes and other markings not acceptable to the jurisdiction in which you are testing.

- A silent, nonprinting, nonprogrammable calculator that is not capable of alphabetic entry.

- Two No. 2 pencils.

The time has finally arrived for you to take your examination. Be sure you allow plenty of time to travel to the testing location; it is important to arrive on time. Some testing agencies will not permit admission if you arrive late. On the other hand, do not arrive too early. Discussing the test with other examinees may raise your anxiety level. Nervousness is contagious! Be sure you have your receipt for your application and your entrance ticket. Be sure you have all materials required by the jurisdiction or testing agency, including study books and a handheld calculator.

When a test is multiple choice, you generally have four alternative answers; therefore, the odds are four to one that you can get the right answer. If you can eliminate two of the alternatives, your chances are now increased to 50/50. This is extremely beneficial when referencing the *Code*, as you can quickly eliminate half of the possibilities leaving you only two to research.

A lighting and branch-circuit panelboard is installed in a dwelling unit bathroom. Which of the following answers is correct?
1. It is not a *NEC®* violation.
2. It is permitted, but it must be suitable for wet locations.
3. The location of panelboards is not covered by the *NEC®*.
4. It is a *NEC®* violation.

We can quickly eliminate 2 and 3. The answer is 4. The *NEC®* 240.24(E) prohibits installing overcurrent devices in dwelling unit bathrooms.

Read the directions carefully. Many mistakes are made merely because of misunderstanding. If you have any questions, discuss them with the proctor before the test time starts. Any discussions after the test starts will take away from the time you have to answer the questions. Allow yourself ample time for each question. Answer the question first before you check any alternatives. This way you can evaluate your answer against any alternatives that you may find in the *NEC®*. Do not spend too much time on any one question. If you find the question extremely difficult and you are unsure of the answer or cannot find the answer in the *NEC®*, skip that question and go back to it only if you have time at the end of the test.

You should never spend more than 3½ minutes per question. Your first choice is usually best. Read the questions very carefully. Make sure you know key words that might change the meaning of the question. Note any negatives, such as "Which of the following

are not, or shall not be permitted?" It may be beneficial to underline key words as you read the question. This has a tendency to channel your thoughts in the right direction.

Sample Question

Tip: Break the question down and get to the root question.

The bathroom in a custom home includes a vanity with double sinks (basins) located on one side of the bathroom. There is an outlet located adjacent to each sink (basin) with ground-fault circuit-interrupter protection (GFCI). On the other side of the bathroom, just opposite to and about 8 feet away, is a makeup vanity (without a sink [basin]) with a receptacle outlet located at the same height as the GFCI outlets across the room.

Would this outlet above this vanity be required to also have GFCI protection?

Answer:
The key words are: bathroom
home (dwelling)
GFCI

The question is, Would an outlet not adjacent to a sink in a bathroom need GFCI protection? The answer is Yes, all receptacles located in a residential bathroom are required to be protected with ground-fault circuit-interrupter (GFCI) protection. The reference is *NEC®* 210.8(A)(1).

This question could be time-consuming because it causes you to look at the location in relation to the sink. *NEC®* 210.52(D) requires a receptacle to be located within 3 feet (900 mm) of the outside edge of each basin (sink). However, the question does not ask that! The question asks whether another receptacle located outside this dimension needs GFCI protection.

Summary

- Dress properly; wear comfortable shoes and loose-fitting clothing. The test facility should be approximately 70°F. Too many clothes or uncomfortable clothes have a tendency to make you drowsy and concentration difficult.

- Plan to arrive at the test site 20 to 30 minutes early. Anxiety will reduce your effectiveness. Do not be late or too early. Make sure ahead of time that you know where the test site is located and whether parking is available.

- Be sure you have your acceptance letter and proper I.D. Verify ahead of time so that you have the identification materials required by the testing agency.

- Do not forget your *NEC®* book, reference books, calculator, and other acceptable/required materials. Pack all these materials the night before in a backpack or briefcase.

- After you have answered all the questions, recheck your work.

1. Make sure that you have not made clerical mistakes.

2. Make sure that your answers are recorded on the correct line on the answer sheet.

3. Go over the extremely difficult questions.

4. Regarding questions on which you know you have guessed—you may want to mark them in some way so that if there is time at the end of the test, you can do additional research.

Good luck!

SAMPLE EXAMINATION
Examination Instructions:

For a positive evaluation of your knowledge and preparation awareness, you must:

1. Locate yourself in a quiet atmosphere (room by yourself).

2. Have with you at least two sharp No. 2 pencils, the 2005 *NEC®*, and a handheld calculator.

3. Time yourself (three hours) with no interruptions.

4. After the test is complete, grade yourself honestly and concentrate your studies on those sections of the *NEC®* in which you missed the questions.

   Caution: Do not just look up the correct answers because the questions in this examination are only an exercise and not actual test questions. Therefore, it is important that you be able to quickly find answers from throughout the *NEC®*.

Directions: Each question is followed by four suggested answers: A, B, C, and D. In each case, select the **one** that best answers the question.

1. Liquidtight flexible metal conduit can be used in which of the following locations?
   A. in areas that are both exposed and concealed
   B. in areas that are subject to physical damage
   C. in connection areas for gasoline dispensing pumps
   D. in areas where ambient temperature is to be greater than 200°C

   Answer: _____   Reference: _____

2. The maximum number of quarter bends allowed in one run of nonmetallic rigid conduit is
   A. 2.
   B. 4.
   C. 6.
   D. 8.

   Answer: _____   Reference: _____

3. The maximum size of flexible metallic tubing that can be used in any construction or installation is
   A. trade size ⅜ (12).
   B. trade size ½ (16).
   C. trade size ¾ (21).
   D. trade size 1 (27).

   Answer: _____   Reference: _____

4. The size of the pull box in the diagram above should not be less than
   A. 12 inches × 14 inches (300 mm × 350 mm).
   B. 16 inches × 18 inches (400 mm × 450 mm).
   C. 22 inches × 22 inches (560 mm × 560 mm).
   D. 24 inches × 24 inches (600 mm × 600 mm).

   Answer: _____   Reference: _____

5. A metal raceway system installed in the washing area of a car wash must be spaced at least how far from the walls?
   A. ⅛ inch (3 mm)
   B. ¼ inch (6.4 mm)
   C. ¾ inch (19 mm)
   D. 1 inch (25 mm)

   Answer: _____   Reference: _____

6. Unless otherwise indicated, busways should be supported at intervals not to exceed how many feet?
   A. 3 (900 mm)
   B. 5 (1.5 m)
   C. 7 (2.1 m)
   D. 10 (3.0 m)

   Answer: _____   Reference: _____

7. Which of the following percentages of conduit fill should be used when four Type THW conductors are being installed?
   A. 55 percent
   B. 53 percent
   C. 40 percent
   D. 31 percent

   Answer: _____   Reference: _____

8. Flexible 18 AWG supply cords of listed appliances are considered adequately protected if the circuit over-current device is set at a maximum value of how many amperes?
   A. 15
   B. 20
   C. 30
   D. 40

   Answer: _____   Reference: _____

9. All of the following sizes of solid aluminum conductors shall be made of an aluminum alloy except
   A. 8 AWG.
   B. 10 AWG.
   C. 12 AWG.
   D. 14 AWG.

   Answer: _____   Reference: _____

10. An insulated bushing is required to be used on a raceway entering a cabinet if the ungrounded conductors entering the raceway are at least
    A. 10 AWG.
    B. 8 AWG.
    C. 6 AWG.
    D. 4 AWG.

    Answer: _____   Reference: _____

11. The allowable distance between supports of nonmetallic-sheathed cable installed in an on-site, constructed, one-family dwelling is a maximum of how many feet?
    A. 2½ (750 mm)
    B. 3 (900 mm)
    C. 4 (1.2 m)
    D. 4½ (1.4 m)

    Answer: _____   Reference: _____

12. It is permissible to use 4 THW CU AWG for service in dwelling units to a maximum load of
    A. 100 amperes.
    B. 95 amperes.
    C. 70 amperes.
    D. 65 amperes.

    Answer: _____   Reference: _____

13. Where conductors of less than 600 volts emerge from the ground, they must be protected by enclosures or raceways that extend from below grade to a point how many feet above the finish grade?
   A. 5 (1.5 m)
   B. 6 (1.8 m)
   C. 7 (2.1 m)
   D. 8 (2.4 m)

   Answer: _____   Reference: _____

14. An installation consisting of six 3-wire cables installed in a 6-foot (1.8 m) nonmetallic wireway is being installed in a light commercial building. Which of the following statements is (are) not true?
   A. The six multiwire cables are not required to be laced together in a neat workmanlike manner.
   B. The sum of the cross-sectional areas of all contained current-carrying conductors shall not exceed 20 percent of the interior area of the wireway.
   C. The derating factors specified in Article 310, Note 8 to the 0- to 2000-volt tables is not applicable because there are less than 30 current-carrying conductors at 20 percent fill.
   D. The wireway may not be installed where subject to physical damage.

   Answer: _____   Reference: _____

15. The service rating for a residence that uses 2 AWG THW copper service-entrance conductors is
   A. 95 amperes.
   B. 110 amperes.
   C. 115 amperes.
   D. 125 amperes.

   Answer: _____   Reference: _____

16. Surge arrester grounding conductors that run in metal enclosures should be
   A. bare.
   B. insulated.
   C. bonded on one end of the enclosure only.
   D. bonded at both ends of the enclosure.

   Answer: _____   Reference: _____

17. The number of overcurrent devices in a single cabinet of a lighting and appliance panelboard should not exceed
   A. 30.
   B. 36.
   C. 42.
   D. 48.

   Answer: _____   Reference: _____

18. When it is impractical to locate a service head directly above the point of attachment, the maximum allowable placement distance from the point of attachment is how many feet?
    A. 1 (300 mm)
    B. 2 (600 mm)
    C. 3 (900 mm)
    D. 4 (1.2 m)

    Answer: _____   Reference: _____

19. An underground service to a small, controlled water heater is to be installed. The single branch-circuit used for this service requires that copper conductors be at least
    A. 12 AWG.
    B. 10 AWG.
    C. 8 AWG.
    D. 6 AWG.

    Answer: _____   Reference: _____

20. Service conductors can have less than 3 feet (900 mm) of clearance from which of the following?
    A. doors
    B. tops of windows
    C. fire escapes
    D. porches

    Answer: _____   Reference: _____

21. Service conductors that pass over rooftops should have a vertical clearance of not less than how many feet?
    A. 3 (900 mm)
    B. 7 (2.1 m)
    C. 8 (2.4 m)
    D. 10 (3.0 m)

    Answer: _____   Reference: _____

22. The interrupting rating shall be marked on all circuit breakers rated more than
    A. 1000 amperes.
    B. 2000 amperes.
    C. 2500 amperes.
    D. 5000 amperes.

    Answer: _____   Reference: _____

23. Illumination is mandatory for service equipment or panelboards in a dwelling unit if the service to the unit exceeds how many amperes?
    A. 100
    B. 200
    C. 400
    D. required for all services regardless of amperage

    Answer: _____   Reference: _____

24. Ground-fault protection is required for _____-ampere, 480-volt, 4-wire, 3-phase solidly grounded wye services.
    A. 800
    B. 1000
    C. 1200
    D. above 2000

    Answer: _____   Reference: _____

25. Which of the following is the correct connection sequence for a 3-phase, delta-connected motor to be connected to 240-volt supply?

L1	*L2*	*L3*
A. To 1 and 7	To 2 and 8	To 3 and 9; 4, 5, and 6
B. To 1, 7, and 6	To 2, 8, and 4	To 3, 9, and 5
C. To 1, 7, and 4	To 2, 8, and 5	To 3, 9, and 6
D. To 1	To 2	To 3; 7 and 4; 8 and 5; 9 and 6

    Answer: _____   Reference: _____

26. Impedance is present in which of the following circuits?
    A. ac only
    B. dc only
    C. resistance only
    D. both ac and dc

    Answer: _____   Reference: _____

27. Excluding fuses and exceptions, how many overcurrent protection devices, such as trip coils, relays, or thermal cutouts, are required on a 3-phase motor?
    A. 5
    B. 3
    C. 2
    D. 0

    Answer: _____   Reference: _____

28. All of the following are induction-type motors except
    A. wound-rotor.
    B. split-phase.
    C. capacitor.
    D. universal.

    Answer: _____   Reference: _____

Feeder

A   B   C

10 hp   15 hp   20 hp

Autotransformer stg. squirrel cage	
28 A	No code
230V	Temp rise 40° C
10 hp	3 PH

Full voltage start squirrel cage	
42 A	A
230V	Temp rise 40° C
15 hp	3 PH

Full voltage start wound cage	
54 A	No code
230V	Temp rise 40° C
20 hp	3 PH

29. In the diagram above, the size of THW feeder conductors should not be less than
   A. 4 AWG.
   B. 2 AWG.
   C. 1 AWG.
   D. 1/0 AWG.

Answer: _____   Reference: _____

30. In the diagram on page 12, the size of Type THW conductors for branch-circuit A should not be less than
    A. 10 AWG.
    B. 8 AWG.
    C. 6 AWG.
    D. 4 AWG.

    Answer: _____   Reference: _____

31. In the diagram on page 12, the size of Type THW conductors for branch-circuit C should not be less than
    A. 8 AWG.
    B. 6 AWG.
    C. 4 AWG.
    D. 3 AWG.

    Answer: _____   Reference: _____

32. The attachment plug and receptacle can be used as the controller for a portable motor that has a maximum horsepower rating of
    A. ⅓ horsepower.
    B. ½ horsepower.
    C. 1 horsepower.
    D. 2 horsepower.

    Answer: _____   Reference: _____

33. Snap switches can be grouped or ganged in outlet boxes if voltages between adjacent switches do not exceed how many volts?
    A. 200
    B. 300
    C. 400
    D. 500

    Answer: _____   Reference: _____

34. Each doorway that leads into a transformer vault from the building interior should have a tight-fitting door with a minimum fire rating of how many hours?
    A. 1
    B. 2
    C. 3
    D. 5

    Answer: _____   Reference: _____

35. The conduit run between warm and cold locations should be
    A. sealed.
    B. sleeved.
    C. insulated.
    D. provided with drains.

    Answer: _____   Reference: _____

36. Except for points of support, recessed portions of luminaire (lighting fixture) enclosures should be spaced at least how many inches from combustible material?
    A. ¼ (6 mm)
    B. ½ (13 mm)
    C. 1 (25 mm)
    D. 3 (75 mm)

    Answer: _____   Reference: _____

37. Surface-mounted fluorescent luminaires (lighting fixtures) that contain a ballast and are to be installed on combustible, low-density cellulose fiberboard should be spaced not less than how many inches from the surface of the fiberboard?
    A. ½ (13 mm)
    B. 1 (25 mm)
    C. 1½ (38 mm)
    D. 2 (50 mm)

    Answer: _____   Reference: _____

38. The minimum amount of on-site fuel for an emergency generator should be sufficient for not less than
    A. 2 hours at full demand.
    B. 4 hours at 75 percent demand.
    C. 6 hours at 50 percent demand.
    D. 8 hours at 25 percent demand.

    Answer: _____   Reference: _____

39. Which of the following statements about dimmers installed in theaters is (are) true?
    I. Dimmers installed in ungrounded conductors must have overcurrent protection not greater than 125 percent of the dimmer rating.
    II. The circuit-supplying autotransformer type dimmers must not exceed 150 volts between conductors.
    A. I only
    B. II only
    C. both I and II
    D. neither I nor II

    Answer: _____   Reference: _____

Remote panelboard

Gasoline dispenser

Diesel fuel dispenser

Gasoline dispenser

Light pole

Trade size 1 (27 mm)
Rigid metal conduit

40. In the diagram above, how many conduit seals are required?
   A. 2
   B. 4
   C. 6
   D. 8

   Answer: _____   Reference: _____

41. An area must be classed as a Class II hazardous location if it contains which of the following?
   A. combustible dust
   B. flammable gases
   C. ignitable liquid
   D. ignitable vapors

   Answer: _____   Reference: _____

42. If the wiring is to be placed 18 inches (450 mm) above the floor, which of the following wiring methods is prohibited for a commercial automotive repair shop?
   A. metal-clad cable
   B. electrical metallic tubing
   C. rigid nonmetallic conduit
   D. nonmetallic-sheathed cable

   Answer: _____   Reference: _____

43. The minimum number of receptacles in a patient bed location of a hospital's general care area should be
    A. 1 duplex.
    B. 2 duplex.
    C. 3 duplex.
    D. 3 single.

    Answer: _____   Reference: _____

44. Which of the following statements about lightning protection is (are) true?
    I. If lightning protection is required for an irrigation machine, a driven ground rod should be connected to the machine at a stationary point.
    II. Lightning rods should be spaced at least 6 feet (1.8 m) away from noncurrent-carrying metal parts of electrical equipment.
    A. I only
    B. II only
    C. both I and II
    D. neither I nor II

    Answer: _____   Reference: _____

45. Which of the following will effectively ground a 230-volt, single-phase, residential air-conditioning unit that replaces an older unit supplied by a 2-wire, 230-volt circuit?
    I. a new circuit containing a grounding conductor
    II. an 8-foot (2.5 m) ground rod installed at the unit
    A. I only
    B. II only
    C. both I and II
    D. neither I nor II

    Answer: _____   Reference: _____

46. When connected to a rod, pipe, or plate grounding electrode, grounding electrode copper conductors need not be sized larger than
    A. 8 AWG.
    B. 6 AWG.
    C. 4 AWG.
    D. 2 AWG.

    Answer: _____   Reference: _____

47. Which of the following is the most acceptable grounding electrode?
    A. a driven steel approved ground rod
    B. well casing
    C. building foundation steel
    D. an underground cold water metal piping system

    Answer: _____   Reference: _____

48. Which of the following terms can be used in place of "resistance" in the phrase "resistance is to ohms"?
    A. reactance
    B. inductance
    C. impedance
    D. capacitance

    Answer: _____   Reference: _____

49. The *NEC®* requires a minimum of which of the following for the kitchen small-appliance load in dwelling units?
    A. two 20-ampere circuits
    B. two 15-ampere circuits
    C. one 20-ampere circuit
    D. one 15-ampere circuit

    Answer: _____   Reference: _____

Questions 50 and 51 refer to a 120-gallon water heater with 220-volt, 4000-watt units, and an interlocking thermostat.

50. Installation of this water heater will require branch-circuit conductors of size
    A. 8.
    B. 10.
    C. 12.
    D. 14.

    Answer: _____   Reference: _____

51. The maximum branch-circuit overcurrent protection for this water heater is
    A. 20 amperes.
    B. 25 amperes.
    C. 30 amperes.
    D. 40 amperes.

    Answer: _____   Reference: _____

Questions 52 and 53 refer to a 230-volt circuit that is 500 feet long, displays a resistance of 1 ohm, and supplies a load of 10 amperes.

52. What is the wattage of the load in this circuit?
    A. 2000
    B. 2200
    C. 2300
    D. 2400

    Answer: _____   Reference: _____

53. How much wattage is lost in this circuit?
    A. 100
    B. 120
    C. 240
    D. 220

    Answer: _____   Reference: _____

54. Ground-fault protection of equipment is a requirement for solidly grounded wye services that are rated at more than 150 volts-to-ground but do not exceed 600 volts, phase-to-phase, for each service disconnecting means rated at 1000 amperes or more. The maximum amperage for this ground-fault protection relay should be set at
    A. 1000 amperes.
    B. 1200 amperes.
    C. 1500 amperes.
    D. 2000 amperes.

    Answer: _____   Reference: _____

Number	Size (kcmil)	Type	Function
2	500	THW	Phases A and B
1	400	THW	Neutral
1	300	THW	Phase C
1	3 AWG	THW	Ground Conductor

55. The minimum size intermediate metal conduit required for installation of the conductors above should not be less than
    A. trade size 2 (53).
    B. trade size 2½ (63).
    C. trade size 3 (78).
    D. trade size 3½ (91).

    Answer: _____   Reference: _____

Questions 56 and 57 refer to three 3-phase induction motors, rated respectively at 28, 16, and 12 amperes.

56. The motors would require a feeder rated at how many amperes?
    A. 56
    B. 63
    C. 70
    D. 112

    Answer: _____   Reference: _____

57. The motors are started at line voltage and are protected with time-delay fuses. The maximum feeder time-delay fuse should be rated at how many amperes?
    A. 70
    B. 80
    C. 90
    D. 100

    Answer: _____ Reference: _____

58. The combined load of several 240-volt, fixed space heaters on a 20-ampere circuit should not exceed how many kilowatts?
    A. 2.4
    B. 2.6
    C. 3.8
    D. 4.8

    Answer: _____ Reference: _____

59. If three resistors with values of 5, 10, and 15 ohms, respectively, are connected in parallel, the combined resistance of the units will be
    A. 1.63 ohms.
    B. 2.73 ohms.
    C. 20.0 ohms.
    D. 30.0 ohms.

    Answer: _____ Reference: _____

60. Conductors that supply an 8-kilowatt, 230-volt, fixed electric space heater consisting of resistance elements should have a kilowatt rating of not less than
    A. 6.4.
    B. 8.
    C. 10.
    D. 12.

    Answer: _____ Reference: _____

61. The *NEC*® covers the installation of electrical conductors and equipment in all of the following locations except
    A. public and private buildings.
    B. floating dwelling units.
    C. buildings used by a utility for warehousing.
    D. centers of transmission and distribution of electrical energy.

    Answer: _____ Reference: _____

62. *Refer to the figure above.* Three 35-ampere resistive balanced resistive loads are connected to a 240-volt, 3-phase, 3-wire circuit. What is the total power of this circuit in kilowatts?
    A. 14.6 kW
    B. 8.4 kW
    C. 7.3 kW
    D. 49.9 kW

    Answer: _____   Reference: _____

63. The most recent editions of the *NEC*® are revised every
    A. year.
    B. 2 years.
    C. 3 years.
    D. 4 years.

    Answer: _____   Reference: _____

64. Galvanized rigid conduit is used to protect conductors in a cable tray. Where the conductors enter and leave the conduit within the cable tray,
    A. junction boxes are required even if the conduit is bushed.
    B. junction boxes are not required if the conduit is bushed.
    C. *no* bushings are required if the conduit is smaller than 1 inch.
    D. *no* bushings are required if the conductors are smaller than #4 AWG copper.

    Answer: _____   Reference: _____

65. Surface-type cabinets for electrical equipment in damp or wet locations shall be mounted so at least _____ inch air space between the wall or other supporting surface exists.
    A. ⅛ (3.1 mm)
    B. ¼ (6.3 mm)
    C. ⅜ (10 mm)
    D. ½ (13 mm)

    Answer: _____   Reference: _____

66. Generally, rigid metal conduit is required to be fastened in place within _____ foot/feet of each box, outlet, cabinet, or conduit termination.
    A. 1 (300 mm)
    B. 2 (600 mm)
    C. 3 (900 mm)
    D. 4 (1.2 m)

    Answer: _____   Reference: _____

67. For six size 12 AWG conductors, the minimum trade size octagonal box that can be used is
    A. 4 inches × 1¼ inches (100 mm × 32 mm).
    B. 4 inches × 1½ inches (100 mm × 38 mm).
    C. 4 inches × 2⅛ inches (100 mm × 54 mm).
    D. 4 inches × 2¼ inches (100 mm × 57 mm).

    Answer: _____   Reference: _____

68. A motor controller that is installed with the expectation of its being submerged occasionally for short periods shall be installed in a rated enclosure type No. ____.
    A. 3
    B. 3S
    C. 4X
    D. 6

    Answer: _____   Reference: _____

69. Where recessed high-intensity discharge fixtures are installed indoors and operated by remote ballasts,
    A. thermal protection is not required.
    B. only the fixture requires thermal protection.
    C. only the ballast requires thermal protection.
    D. both the fixture and the ballast require thermal protection.

    Answer: _____   Reference: _____

70. Any pipe or duct system foreign to the electrical installation must not enter a transformer vault. The _____ is not considered foreign to the vault.
    A. vault automatic fire protection
    B. building water main
    C. air duct passing through the vault
    D. roof drain piping

    Answer: _____   Reference: _____

71. A dry-type transformer rated at 480 volts can be installed on a building column and shall not be required to be
    A. accessible.
    B. readily accessible.
    C. rated less than 112½ kVA.
    D. rated less than 600 volts.

    Answer: _____ Reference: _____

72. A capacitor is located indoors. It must be enclosed in a vault if it contains more than a minimum of _____ gallon(s) of flammable liquid.
    A. 1
    B. 2
    C. 3
    D. 5

    Answer: _____ Reference: _____

73. A single-phase hermetic refrigerant motor compressor has a rated load current of 24 amperes and a branch-circuit selection current of 30 amperes. The branch-circuit conductors are copper, Type TW. They operate at 80°F, and they are the only conductors in the conduit to this compressor. The smallest possible branch-circuit conductors must be at least size _____ AWG.
    A. 12
    B. 10
    C. 8
    D. 6

    Answer: _____ Reference: _____

74. A 3-phase, squirrel-cage motor with full-voltage reactor starting has no code letter. The calculated fuse size to protect the branch circuit of the motor would be sufficient for the starting current of the motor. The nontime delay fuse to protect the circuit of this motor shall be sized at _____ percent of full-load current.
    A. 150
    B. 200
    C. 250
    D. 300

    Answer: _____ Reference: _____

75. A dry-type transformer is to be installed indoors. If rated more than _____ kilovolt-amperes, the transformer must be installed in a fire-resistant transformer room.
    A. 25.0
    B. 35.0
    C. 100.5
    D. 112.5

    Answer: _____ Reference: _____

Chapter Two

NATIONAL FIRE PROTECTION ASSOCIATION STANDARDS

The National Fire Protection Association (NFPA) has acted as the sponsor of the *National Electrical Code®* (*NEC®*) as well as many other safety standards. The most widely used electrical code in the world is the *NEC®*. The official designation for the *NEC®* is ANSI/NFPA 70.

NATIONAL ELECTRICAL CODE® (*NEC®*)

The *NEC®* was first developed about 100 years ago by interested industry and governmental authorities to provide a standard for the safe use of electricity. It is now revised and updated every three years. It was first developed to provide the regulations necessary for safe installations and to provide the practical safeguards of persons and property from the hazards arising from the use of electricity. It still provides these regulations and safeguards to the industry today. The *NEC®* is developed by 20 different Code-Making Panels composed entirely of volunteers. These volunteers come from industry, testing laboratories, inspection agencies, engineering, users, government, and the electrical utilities. Suggestions for the content come from various sources, including individuals like you. Anyone can submit a proposal or make a comment on a proposal submitted by another individual.

For more specifics on the *NEC®* process, you can contact the National Fire Protection Association, One Batterymarch Park, Quincy, MA, 02269-9101. Request their free booklet "The NFPA Standards Making System."

The *NEC®* is advisory as far as NFPA and ANSI are concerned but is offered for use in law and regulations in the interest of life and property protection. The name *National Electrical Code®* might lead one to believe that this document is developed by the federal government. This is not so; the *NEC®* only recognizes uses of products. It approves nothing. The *NEC®* has no legal standing until it has been adopted by the *authority having jurisdiction* (AHJ) (see *NEC®* definition Article 100), usually a governmental entity. Therefore, we must first check with the local electrical inspector to see which edition of the *NEC®* has been adopted. Compliance within and proper maintenance will result in an installation essentially free from hazard but not necessarily efficient, convenient, or adequate for good service or future expansion of electrical use. The *NEC®* is not an instruction manual for untrained persons, nor is it a design specification. However, it does offer design guidelines.

NEC® 90.2 defines the scope of this document. It is intended to cover all electrical conductors and equipment within or on public and private buildings or other structures, including mobile homes, recreational vehicles, and floating buildings, and other premises such as yards, carnival, parking, and other lots, and industrial substations. Installations of conductors and equipment that connect to the supply of electricity, other installations of outside conductors on the premises, and installations of optical fiber cables are clearly covered by the *NEC®*. The *NEC®* is not intended to cover installations on ships, watercraft, railway rolling stock, automotive vehicles, underground mines, and surface mobile mining equipment. The *NEC®* is not intended to cover installations governed by the utilities, such as communication equipment, transmission, generation, and distribution installations on right-of-way.

Note: For the complete list of the exemptions and coverage, see 2005 *NEC*® 90.2. Mandatory rules are characterized by the word "shall." Explanatory rules are in the form of fine print notes (FPN). All tables and footnotes are a part of the mandatory language. Material text extracted from other NFPA standards and documents shall not be compromised or violated. Editing is permitted for adapting the extracted text into *NEC*® style. A complete list of all NFPA documents referenced can be found in Annex A. New revisions are identified by a vertical line on the margin where inserted.

To use the *NEC*®, one must first have a thorough understanding of Article 90—The Introduction; Article 100—Definitions; Article 110—Requirements for Electrical Installations; and Article 300—Wiring Methods. The rest of the *Code* can be referred to on an as-needed basis.

New for the 2005 *National Electrical Code*®

The 2005 *NEC*® has some notable changes but not as many or as significant as have been adopted in past editions. Some of the major changes that appear in the 2005 *NEC*® are noted here.

Article 80 has been relocated as a new addendum G. This was a logical move because Article 80 was never intended to be enforceable unless adopted separately.

Handhole enclosures have been added to Article 314. These are not new. They have been used by utilities for many years but are now being recognized in the *NEC*®. They are generally access enclosures without a bottom for access to splices for direct buried conductors.

A new Article 353 for high-density polyethylene (HDPE) conduit. This is a limited-use product that is only permitted outside underground.

Article 366—Auxiliary Gutters and Article 368—Busways have been rewritten and renumbered to conform to the numbering style in Chapter 3.

A new Article 409—Industrial Control Panels, has been added. This covers many types of contactors and control panels commonly fabricated to UL 508.

Adjustable speed drives (variable speed ac) have been added to Article 430. These have become very popular in all industries and needed to be added to the *NEC*®.

A new Article 506 is added to cover the Zone alternative system for hazardous dusts. This article will look similar to Article 505, which was added two editions earlier for hazardous gases and vapors.

Temporary Installations have been relocated from Article 305 in the 1999 *NEC*® to Article 527 in the 2002 *NEC*®, and in the 2005 *NEC*® will be renumbered as Article 590. There are no significant changes to this article.

A new Article 682 has been added to cover bodies of water not covered by Article 680. This will include fish farms and similar bodies of water.

Table 344.24, covering the bending radius allowed for raceways, is relocated to Chapter 9 as a new Table 2.

These are the major differences, but remember that there are hundreds of editorial and substantive changes that are enforceable. As you prepare for your examination, check all applicable *Code* sections so that your answers are accurate.

Article 90—Introduction

Article 90 contains the scope of the *NEC*® and is enforceable. This article contains information that applies to every installation. Every electrician should read this information carefully and understand it.

Chapter 1—General

Chapter 1 includes Article 100—Definitions. These definitions are unique to this document and apply where the term is used in the *NEC*®. Article 110 covers the general requirements for all electrical installations and contains information that applies to every installation. Every electrician should read this information carefully and understand it completely. These requirements apply to every installation regardless of how large or small it is.

Chapter 2—Wiring and Protection

Chapter 2 covers Article 200 through Article 285. These are general requirements that apply to all installations from Article 200, the use and identification of the grounded or neutral conductor, to a new Article 285 covering transient voltage surge suppressors (TVSS). Although the requirements in this chapter are applicable to all installations, the electrician should refer to the material because there are too many requirements to remember.

Chapter 3—Wiring Methods

Chapter 3 covers Article 300 through Article 398. This chapter was completely reorganized for the 2002 cycle and renumbered. Temporary Wiring has been moved to 590 and renamed Temporary Installations.

Article 300 covers the general wiring methods that apply to all of the wiring methods in Chapter 3. Every electrician needs to read this article carefully and understand it because these requirements apply to every installation regardless of how large or small it is. The balance of this chapter is used where applicable. The electrician should use it as reference material because there are too many requirements for anyone to remember.

Chapter 4—Equipment for General Use

Chapter 4 covers Article 400 through Article 490. This chapter covers the general requirements for equipment for all installations. Article 404—Switches and Article 408—Switchboards and Panelboards have been moved to Chapter 4. Article 410 now covers only luminaires (lighting fixtures), lampholders, and lamps. The requirements for receptacles, cord connectors, and attachment plugs (caps) are covered in Article 406. A new Article 409 has been added to cover industrial control panels. Although it covers general equipment such as appliances and motors that are used on all installations, the electrician should use this chapter as reference material because there are too many requirements for anyone to remember.

Chapter 5—Special Occupancies

Chapter 5 covers Articles 500 through 555. These chapters do not apply generally—they modify or amend Chapters 1 through 4. They are only applicable in the types of occupancies or portions of buildings or structures that contain areas with special occupancies, such as an area within another building that contains hazardous materials. However, Article 527—Temporary Wiring has been relocated to Chapter 5 as Article 590—Temporary Installations. The electrician should use Chapter 5 as reference material because there are too many requirements for anyone to remember.

Chapter 6—Special Equipment

Chapter 6 covers Article 600 through Article 695. These chapters do not apply generally—they modify or amend Chapters 1 through 4. They are only

2005 NEC®	Title
300	General Wiring Methods
310	Conductors for General Wiring
312	Cabinets, Cutout Boxes and Meter Socket Enclosures
314	Outlet, Device, Pull and Junction Boxes, Conduit Bodies and Fittings
320	Armored Cable: Type AC
322	Flat Cable Assemblies: Type FC
324	Flat Conductor Cable: Type FCC
326	Integrated Gas Spacer Cable: Type IGS
328	Medium Voltage Cable: Type MV
330	Metal-Clad Cable: Type MC
332	Mineral-Insulated Metal-Sheathed Cable: Type MI
334	Nonmetallic-Sheathed Cable: Types NM, NMC, and NMS
336	Power and Control Tray Cable: Type TC
338	Service-Entrance Cable: Types SE and USE
340	Underground Feeder and Branch-Circuit Cable: Type UF
342	Intermediate Metal Conduit: Type IMC
344	Rigid Metal Conduit: Type RMC
348	Flexible Metal Conduit: Type FMC
350	Liquidtight Flexible Metal Conduit: Type LFMC
352	Rigid Nonmetallic Conduit: Type RNC
353	High Density Polyethylene Conduit: Type HDPE
354	Nonmetallic Underground Conduit with Conductors: Type NUCC
356	Liquidtight Flexible Nonmetallic Conduit: Type LFNC
358	Electrical Metallic Tubing: Type EMT
360	Flexible Metallic Tubing: Type FMT
362	Electrical Nonmetallic Tubing: Type ENT
366	Auxiliary Gutters
368	Busways
370	Cablebus
372	Cellular Concrete Floor Raceways
374	Cellular Metal Floor Raceways
376	Metal Wireways
378	Nonmetallic Wireways
380	Multioutlet Assembly
382	Nonmetallic Extensions
384	Strut-Type Channel Raceway
386	Surface Metal Raceways
388	Surface Nonmetallic Raceways
390	Underfloor Raceways
392	Cable Trays
394	Concealed Knob-and-Tube Wiring
396	Messenger Supported Wiring
398	Opening Wiring on Insulators
404	Switches
406	Receptacles, Cord Connectors, and Attachment Plugs (Caps)
408	Switchboards and Panelboards
409	Industrial Control Panels
590	Temporary Installations

applicable where the special equipment covered, such as a sign on a building or a swimming pool, is installed. The electrician should use Chapter 6 as reference material because there are too many requirements for anyone to remember.

Chapter 7—Special Conditions

Chapter 7 covers Article 700 through Article 780. These chapters do not apply generally—they modify or amend Chapters 1 through 4. They are only applicable where the special systems covered, such as limited voltage or emergency systems, are installed. The electrician should use Chapter 7 as reference material when encountering a system covered in this chapter because there are too many requirements for anyone to remember.

Chapter 8—Communications Systems

This chapter covers Article 800, Article 810, Article 820, and Article 830. Chapter 8 stands alone. It is not subject to the requirements of Chapters 1 through 7 unless the requirements are specifically referenced in Chapter 8. These chapters do contain many references, such as requirements for grounding. The electrician should use Chapter 8 only as reference material when installing a communication system. There are too many requirements for anyone to remember.

Chapter 9—Tables

These tables are applicable where referenced elsewhere in the *Code*. Table 344.24 has been relocated as Chapter 9, Table 2.

Annex A–G

Annex A through G (formerly "Appendix") contains informational material that is not mandatory or enforceable. Article 80 in the 2002 *NEC®* has been relocated as a new Addendum G.

The *NEC®* has a new way of numbering the parts to each article and the sections and subsections. The parts, which formerly appeared as A, B, C, and so on, now appear as Roman numerals such as Part I, Part II, Part III, and so on. The sections and subsections have a new look; they are formatted differently. The hyphen between the article number and section designation has been changed to a dot (.). The articles in the *NEC®* subdivide the chapters by a specific subject such as services, branch circuits, grounding, transformers, signs, and the like. The parts are subdivided by sections, and they may be subdivided by subsections or lists.

Example	
Chapter:	**Chapter 2—Wiring and Protection**
Article:	**Article 250—Grounding**
Part:	**IV. Conductors**
Section:	**250.119** Identification of Equipment Grounding Conductors. Unless required.
Level 1:	**(A)** Conductors larger than 6 AWG.
List item:	(1) Stripping the insulation or covering from the entire exposed length . . .
List item:	(2) Coloring the exposed insulation or covering green . . .
List item:	(3) Marking the exposed insulation or covering . . .
Level 2:	**(B)** Multiconductor Cable.
List item:	(1) Stripping the insulation . . .
List item:	(2) Coloring the exposed insulation . . .
List item:	(3) Marking the exposed insulation . . .

LOCAL CODES AND REQUIREMENTS

Although most municipalities, countries, and states adopt the *NEC®*, they may not have adopted the latest edition, and in some cases the authority having jurisdiction (AHJ) may be using an older edition and not the 2005 edition. Most jurisdictions make amendments to the *NEC®* or add local requirements. These may be based on environmental conditions, fire safety concerns, or other local experience. An example of one common local amendment is that all commercial buildings be wired in metal raceways (Rigid, IMC, or EMT). The *NEC®* generally does not differentiate between wiring methods in residential, commercial, or industrial installations; however, many local jurisdictions do. Metal wiring methods, especially in fire zones, are another common amendment. Some major cities have developed their own electrical code—for example, Los Angeles, New York, Chicago, and several metropolitan areas. In addition to the *NEC®* and all local amendments, the designer and installer must also comply with the local electrical utility rules. Most utilities have specific requirements for installing the service to the structure. There have been many unhappy designers and installers who have learned about special jurisdictional requirements after making the installation, thus incurring costly corrections at their own expense.

If you are preparing for a state or city examination, you must check with the AHJ and see exactly what the examination content is based on. If it is a locally developed examination, the content may vary widely. If it is a nationally developed examination, it will generally be based on the latest edition of the *NEC*® (2005 *NEC*®). However, some jurisdictions have supplemental examinations that cover their unique amendments, and in some jurisdictions a practical hands-on examination is given in addition to the written portion. (See Chapter One of this book for information related to some of these unique requirements.)

APPROVED TESTING LABORATORIES

Underwriters Laboratories (UL) has long been the major product testing laboratory in the electrical industry. In addition to the testing, UL is a longtime developer of product standards. By producing these standards and contracting for follow-up service after undergoing a listing procedure, a manufacturer is authorized to apply the UL label or to mark their product. (See Figure 2-1, Figure 2-2, and Figure 2-3.) Underwriters Laboratories is not the only testing laboratory evaluating electrical products. In the last decade, numerous other electrical testing laboratories have arrived and are being officially recognized; for example, Electrical Testing Laboratories (ETL), Applied Research Laboratories (APL), and Canadian Standards (CSA). Some jurisdictions evaluate and approve laboratories; others accept them based on reputation. OSHA is now evaluating and approving testing laboratories. It is the responsibility of the entity responsible for specifying the materials to verify that the product has been evaluated by a testing laboratory acceptable by the AHJ where the installation is being made. It is the installer's responsibility that the product is installed in accordance with the products listing (*NEC*® 110.3(B)).

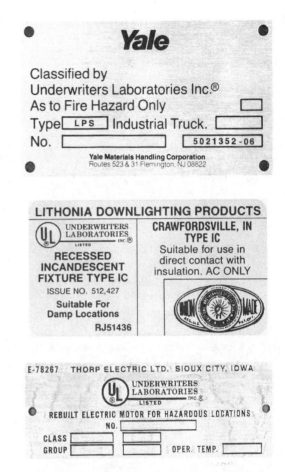

Figure 2–2(A) Examples of testing laboratories' labels as they appear on equipment. (*Reprinted with permission by Underwriters Laboratories, Inc.*®)

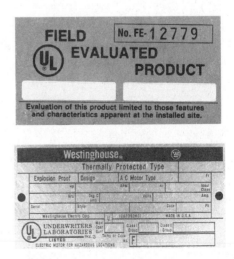

Figure 2–2(B) Examples of testing laboratories' labels as they appear on equipment. (*Reprinted with permission by Underwriters Laboratories, Inc.*®)

Figure 2–1 Common testing laboratory trademarks. (*Reprinted with permission by Underwriters Laboratories, Inc.*®)

IDENTIFICATION OF UL LISTED PRODUCTS THAT ARE ALSO UL CLASSIFIED IN ACCORDANCE WITH INTERNATIONAL PUBLICATIONS

Underwriters Laboratories Inc. (UL) provides a service for the Classification of products that not only meet the appropriate requirements of UL but also have been determined to meet appropriate requirements of the applicable international publication(s). For those products which comply with both the UL requirements and those of an international publication(s), the traditional UL Listing Mark and a UL Classification Marking, as described below, may appear on the product as a combination Listing and Classification Marking.

The combination of the UL Listing Mark and Classification Marking may appear as authorized by Underwriters Laboratories Inc.

LISTED
(Product Identity)
(Control Number)

ALSO CLASSIFIED BY UNDERWRITERS LABORATORIES INC.® IN ACCORDANCE WITH IEC PUBLICATION _____

LISTED
(Product Identity)
(Control Number)

ALSO CLASSIFIED BY UNDERWRITERS LABORATORIES INC.® IN ACCORDANCE WITH IEC PUBLICATION _____

UND. LAB. INC. ®   LISTED
(Product Identity)
(Control Number)

ALSO CLASSIFIED BY UNDERWRITERS LABORATORIES INC.® IN ACCORDANCE WITH IEC PUBLICATION _____

Underwriters   LISTED
Lab. Inc. ®   (Product Identity)
(Control Number)

ALSO CLASSIFIED BY UNDERWRITERS LABORATORIES INC.® IN ACCORDANCE WITH IEC PUBLICATION _____

(Product Identity)
CLASSIFIED BY
UNDERWRITERS LABORATORIES INC.®
IN ACCORDANCE WITH
IEC PUBLICATION _____
(Control Number)

OR

(Product Identity)
CLASSIFIED BY
UNDERWRITERS LABORATORIES INC.®
IN ACCORDANCE WITH
IEC PUBLICATION _____
(Control Number)

LOOK FOR THE CLASSIFICATION MARKING

(Product Identity)
(Control Number)

ALSO CLASSIFIED BY UNDERWRITERS LABORATORIES INC.® IN ACCORDANCE WITH IEC PUBLICATION _____

(Product Identity)
(Control Number)

ALSO CLASSIFIED BY UNDERWRITERS LABORATORIES INC.® IN ACCORDANCE WITH IEC PUBLICATION _____

UNDERWRITERS LABORATORIES INC. ®   (Product Identity)
LISTED   (Control Number)

ALSO CLASSIFIED BY UNDERWRITERS LABORATORIES INC.® IN ACCORDANCE WITH IEC PUBLICATION _____

LOOK FOR THE COMBINATION LISTING MARK AND CLASSIFICATION MARKING

IDENTIFICATION OF PRODUCTS CLASSIFIED TO INTERNATIONAL PUBLICATIONS ONLY

Underwriters Laboratories Inc. (UL) provides a service for the classification of products that have been determined to meet the appropriate requirements of the applicable international publication(s). For those products which comply with the requirements of an international publication(s), the Classification Marking may appear in various forms as authorized by Underwriters Laboratories Inc. Typical forms which may be authorized are shown below, and include one of the forms illustrated, the word "Classified", and a control number assigned by UL. The product name as indicated in this Directory under each of the product categories is generally included as part of the Classification Marking text, but may be omitted when in UL's opinion, the use of the name is superfluous and the Classification Marking is directly and permanently applied to the product by stamping, molding, ink-stamping, silk screening or similar processes.

Separable Classification Markings (not part of a name plate and in the form of decals, stickers or labels) will always include the four elements: UL's name and/or symbol, the word "Classified", the product category name, and a control number.

The complete Classification Marking will appear on the smallest unit container in which the product is packaged when the product is of such a size that the complete Classification Marking cannot be applied to the product or when the product size, shape, material or surface texture makes it impossible to apply any legible marking to the product. When the complete Classification Marking cannot be applied to the product, no reference to Underwriters Laboratories Inc. on the product is permitted.

Figure 2–3 Identification of marks listed and classified products in accordance with national and international publications. (*Reprinted with permission by Underwriters Laboratories, Inc.®*)

QUESTION REVIEW

Each lesson is designed purposely to require the student to apply the entire *NEC*® text, and not specific chapters or articles. It has been found that when studying for a timed, open-book examination, the student must gain proficiency in the Table of Contents, the Index, and the ability to move quickly from cover to cover to find the correct answer to each question in a timely fashion.

1. Is liquidtight flexible nonmetallic conduit permitted to be used in circuits in excess of 600 volts?

   Answer: _____

   _____

   _____

   _____

   Reference: _____

2. Who is the authority having jurisdiction (AHJ)?

   Answer: _____

   _____

   _____

   _____

   Reference: _____

3. Name four types of installations not covered by the *NEC*®.

   Answer: _____

   _____

   _____

   _____

   Reference: _____

4. How is explanatory information characterized in the *NEC*®?

   Answer: _____

   _____

   _____

   _____

   Reference: _____

5. What section of the *NEC*® requires equipment to be installed in accordance with manufacturer's instructions?

Answer: _____

_____

_____

_____

Reference: _____

6. When does the *NEC*® become the legal document?

Answer: _____

_____

_____

_____

Reference: _____

7. Are all installations by utilities exempt from the *NEC*®? Explain.

Answer: _____

_____

_____

_____

Reference: _____

8. What does the vertical line in the margin of the *NEC*® indicate?

Answer: _____

_____

_____

_____

Reference: _____

9. What is the latest amended edition of the *NEC*® used in any jurisdiction? When was the latest edition of the *NEC*® published?

Answer: _____

_____

_____

_____

Reference: _____

10. Who set up the first meeting that resulted in the *NEC*® and when and where was it held?

    Answer: _____

    _____

    _____

    _____

    Reference: _____

11. Can Type NM cable be used for a 120/240-volt branch circuit as temporary wiring in a building under construction where the cable is supported on insulators at intervals of not more than 10 feet (3 m)?

    Answer: _____

    _____

    _____

    _____

    Reference: _____

12. Which chapter of the *NEC*® is independent of all other chapters?

    Answer: _____

    _____

    _____

    _____

    Reference: _____

13. Is the *NEC*® considered a training manual?

    Answer: _____

    _____

    _____

    _____

    Reference: _____

14. Are all mining facilities exempt from the *NEC*®?

    Answer: _____

    _____

    _____

    _____

    Reference: _____

15. A comment often made by those involved in electrical design or installation states that the *NEC*® is a minimum standard. Where in the *NEC*® does it state that it is the true minimum permitted for electrical installations?

    Answer: _____

    _____

    _____

    _____

    Reference: _____

16. As the inspector was making an inspection and looked at the size of overcurrent device and conductors supplying a motor, the inspector asked, "Is this motor circuit rated for continuous duty?" What is a continuous load?

    Answer: _____

    _____

    _____

    _____

    Reference: _____

17. When you arrive in a nearby city to make an installation, the electrical inspector informs you that the city has not adopted the last two editions of the *NEC*®. Which edition would that city be enforcing?

    Answer: _____

    _____

    _____

    _____

    Reference: _____

18. In wiring a small residence, the owners inform you that they wanted a doorbell mounted on the front and back of the house. What class wiring would this small 24-volt doorbell circuit be wired in and what article governs that wiring method?

    Answer: _____

    _____

    _____

    _____

    Reference: _____

19. We have been asked to bid a nursing home where patients will have varied degrees of mobility. Some will come and go freely, cook their own meals, and do their own housekeeping; others may require meals to be prepared and minor medical attention, such as someone ensuring that they take their medicine regularly. Other patients may be bedridden and require oxygen or a doctor's care. Which article of the *NEC*® covers a nursing home facility of this type?

Answer: _____

_____

_____

_____

Reference: _____

20. In establishing a grounding electrode on a new installation, the owner says that although an underground metal water pipe exists, the owner prefers that it not be used and that a standard 8-foot (2.44 m) ground rod be used instead. Which section of the *NEC*® governs grounding electrodes?

Answer: _____

_____

_____

_____

Reference: _____

21. In wiring the kitchen of a residential home, the inspector informs you that the cord on the disposal is too long. Where are the requirements for the cord and attachment plug for a home disposal?

Answer: _____

_____

_____

_____

Reference: _____

22. In working in an industrial facility, you have been informed that some control circuitry will be installed to cut down on noise interference. A fiber-optic cable and fiber optics will be used. Are fiber optics covered in the *NEC*®?

Answer: _____

_____

_____

_____

Reference: _____

23. You recently have been asked to wire an irrigation pump on a nearby farm. On arriving, you find that the farmer has a 3-phase irrigation pump; however, the utility can only serve single-phase power. The farmer tells you he has heard about a phase converter that can be purchased at a relatively low cost and can convert single-phase power to 3-phase power. Which section of the *NEC®* would cover an installation of this type?

Answer: _____

_____

_____

_____

Reference: _____

24. In making an installation, you are asked to connect deicing equipment to the rain gutters on a house. Which section of the *NEC®* covers these requirements?

Answer: _____

_____

_____

_____

Reference: _____

25. In making an installation at a new golf course clubhouse, several ornamental pools are to be installed in and around the golf course with electrical lighting in the pools and around the pools. Which section of the *NEC®* covers this installation? What part of that article covers this area?

Answer: _____

_____

_____

_____

Reference: _____

Chapter Three

OBJECTIVES

After completing this chapter, you should be able to:

- Understand concepts of basic mathematics and mathematical terms
- Apply basic mathematics to electricity
- Explain the basic authority of the *NEC®*
- Perform simple calculations relating to mathematical formulae
- Perform simple calculations relating to Ohm's law

BASIC ELECTRICAL MATHEMATICS REVIEW

In this chapter, we refresh our knowledge of the basic mathematical calculations needed to perform the simple day-to-day mathematical tasks required in our work as electrical contractors or electricians. Many tasks reviewed in this chapter may appear simple. However, if you have difficulty in solving any of these problems, then it may be necessary for you to obtain a more complete study guide on basic electrical mathematics. This chapter is not only a mathematical teaching aid but also a refresher to see that your mathematical skills are sufficient to solve the problems that you will encounter in most electrical examinations.

CALCULATOR MATH

Many of us were taught and still believe that mathematical study should be done manually. But in today's world, with the introduction of handheld calculators into the industrialized world in the past 35 years, it would be somewhat silly to depend on all hand calculations. Most testing jurisdictions permit the use of handheld, silent, battery-powered, nonprogrammable calculators to use during the test. Therefore, it is advantageous to work the problems in this chapter using the same calculator that you will take to the examination. Many calculators are on the market, priced from advertising giveaways to complex scien-

tific notation calculators. However, for your purposes, any quality handheld calculator will do. It should have the standard engineering functions, but it is not necessary to have one with scientific capabilities. The button arrangement should be good quality. It should be battery-powered, not solar-powered, because the room lighting might be inadequate for solar power. You should have a memory function and most modern mathematical functions, including the basic keys, input, error correction, combining operations, calculator hierarchy (calculations with a constant), roots, powers, reciprocal, factorial, percents, natural logarithm and natural antilogarithm, trigonometric functions, and error indication, accuracy and rounding, good memory usage for storing, recalling, and memory exchange, conversion factors from English to metric, and temperature conversions. A calculator of this quality is available in most discount stores for under $20.00. Get the calculator that you plan to use and familiarize yourself with it so that you are familiar with its operations before the examination.

The Basic Keys

| + | − | × | ÷ | = |

Example: If you have $300.00 in your checking account, and you write checks for $6.00, $35.00, $53.00, and $125.25 and make a deposit for $200.00, what is your balance?

$300 − 6 − 35 − 53 − 125.25 + 200 = $280.75 balance

Example: If 250 feet of Type NM Cable costs $31.75, how much does 190 feet cost?

$31.75 ÷ 250 = .127 × 190 = $24.13 cost

Caution: Do not rely on programmable calculators to verify calculations or for preparation for the examination. They are unacceptable and are not permitted.

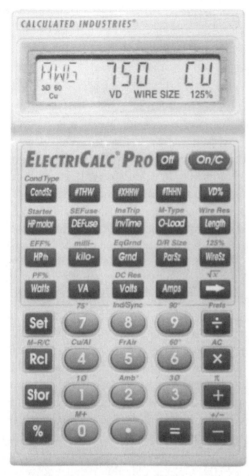

Figure 3–1 An example of a handheld calculator suitable for quick field calculation. (*Courtesy of Calculated Industries, Inc.*)

Although programmable calculators are not permitted in most testing organizations, there are good programmable handheld calculators on the market as shown in Figure 3–1 and Figure 3–2. Although these are unacceptable in an examination, they can provide a convenient tool for the contractor or electrician or those studying to improve their skills in the electrical industry.

WORKING WITH FRACTIONS

Assuming that your basic addition, subtraction, multiplication, and division skills are still sharp, we begin with a brief review of fractions. A fraction is a part of a number. For instance, if we have a pie, we have one whole pie. If we cut that pie into four pieces, then each piece is one fourth of a pie. The four pieces together equal four fourths, or one. If we cut that whole pie into five pieces, then each piece would be one fifth of the pie. The five pieces together would be one pie.

If we have two pieces of a pie that has been cut into four pieces, we would have two fourths. This fraction can be expressed in two different ways—by ²⁄₄ or ½. To reduce a fraction is to change it into another equal fraction. To do this, you divide the numerator (top number) and the denominator (bottom number) by the same number. For instance, to reduce ²⁄₆, divide the 2 and the 6 by 2. Then the numerator would be 2 divided by 2, or 1. The denominator would be 6 divided by 2, which equals 3. Therefore, the reduced fraction is ⅓; ²⁄₆ = ⅓.

Example: $\dfrac{2}{6} = \dfrac{2/2}{6/2} = \dfrac{1}{3}$

To reduce larger fractions, such as ²⁴⁄₉₆, to the lowest form, we can divide both the 24 and the 96 by 24 (to get ¼) or we can divide the 24 and the 96 by 8 (which gives ³⁄₁₂) and then divide the 3 and the 12 by 3, which gives ¼, the answer.

Example: $\dfrac{24}{96} = \dfrac{24/24}{96/24} = \dfrac{1}{4}$

Reducing Mixed Numbers to Improper Fractions

In a proper fraction, the numerator is always smaller than the denominator, such as ⅓, ⅙, ¹⁄₁₂. Mixed numbers are made of two numbers, a whole number and a fraction, for example: 2½, 5½, 3⁹⁄₁₆, numbers that we use every day in our work. To change these to improper fractions, the numerator will always be larger than the denominator, such as ¹²⁄₃ or ⁹⁄₆ or ²⁴⁄₁₀. For example, to change 2½ to an improper fraction, the 2 will be a certain number of halves (2 = ⁴⁄₂), which we then add to the fraction we are given (½). Thus, 2½ = ⁴⁄₂ + ½ = ⁵⁄₂ .

Example: $2\dfrac{1}{2} = \dfrac{2 \times 2 + 1}{2} = \dfrac{5}{2}$

To change a mixed number to an improper fraction, multiply the denominator of the fraction by the whole number and add the numerator of the fraction. Place this answer over the denominator to make an improper fraction.

Example: $5\dfrac{1}{2} = \dfrac{2 \times 5 + 1}{2} = \dfrac{11}{2}$

Figure 3–2 Examples of some of the types of multifunction tasks that can be quickly solved. (*Courtesy of Calculated Industries, Inc.*)

Changing Improper Fractions to Mixed Numbers

To change an improper fraction, such as $\frac{13}{3}$, to a mixed number, divide the numerator by the denominator (13 divided by 3). Any remainder is placed over the denominator (13 divided by 3 = 4 with a remainder of 1). The resulting whole number and proper fraction form a mixed number. Thus, $\frac{13}{3} = 4 + \frac{1}{3} = 4\frac{1}{3}$.

Example: $\dfrac{13}{3} = 4 + \dfrac{1}{3} = 4\dfrac{1}{3}$

Multiplication of Fractions

To multiply fractions, place the multiplication of the numerators over the multiplication of the denominators and reduce to lowest terms. For example: ½ × ¾. Answer: ½ × ¾ means that $1 \times 3 = 3$ and $2 \times 4 = 8$; thus, the answer is ⅜. Example: ½ × 5. Answer: ½ × 5 means that $1 \times 5 = 5$ and $2 \times 1 = 2$, giving an improper fraction of 5⁄2, which then has to be reduced to 2½.

Examples: $\dfrac{1}{2} \times \dfrac{3}{4} = \dfrac{1 \times 3}{2 \times 4} = \dfrac{3}{8}$

$$\dfrac{1}{2} \times 5 = \dfrac{1 \times 5}{2 \times 1} = \dfrac{5}{2} = 2\dfrac{1}{2}$$

To cancel the numerator and denominator means to divide both numerator and denominator by the same number. As an example, when multiplying ⅜ × ⁴⁄9, notice that the 3 on top and the 9 on the bottom can both be divided by 3. Cross out the 3 and write a 1 over it. Cross out the 9 and write a 3 under it. Also, the 4 on the top and the 8 on the bottom can both be divided by 4. Cross out the 4 and write a 1 over it. Cross out the 8 and write a 2 under it. The solution then is 3 reduced to 1×4 reduced to 1 (1×1) over 8 reduced to 2×9 reduced to 3 (2×3), thus giving ⅙.

Example:

$$\dfrac{3}{8} \times \dfrac{4}{9} = \dfrac{\overset{1}{\cancel{3}} \times \overset{1}{\cancel{4}}}{\underset{2}{\cancel{8}} \times \underset{3}{\cancel{9}}} = \dfrac{1}{6}$$

WORKING WITH DECIMALS

A decimal is a fraction in which the denominator is not written. The denominator is 1. Many answers to electrical problems will be fractions, such as ¼ of an ampere, ½ of an ampere, ½ of a volt, ⅓ of a volt, and the like. These will be correct mathematical answers, but they will be worthless to an electrician. The electrical measuring instruments give values expressed in decimals, not in fractions. In addition, the manufacturers of components give the values of parts in terms of decimals, such as 3.4 amperes, and the like, on the nameplate of a motor. Suppose we worked out a problem and found that the current in the circuit should be ⅛ of an ampere, and using the ampmeter, we test the circuit and find that .125 ampere flows. Is our circuit correct? How will we know? How can we compare

⅛ to .125? The easiest way is to change the fraction ⅛ to its equivalent decimal and then compare the decimals. To change a fraction into a decimal, divide the numerator by the denominator, (1 divided by 8 = .125).

Example: $\quad \dfrac{1}{8} = \dfrac{1}{8.000} = .125$

Converting Fractions to Decimals
Divide the numerator (top number) by the denominator (bottom number).

$$\begin{aligned} \tfrac{1}{2} &= 1 \div 2 = 0.5 \\ \tfrac{3}{4} &= 3 \div 4 = 0.75 \\ \tfrac{7}{8} &= 7 \div 8 = 0.875 \end{aligned}$$

$$9\tfrac{3}{8} = \dfrac{9 \times 8 + 3}{8} = \dfrac{75}{8} = 75 \div 8 = 9.375$$

Converting Percentages to Decimals
Move decimal two places to the left.

$$\begin{aligned} 100\% \ &= 1.00 \\ 50\% \ &= 0.50 \\ 37\tfrac{1}{2}\% &= 0.375 \end{aligned}$$

WORKING WITH PERCENTAGES

To apply the rules of the *NEC®*, you will need to know how to work with percentages—for example, **430.110 (A) General.** *The disconnecting means for motor circuits rated 600 volts, nominal, or less shall have an ampere rating of at least 115 percent of the full-load current rating of the motor.* [Reprinted with permission from NFPA 70-2005.]

Question: A 480-volt motor has a full-load current of 28 amperes. What size disconnecting means is required?

   A. 20 amperes
   B. 25 amperes
   C. 30 amperes
   D. 40 amperes

Because the motor load is 28 amperes, you can quickly eliminate answers A and B.

To solve: 28 × 115% = 28 × 1.15 = 32.2 = 40 amperes disconnecting means required.

Question: A custom home (single-family dwelling) has branch-circuit general illumination load of 200,000 VA (volt-amperes). What demand factor must be applied to this general illumination load?

- A. 200,000
- B. 43,950
- C. 105,000
- D. 63,950

Answer: D

To solve:

First 3000 at 100% = 3000 × 1   = 3000 VA

3001 to 120,000 = 117,000 at 35%
    = 117,000 × .35   = 40,950 VA

Over 120,000 = 80,000 at 25%
    = 80,000 × .25   = 20,000 VA

Total 200,000 VA lighting load demand   63,950 VA

220.42 General Lighting. *The demand factors specified in Table 220.42 shall apply to that portion of the total branch-circuit load calculated for general illumination. They shall not be applied in determining the number of branch circuits for general illumination.* [Reprinted with permission from NFPA 70-2005.]

WORKING WITH SQUARE ROOT

Roots are the opposite of powers. A square root is the opposite of the number to the second power. The sign of the square root is $\sqrt{}$. To find a square root, find a number that when multiplied by itself is the number inside the square root sign. For example, to find the value of the square root of 36, ask yourself what number times itself is 36. The answer is 6, so 6 is the square root of 36.

If you can find the square root of a number with the method that uses averages, suppose that you did not know that the square root of 144 is 12. When you divide the number by its square root, the answer is the square root. If you cannot find this answer, then guess as close as possible. A good guess for 144 would be 10, because 10 × 10 = 100, which is close to 144. Divide 144 by 10 and the answer is 14 + a remainder. Average the guess, 10, and the answer to the division problem, 14. 10 + 14 = 24, ²⁴⁄₂ = 12, which is the correct answer.

Follow these steps to find the square root of the larger number: Guess the answer; divide the guess into the large number; average the guess and the answer to the division problem; and check. For example, find the value of the square root of 1024.

Table 220.42 Lighting Load Demand Factors

Type of Occupancy	Portion of Lighting Load to Which Demand Factor Applies (Volt-Amperes)	Demand Factor (Percent)
Dwelling units	First 3000 or less at	100
	From 3001 to 120,000 at	35
	Remainder over 120,000 at	25
Hospitals*	First 50,000 or less at	40
	Remainder over 50,000 at	20
Hotels and motels, including apartment houses without provision for cooking by tenants*	First 20,000 or less at	50
	From 20,001 to 100,000 at	40
	Remainder over 100,000 at	30
Warehouses (storage)	First 12,500 or less at	100
	Remainder over 12,500 at	50
All others	Total volt-amperes	100

*The demand factors of this table shall not apply to the calculated load of feeders or services supplying areas in hospitals, hotels, and motels where the entire lighting is likely to be used at one time, as in operating rooms, ballrooms, or dining rooms.

Reprinted with permission from NFPA 70-2005.

Step 1. Guess: in the list of square roots, 30 × 30 = 900. This is too small, but it is easy to divide by.

Step 2. Divide 1024 by 30. The answer is 34, with a remainder.

Step 3. Find the average of 30 and 34. 30 + 34 = 64; ⁶⁴⁄₂ = 32.

Step 4. Check: multiply 32 × 32. The answer is 1024. Thus, 32 is the square root of 1024. When you use this method to find square roots, always guess a number that ends in 0. It is easier and faster to divide by these numbers. If the average is not the square root of the number, use the average as a new guess and try again.

Finding the Reciprocal of a Number

There are instances in the *NEC®* where it is easier to find the reciprocal of a number and then make the calculation, rather than use the number in the

(continued)

calculation. For example, the reciprocal of 125% is 80%, and the reciprocal of 1.73 is .578, which may be rounded off to .58 for quick calculations.

To find a reciprocal of a number, divide the number into 1.

$$1 \div 1.73 = .5780 \text{ or } .58$$

If you are finding the reciprocal of a percentage, you must first convert the percent to a decimal.

$$125\% = 1 \div 1.25 = 0.8 \text{ or } 80\%$$

(E) Single Non–motor-Operated Appliance. *If the branch-circuit supplies a single non–motor-operated appliance, the rating of overcurrent protection shall*
(1) Not exceed that marked on the appliance;
(2) Not exceed 20 amperes if the overcurrent protection rating is not marked and the appliance is rated 13.3 amperes or less; or
(3) Not exceed 150 percent of the appliance rated current if the overcurrent protection rating is not marked and the appliance is rated over 13.3 amperes. Where 150 percent of the appliance rating does not correspond to a standard overcurrent device ampere rating, the next higher standard rating shall be permitted. [Reprinted with permission from NFPA 70-2005.]

Question: A quick recovery electrical water heater is rated at 4500 watts 240 volts. Is a 30-ampere circuit breaker permitted to protect this appliance?

Answer: Yes, 4500 ÷ 240 = 18.75 × 150% = 28.1 = the next higher standard circuit breaker rating is 30 amperes.

You may first think a 20-ampere circuit breaker would be acceptable for a 18.75-ampere load. However, 422.10(A) requires the branch-circuit rating to be at least 125% of the marked rating. A 20-ampere circuit breaker is permitted to protect the reciprocal of 125% or 80% of the circuit breaker rating. Thus 80% × 20 = 16 amperes.

A 25-ampere circuit breaker would be acceptable. However, a 25-ampere circuit breaker is not a common size and may not be readily available.

POWERS

A power is a product of a number multiplied by itself one or more times. For an example, what is the value of 3^2? The exponent is 2; write 3 two times, or $3^2 = 3 \times 3$; $3^2 = 9$. What is the value of 5^3? The exponent is 3;

write 5 three times; in other words, $5^3 = 5 \times 5 \times 5 = 125$. These powers are terms often used in the electrical field and should be known.

OHM'S LAW

E = voltage; I = current (amperes); R = resistance

$$E = IR; \quad I = \frac{E}{R}; \quad R = \frac{E}{I}$$

As an example, all numbers are relative. For example, if E = 12, I = 3, and R = 4, then 3 × 4 = 12, $\frac{12}{3} = 4$, and $\frac{12}{4} = 3$. All numbers are relative. Remember, in any electrical circuit, the voltage force is the current through the conductor against its resistance. The resistance tries to stop the current from flowing. The current that flows in the circuit depends on the voltage and the resistance. The relationship among these three quantities is described by Ohm's law. Ohm's law applies to an entire circuit or any component part of a circuit.

Voltage Drop Calculations

We have discussed the methods for finding the resistance by using Ohm's law. Now we can find the voltage drop for circuit loads using this same formula.

$$E = I \times R$$

Voltage Drop Formula

Single-phase voltage drop = amperes × length × resistance × (2)

Voltage drop % = voltage drop × 100/volts

Three-phase voltage drop—calculate as a single-phase circuit and multiply resultant by .866.

$$V_{(D)} = I \times 2L \times R$$

I = Amperes
L = Length of circuit
R = Resistance values from *NEC®* Chapter 9, Table 8
Resistance = R × [1 + a (temperature − 75)]
a = 0.00323 for copper, 0.000330 for aluminum
Temperature = ambient temperature

Voltage drop can be calculated with the known elements of the power circuit. The resistance of conductors commonly used for electrical power conductors can be found in *NEC®* Chapter 9, Table 8.

P = watts
I = amps
R = ohms
E = volts

$$I = \sqrt{\frac{P}{R}}$$

$$I = \frac{P}{E}$$

$$I = \frac{E}{R}$$

In this chart, the values in the inner circle are equal to the values in the corresponding segments of the outer circle.

Example: What is the voltage drop for a 120-volt, 20-ampere, single-phase circuit supplying a 12-ampere load located 125 feet (37.5 m) away from the source of supply? The circuit conductors have a combined resistance of 0.35 ohm.

    A. 21 volts
    B. 42 volts
    C. 2.1 volts
    D. 4.2 volts

Answer: D

What is the question? What is the voltage drop? **Ohm's law works!**

$E = IR$ $I = 12$ amperes $R = 0.35$ ohm
$E = 12 \times 0.35$
$E = 4.2$ volts

Note that this is a 3½ percent voltage drop. 210.19 FPN No. 4 recommends that branch circuits do not exceed a 3 percent drop. For maximum efficiency, the circuit conductors should be increased in size.

NEC® 2005 210.19(A)(1) FPN No. 4: *Conductors for branch circuits as defined in Article 100, sized to prevent a voltage drop exceeding 3 percent at the*

*farthest outlet of power, heating, and lighting loads, or combinations of such loads, and where the maximum total voltage drop on both feeders and branch circuits to the farthest outlet does not exceed 5 percent, provides reasonable efficiency of operation. See FPN No. 2 of 215.2(A)(3) for voltage drop on feeder conductors.**

The *NEC®* references prescribe voltage drop values or percentages for feeder conductors and branch-circuit conductors; however, it should be noted these appear in a fine print note (FPN) and are only suggested and not mandatory requirements. The *NEC®* 215.2(A)(3) FPN 2 recommends voltage drop does not exceed 3 percent of the farthest outlet of power, heating, or lighting loads or combinations of such loads, and the maximum voltage drop for both feeders and branch circuits to the farthest outlet is not to exceed 5 percent and will provide reasonable efficiency of operation.

Example: A dc circuit has 2 amperes of current flowing in it. A voltmeter reads 10 volts line to line. How much resistance is in the circuit? Answer: 5 ohms.

* Reprinted with permission from NFPA 70-2005.

$$R = \frac{E}{I} \quad R = \frac{10}{2} \quad R = 5 \text{ ohms}$$

Example: If the voltage is 100 volts and the resistance is 25 ohms, what amount of current will flow in the circuit? Answer: 4 amperes.

$$I = \frac{E}{R} \quad I = \frac{100}{25} \quad I = 4 \text{ amperes}$$

Example: If the potential across a circuit is 120 volts and the current is 6 amperes, what is the resistance? Answer: 20 ohms.

$$R = \frac{E}{I} \quad R = \frac{120}{6} \quad R = 20 \text{ ohms}$$

A series circuit may be defined as a circuit in which the resistive elements are connected in a continuous run (end to end). It is evident that because the cir-

cuit has no branches, the same current flows in each resistance. The total potential across the circuit equals the sum of potential drops across each resistance.

Example:
$$E_1 = IR_1$$
$$E_2 = IR_2$$
$$E_3 = IR_3$$
$$E = E_1 + E_2 + E_3$$
$$R = R_1 + R_2 + R_3$$

The total potential of the circuit is:

$$E = 1R_1 + 1R_2 + 1R_3 = I(R_1 + R_2 + R_3)$$

$$I = \frac{E}{R_1 + R_2 + R_3}$$

$R_1 = 10$ ohms, $R_2 = 5$ ohms, and $R_3 = 15$ ohms.

What amount of voltage must flow to force 0.5 amperes through the circuit?

$$R_T = 5 + 10 + 15 = 30 \text{ ohms}$$

Hence, $E_T = 0.5 \times 30 = 15$ volts

What is the voltage drop across each resistance?

$$E_1 = 0.5 \times 10 = 5.0 \text{ volts}$$
$$E_2 = 0.5 \times 5 = 2.5 \text{ volts}$$
$$E_3 = 0.5 \times 15 = \underline{7.5 \text{ volts}}$$
$$E_T = 15 \text{ volts}$$

When the resistance and current are known, what is the formula to determine the voltage?

The current is 2 amperes, the resistance is 15 ohms, 10 ohms, and 30 ohms.

$$E = IR \qquad E = 2(15 + 10 + 30)$$
$$E = 2 \times 55 \qquad E = 110 \text{ volts}$$

In the industry today most circuits are connected in parallel. In a parallel circuit the voltage is the same across each resistance (load).

The total current is equal to the sum of all the currents.

$$E = I_1R_1 = I_2R_2 = I_3R_3 = I_4R_4$$

and

$$I = I_1 + I_2 + I_3$$

Individual Resistances =

$$I_1 = \frac{E}{R_1} \quad I_2 = \frac{E}{R_2} \quad I_3 = \frac{E}{R_3} \quad I_4 = \frac{E}{R_4}$$

Hence: $I_1 = \dfrac{E}{R_1} + I_2 = \dfrac{E}{R_2} + I_3 = \dfrac{E}{R_3} + I_4 = \dfrac{E}{R_4}$

or $I = E\left(\dfrac{1}{R_1} + \dfrac{1}{R_2} + \dfrac{1}{R_3} + \dfrac{1}{R_4}\right)$

and several resistances in parallel is

$$\frac{1}{R} = \frac{1}{R_1} + \frac{1}{R_2} + \frac{1}{R_3} + \frac{1}{R_4}$$

The sum of the resistances in parallel always will be smaller than the smallest resistance in the circuit where only two resistances are in parallel.

$$R = \frac{R_1 \times R_2}{R_1 + R_2} \text{ ohms}$$

Power Factor

Power factor is a phase displacement of current and voltage in an ac circuit. The cosine of the phase angle of displacement is the power factor. The cosine is multiplied by 100 and is expressed as a percentage. A cosine of 90° is 0; therefore, the power factor is 0 percent. If the angle of displacement were 60°, the cosine of which is .5, the power factor would be 50 percent. This is true whether the current leads or lags the voltage. Power is expressed in dc circuits and ac circuits that are purely resistive in nature. Where these circuits contain only resistance, P (watts) = E × I. In ac circuits that contain inductive or capacity reactances, VA (volt-ampere) = E × I. In a 60-cycle ac circuit, if the voltage is 120 volts, the current is 12 amperes, and the current lags the voltage by 60°, find (a) the power factor, (b) the power and voltage-amperes (VA), and (c) the power in watts. The cosine of 60 is .5; therefore, the power factor is 50 percent. 120 × 12 = 1440 volt-amperes, which is called the apparent power. 120 × 12 × .5 = 720 watts, which is called the true power. Power factor is important. There is an apparent power of 1440 VA and a true power of 720 watts. There also are 12 amperes of line current and 6 amperes of in-phase or effective current. This means that all equipment from the source of supply to the power consumption device must be capable of handling a current of 12 amperes, when actually the device is only using the current of 6 amperes. A 50 percent power factor was used intentionally to make the results more pronounced. The I²R loss is based on the 12-ampere current, while only 6 amperes are effective. Power factor can be measured by a combined use of a voltmeter, ammeter, and watt-meter, or by the use of a power factor meter. When using the three meters, volt-ampere meter and watt-meter, all connected properly in the circuit, the readings of the three meters are taken simultaneously under the same load conditions and calculated as follows: power factor = true power (watts)/apparent power (which is volt-amperes). Power factor = W/EI.

```
1000 watts = 1 KW
1000 watts operating for 1 hour = 1 KWH used
3413 BTU = 1 KW
3.413 BTU = 1 watt
Fahrenheit = (9/5 × C) + 32
Celsius = 5/9 × (F − 32)
P. F. = KW / KVA
KVA = KW / P. F.
```

PRACTICE EXAM

1. 0.875 is the equivalent of which of the following?
   A. 9/32
   B. 15/16
   C. 7/8
   D. 5/8

2. How much current will a 240-volt circuit with 20-ohm load draw?
   A. 24 amperes
   B. 2.4 amperes
   C. 1.2 amperes
   D. 12 amperes

3. 4.75 is the equivalent of which of the following improper fractions?
   A. 23/4
   B. 17/5
   C. 15/3
   D. 38/8

4. A 10 kW load is increased by 12 percent every summer. What is the VA in the summer?
   A. 11.2
   B. 1120
   C. 11,200
   D. 112,000

5. What is the circumference of a circle with a 1-inch diameter?
   A. 1 inch
   B. 1.2 inches
   C. 3.14 inches
   D. 2½ inches

6. What is the radius of a circle with a 1-inch diameter?
   A. ½ inch
   B. 2½ inches
   C. 1 inch
   D. 1½ inches

7. What is the resistance of a 20-ampere, 120-volt circuit?
   A. 5 ohms
   B. .5 ohm
   C. .6 ohm
   D. 6 ohms

8. One meter equals how many feet?
   A. 32 inches
   B. 3.28 feet
   C. 2.88 feet
   D. 1½ feet

9. One meter equals how many millimeters?
   A. 100
   B. 10
   C. 1000
   D. 10,000

10. According to the *NEC®*, 10 feet converted to metric (soft conversion) equals
    A. 3 m.
    B. 3.05 m.
    C. 30 m.
    D. 300 m.

11. According to the *NEC®*, 3 feet converted to metric (hard conversion) equals
    A. .3 m.
    B. 300 mm.
    C. 900 mm.
    D. 9 m.

12. How many amperes does a 3-phase, 480-volt, 24-kW load draw? Round off to a whole number.
    A. 29
    B. 50
    C. 290
    D. 5

13. ¾ can be also shown as what percent?
    A. .75%
    B. 75%
    C. ¾%
    D. none of the above

14. Unless specifically permitted in the *NEC®*, 12 AWG conductors, Type THWN overcurrent protection shall not exceed
    A. 16 amperes.
    B. 25 amperes.
    C. 20 amperes.
    D. 30 amperes.

15. The reciprocal of 125 percent is _____.
    A. 0.8
    B. .58
    C. 1.73
    D. 1.25

QUESTION REVIEW

Directions: Reduce the following fractions to their lowest terms.

1. $3/6$ = _____

2. $4/16$ = _____

3. $4/40$ = _____

4. $40/100$ = _____

5. $6/8$ = _____

6. $15/20$ = _____

7. $18/36$ = _____

8. $9/15$ = _____

9. $30/50$ = _____

10. $4/20$ = _____

11. $5/20$ = _____

12. $6/24$ = _____

Directions: Change the following mixed numbers to improper fractions.

1. $1\frac{1}{2}$ = _____

2. $3\frac{1}{4}$ = _____

3. $5\frac{3}{8}$ = _____

4. $4\frac{1}{4}$ = _____

5. $7\frac{1}{8}$ = _____

6. $2\frac{1}{6}$ = _____

7. $5\frac{3}{4}$ = _____

8. $8\frac{1}{3}$ = _____

9. $6\frac{4}{7}$ = _____

10. $3\frac{5}{8}$ = _____

Directions: Change the following improper fractions to mixed numbers.

1. $\frac{9}{2}$ = _____

2. $\frac{12}{5}$ = _____

3. $\frac{64}{5}$ = _____

4. $\frac{26}{8}$ = _____

5. $\frac{29}{3}$ = _____

6. $\frac{31}{2}$ = _____

7. $\frac{5}{3}$ = _____

8. $\frac{13}{3}$ = _____

Directions: Multiply the following whole numbers and fractions. Give the answer in proper reduced form.

1. $8 \times \frac{1}{2}$ = _____

2. $\frac{1}{2} \times \frac{3}{4}$ = _____

3. $\frac{1}{5} \times \frac{1}{8}$ = _____

4. $9 \times \frac{1}{2}$ = _____

5. $\frac{3}{4} \times 7$ = _____

6. $\frac{1}{2} \times 5$ = _____

7. $\frac{3}{4} \times 12$ = _____

8. $\frac{1}{7} \times 15$ = _____

Directions: Convert the following fractions to decimal equivalents.

1. $\frac{1}{4}$ = _____

2. $\frac{5}{8}$ = _____

3. $\frac{3}{4}$ = _____

4. $\frac{13}{15}$ = _____

5. $\frac{3}{8}$ = _____

Directions: Resolve to whole numbers.

1. 2^4 = _____

2. 9^2 = _____

3. 50^2 = _____

4. 1^5 = _____

5. 10^3 = _____

Directions: Find the square root of the following numbers.

1. Square root of 25   = _____

2. Square root of 81   = _____

3. Square root of 49   = _____

4. Square root of 3136   = _____

5. Square root of 10,201 = _____

6. Square root of 841   = _____

7. Square root of 6064   = _____

Directions: Solve the following problems.

1. A 20-ampere load is fed with two conductors that have a *combined* resistance of 0.3 ohm. If the source voltage is 120 volts dc, the calculated voltage drop on this circuit is _____ percent.

Answer: _____

_____

_____

Reference: _____

2. Refer to the above drawing.

    I(LINE A) = I(LINE B) = I(LINE C) = 10 amperes.

The power factor in the circuit in the figure above is _____ .

Answer: _____

_____

_____

_____

Reference: _____

3. A 120-volt branch circuit has only six 100-watt, 120-volt incandescent luminaires (fixtures) connected to it. The current in the home run of this circuit is _____ amperes.

Answer: _____

_____

_____

_____

Reference: _____

4. Refer to the above drawing.

    With a resistance at R_1 of 4 ohms and a resistance at R_2 of 2 ohms, the total resistance for the above circuit is _____ ohms.

Answer: _____

_____

_____

_____

Reference: _____

5. A 1000-watt, 120-volt lamp uses electrical energy at the same rate as a/an _____ ohm resistor.

Answer: _____

_____

_____

_____

Reference: _____

6. In the above diagram, eight equal resistors are connected in series to a 48-volt source. The voltage drop across each resistor is _____ volts.

Answer: _____

_____

_____

_____

Reference: _____

The above diagram applies to Questions 7 through 14.

7. What is the resistance of R_1 and R_2?

Answer: _____

_____

_____

_____

Reference: _____

8. What is the resistance of $R_3 + R_1$ and R_2?

    Answer: _____

    _____

    _____

    _____

    Reference: _____

9. What is the resistance of R_4 and R_5?

    Answer: _____

    _____

    _____

    _____

    Reference: _____

10. What is the total resistance in the circuit?

    Answer: _____

    _____

    _____

    _____

    Reference: _____

11. What is the voltage drop at R_1 and R_2?

    Answer: _____

    _____

    _____

    _____

    Reference: _____

12. What is the voltage drop at R_3?

    Answer: _____

    _____

    _____

    _____

    Reference: _____

13. What is the voltage drop at R_4 and R_5?

    Answer: _____

    _____

    _____

    _____

    Reference: _____

14. What is the total current in this circuit?

    Answer: _____

    _____

    _____

    _____

    Reference: _____

Chapter Four

OBJECTIVES

Studying this chapter along with the 2005 *NEC®* will give the reader a basic introduction to the Code and an understanding of how to start using the 2005 *NEC®* efficiently.

After completing this chapter, you should know:

*   The unique style the *NEC®* is written in
*   The types of installations covered and exempted
*   The application of Chapters 1–4 generally
*   The application of Chapters 1–5 to supplement and modify the general requirements
*   The application of Chapter 8 independently
*   The application of the tables in Chapter 9
*   The Annex for information and examples
*   The importance of thoroughly understanding contents of
    1. Article 90—Introduction
    2. Article 100, definitions of words and terms unique to the *NEC®*
    3. Article 110, general requirements that apply throughout
    4. Article 300, general wiring applications that apply to all wiring methods

INTRODUCTION TO THE *NATIONAL ELECTRICAL CODE®* (*NEC®*)

The *National Electrical Code®* is a reasonable document, developed and written in a reasonable fashion. It is a design manual, but it is not a design specification. It is not the minimum, but there are many conveniences that we take for granted that are not required by this document. There are also many safety considerations that are not required by this text, such as smoke detectors and other safety and detection devices that are being developed every day. See NFPA 72 the National Fire Alarm Code for requirements related to smoke detectors and fire alarm equipment.

The first page in Article 90, 90.1(C), says it is not an instruction manual for untrained persons. You are going to find that that statement explains much of the confusion about this document. Most engineering schools, vocational schools, and apprentice programs do not have training programs on the proper use of this document as an installation and design standard. Open your 2005 *NEC®* and look up *NEC®* 90.1(C). You will need the 2005 *National Electrical Code®* (*NEC®*) throughout your study of this manual. There are several *NEC®* handbooks available. The NFPA *National Electrical Code® Handbook* contains the exact *NEC®* text and commentary. If you are using that text, remember that the commentary is only the author's opinion and may not be totally correct. Therefore, when studying in a classroom or for an examination, the *NEC®* text will be that from which questions are developed.

KNOW YOUR *CODE* BOOK

To use the *NEC®*, we must first have a thorough understanding of

*   the Table of Contents
*   the Index
*   Article 90—Introduction
*   Article 100—Definitions
*   Article 110—Requirements for Electrical Installations
*   Article 300—Wiring Methods

We will refer to the rest of the *Code* on an as-needed basis; but these are general articles, and we use them continually as applicable.

You must study these parts of the *Code* thoroughly several times until you have a complete understanding of the content in these articles if you expect to successfully pass your examination.

Introduction

Let us turn to Article 90. We often pass over the Introduction of many books that we read or study. The introduction is an important part of any book. It is the application guidelines. Did you read the Introduction to this manual? I repeat! Read Article 90 several times. It is short, only about four pages, but it is a very important part of the document, especially for someone about to embark on the study of the *National Electrical Code®*.

Article 90 covers the foundation requirements of the *NEC®*, the purpose of which is the safeguarding of people and property from the hazards that can occur while using electricity. The *Code* contains rules that, when complied with, along with proper maintenance, will result in an installation that is essentially free from hazards. But it may not be efficient, convenient, or adequate for good service. The requirements are minimal and generally do not provide capacity for future expansion of the electrical system.

> *90.1 FPN: Hazards often occur because of overloading of wiring systems by methods or usage not in conformity with this Code. This occurs because initial wiring did not provide for increases in the use of electricity. An initial adequate installation and reasonable provisions for system changes provide for future increases in the use of electricity.*
>
> Reprinted with permission from NFPA 70-2005.

The *NEC®* is not a design specification, but it is a design manual. For example, specific manufacturers' products are not specified; however, no responsible designer would begin to design an electrical system without considering minimal *Code* requirements such as load calculations, service size and location, or grounding.

> The *NEC®* is not an instruction manual for untrained persons.

Contained in the Introduction is the scope, and it is imperative that the user know the bounds of the *NEC®* (90.2(A)).

> **90.2 Scope. (A) Covered.** *This* Code *covers the installation of electrical conductors, equipment, and raceways; signaling and communications conductors, cables, equipment, and raceways; and optical fiber cables and raceways for the following:*
>
> (1) *Public and private premises, including buildings, structures, mobile homes, recreational vehicles, and floating buildings*
> (2) *Yards, lots, parking lots, carnivals, and industrial substations*

> *FPN to (2): For additional information concerning such installations in an industrial or multibuilding complex, see ANSI C2-2002, National Electrical Safety Code.*
>
> (3) *Installations of conductors and equipment that connect to the supply of electricity*
> (4) *Installations used by the electric utility, such as office buildings, warehouses, garages, machine shops, and recreational buildings, that are not an integral part of a generating plant, substation, or control center.*
>
> Reprinted with permission from NFPA 70-2005.

This *Code* does not cover (90.2(B)):

> **(B) Not Covered.** *This* Code *does not cover the following:*
>
> (1) *Installations in ships, watercraft other than floating buildings, railway rolling stock, aircraft, or automotive vehicles other than mobile homes and recreational vehicles*
>
> *FPN: Although the scope of this* Code *indicates that the* Code *does not cover installations in ships, portions of this* Code *are incorporated by reference into Title 46,* Code of Federal Regulations, *Parts 110–113.*
>
> (2) *Installations underground in mines and self-propelled mobile surface mining machinery and its attendant electrical trailing cable*
> (3) *Installations of railways for generation, transformation, transmission, or distribution of power used exclusively for operation of rolling stock or installations used exclusively for signaling and communications purposes*
> (4) *Installations of communications equipment under the exclusive control of communications utilities located outdoors or in building spaces used exclusively for such installations*
> (5) *Installations under the exclusive control of an electric utility where such installations*
>   a. *Consist of service drops or service laterals, and associated metering, or*
>   b. *Are located in legally established easements, rights-of-way, or by other agreements either designated by or recognized by public service commissions, utility commissions, or other regulatory agencies having jurisdiction for such installations, or*
>   c. *Are on property owned or leased by the electric utility for the purpose of communications, metering, generation, control,*

transformation, transmission, or distribution of electric energy.

FPN to (4) and (5): Examples of utilities may include those entities that are typically designated or recognized by governmental law or regulation by public service/utility commissions and that install, operate, and maintain electric supply (such as generation, transmission, or distribution systems) or communication systems (such as telephone, CATV, Internet, satellite, or data services). Utilities may be subject to compliance with codes and standards covering their regulated activities as adopted under governmental law or regulation. Additional information can be found through consultation with the appropriate governmental bodies, such as state regulatory commissions, Federal Energy Regulatory Commission, and Federal Communications Commission.

Reprinted with permission from NFPA 70-2005.

90.3 Code Arrangement. *This* Code *is divided into the introduction and nine chapters, as shown in* [the figure below]. *Chapters 1, 2, 3, and 4 apply generally; Chapters 5, 6, and 7 apply to special occupancies, special equipment, or other special conditions. These latter chapters supplement or modify the general rules. Chapters 1 through 4 apply except as amended by Chapters 5, 6, and 7 for the particular conditions.*

Chapter 8 covers communications systems and is not subject to the requirements of Chapters 1 through 7 except where the requirements are specifically referenced in Chapter 8.

Chapter 9 consists of tables.

Annexes are not part of the requirements of this Code, *but are included for informational purposes only.*

Reprinted with permission from NFPA 70-2005.

Included in 90.3 is the arrangement of the *Code* and hierarchy. The *Code* is divided into an Introduction, nine chapters, and annexes A, B, and C.

- Chapters 1, 2, 3, and 4 apply generally and are applicable to all installations—branch circuitry, motors, appliances, lighting.
- Chapters 5, 6, and 7 apply to special occupancies—places of assembly, hazardous locations, agricultural buildings, mobile homes, health care facilities; special equipment—swimming pools, signs, elevators, welders; or other special conditions—emergency systems, fire alarms, power-limited circuits. Although the first four still apply, these chapters augment or modify the general rules for the particular conditions.

Example: Article 514 covers the hazardous requirements for wiring a service station. It does not cover other areas within that service station facility. It does not cover other electrical needs, such as service, the grounding, the branch-circuits, or the feeders, and more importantly it does not cover other environmental concerns, such as corrosion or wet locations. The designer must take all these concerns and any others that apply and design the installation so that it will provide good service, have adequate capacity, and provide all the safety needed for the normal operation throughout the life of the facility.

- Chapter 8 is a stand-alone chapter and covers communications systems; it is independent of

NEC® Code Arrangement

Applies generally to all electrical installations	Chapter 1—General Chapter 2—Wiring and Protection Chapter 3—Wiring Methods and Materials Chapter 4—Equipment for General Use
Supplements or modifies Chapters 1 through 4	Chapter 5—Special Occupancies Chapter 6—Special Equipment Chapter 7—Special Conditions
Chapter 8 is not subject to the requirements of Chapters 1 through 7 except where the requirements are specifically referenced in Chapter 8	Chapter 8—Communications Systems
Chapter 9 as referenced in Chapters 1 through 8	Chapter 9 consists of tables.
Informational only; not mandatory	Annex A through Annex G

the other chapters except where they are specifically referenced.

- Chapter 9 consists of tables and examples.

- Annex A lists extract documents; see 90.3 for an explanation.

- Annex B is for information only and contains formula application information.

- Annex C contains optional wire fill tables.

The Introduction lets us know that the *Code* is intended as suitable for mandatory application by those adopting it—that is, governmental bodies exercising legal jurisdiction over electrical installations and for use by insurance inspectors. It also places the responsibility for enforcement on the authority having jurisdiction (AHJ).

NEC® 90.5 gives the style of the document and covers the mandatory rules, which are characterized by the word "**shall**" and explains that explanatory material appears in the form of fine print notes, which are characterized by the designation "**FPN**."

90.6 Formal Interpretations. *To promote uniformity of interpretation and application of the provisions of this* Code, *formal interpretation procedures have been established and are found in the NFPA Regulations Governing Committee Projects.*

Reprinted with permission from NFPA 70-2005.

Warning: The commentary found in the NFPA *National Electrical Code Handbook* is **not** a formal interpretation of the *NEC*®.

NEC® 90.7 states that factory-installed internal wiring or the construction of equipment need not be inspected at the time of installation of the equipment, except to detect alterations or damage, if the equipment has been listed by a qualified electrical testing laboratory that is recognized and that requires suitability for installation in accordance with this *Code*.

See Examination, Identification, Installation, and Use of Equipment, 110.3.
See definition of "Listed," Article 100.

90.9(A) Units of Measurement

For the purpose of this Code, *metric units of measurement are in accordance with the modernized metric system known as the International System of Units (SI).*

Reprinted with permission from NFPA 70-2005.

Values of measurement in the *Code* text will be followed by an approximate equivalent value in SI units. Tables will have a footnote for SI conversion units used in the table.

Conduit size, wire size, horsepower designation for motors, and trade sizes that do not reflect actual measurements—for example, box sizes—will not be assigned dual designation SI units.

NEC® Units of Measurement (90.9)

- In the *NEC*®, metric units of measurement are in accordance with the modernized metric system known as the International System of Units (SI).

- The SI units appear first in the 2002 *NEC*®, followed by the inch-pound units in parentheses. The conversion from the inch-pound units to SI units is based on hard conversion, except as provided here.

    (C)(1) Trade Sizes. Where the actual measured size of a product is not the same as the nominal size, trade size designators shall be used rather than dimensions. Trade practices shall be followed in all cases.

Example:

Table 300.1(C) Metric Designator and Trade Sizes

Metric Designator	Trade Size
12	3/8
16	1/2
21	3/4
27	1
35	1 1/4
41	1 1/2
53	2
63	2 1/2
78	3
91	3 1/2
103	4
129	5
155	6

Note: The metric designators and trade sizes are for identification purposes only and are not actual dimensions.

(C)(2) **Extracted Material.** *Where material is extracted from another standard, the context of the original material shall not be compromised or violated. Any editing of the extracted text shall be confined to making the style consistent with that of the NEC.*

(C)(3) **Industry Practice.** *Where industry practice is to express units in inch-pound units, the inclusion of SI units shall not be required.*

Example: **Extracted Text to Article 516.3 from NFPA 33 and NFPA 34**

(4) For open dipping and coating operations, all space within a 1.5-m (5-ft) radial distance from the vapor sources extending from these surfaces to the floor. The vapor source shall be the liquid exposed in the process and the drainboard, and any dipped or coated object from which it is possible to measure vapor concentrations exceeding 25 percent of the lower flammable limit at a distance of 305 mm (1 ft), in any direction, from the object.

(5) Sumps, pits, or belowgrade channels within 7.5 m (25 ft) horizontally of a vapor source. If the sump, pit, or channel extends beyond 7.5 m (25 ft) from the vapor source, it shall be provided with a vapor stop or it shall be classified as Class I, Division 1 for its entire length.

Reprinted with permission from NFPA 70-2005.

(C)(4) **Safety.** *Where a negative impact on safety would result, soft conversion shall be used.*

Example: **NEC® 110.31 (A) Live Parts (over 600 volts) Guarded Against Accidental Contact.**
The distance from the fence to live parts shall be not less than the following:

601 to 13,799 volts	10 ft (3.05 m)
13,800 to 230,000 volts	15 ft (4.57 m)
Over 230,000 volts	18 ft (5.49 m)

(D) **Compliance.** *Conversion from inch-pound units to SI units shall be permitted to be an approximate conversion. Compliance with the numbers shown in either the SI system or the inch-pound system shall constitute compliance with this* Code. *

* Reprinted with permission from NFPA 70-2005.

NEC® 90.9 Units of Measurement
FPN No. 1: Hard conversion is considered a change in dimensions or properties of an item into new sizes that might or might not be interchangeable with the sizes used in the original measurement. Soft conversion is considered a direct mathematical conversion and involves a change in the description of an existing measurement but not in the actual dimension.
FPN No. 2: SI conversions are based on IEEE/ASTM SI 10-1997, Standard for the Use of the International System of Units (SI): The Modern Metric System.

Reprinted with permission from NFPA 70-2005.

As you can see, Article 90 is very important, and one could never apply the *Code* correctly without first having a thorough understanding of the Introduction, Article 90.

General Requirements

Article 100 contains definitions unique to this document. You must continually refer to these definitions as you apply the requirements set out in each section, article, or table. These terms are unique and essential to this document; that is why they have been included.

Definition Hierarchy

- A word or term unique to this document (*NEC®*) used one time will appear in the location in which it is used.
- A word or term unique to this document (*NEC®*) used two or more times will appear in Article 100.
- A word or term that is unique to the electrical industry but not unique to the *NEC®*, such as diode or transistor, will be defined in the *IEEE Dictionary*.
- All other words are defined by *Webster's Dictionary*.

Article 110 covers the general requirements. *NEC®* 110.3 tells us that we must select equipment and material suitable for the installation, environment, and/or application, and that it must be installed in accordance with any instructions included with its listing or labeling. *NEC®* 110.9 and 110.10 give us some very strong mandatory requirements on interrupting rating and circuit impedance that we cannot ignore. If the equipment is intended to break current, it must be rated to interrupt the available current at fault conditions, and the system voltage and the circuit components including the impedance, shall be coordinated so that the circuit protective device will clear any fault without excessive damage to the equipment. *NEC®* 110.8 and 110.12 tell us that only recognized methods are permitted and that we are required to make our installation in a neat and professional manner. *NEC®* 110.14 gives us mandatory requirements for making our splices and terminations. *NEC®* 110.14(C) tells us that the terminations must be considered when determining the ampacity of a circuit. Read this *NEC®* section very carefully; similar requirements have appeared in the UL *Electrical Constructions Material Directory* (Green Book) and UL *General Information for Electrical Construction, Hazardous Locations, and Electrical Heating Equipment Directory* (White Book).

NEC® 110.26 tells us that we are required to carefully select the proper location for our 600 volts or less equipment (for equipment rated at over 600 volts, the *Code* references are *NEC®* 110 Parts III and IV 110.30 through 110.59) so that proper work space can be maintained. This often requires us to check the building and mechanical plans so that encroachment can be avoided. To discover this encroachment in the last stages of the construction can be and often is very costly to one or more of the craft contractors. Other sections in Chapter 1 are equally important and must be referred to continually every time the *NEC®* is applied.

The title to Article 110 Part 5 has changed; however, the requirements have not changed. The new expanded title—Manholes and Other Electric Enclosures Intended for Personnel Entry, All Voltages—clarifies that all enclosures that are intended for personnel entry shall be of sufficient size to maintain safe work space and clearances from live parts.

Wiring and Protection

Now that we have thoroughly studied the general requirements, the next step is to take a look at the wiring and protection requirements. Let us turn to Chapter 2 of the *National Electrical Code®*, Article 210—Branch Circuits.

Note: Many of the requirements in this article apply only to dwellings; read the pertinent sections very carefully before applying the requirements.

Note: Although, generally, extra circuits are not necessary, careful arrangement of the required branch-circuits is necessary for compliance with 210.11(A), (B), and (C).

You will also find specific branch-circuit requirements in other articles of the *NEC®*, such as Article 430 for Motors, Article 600 for Signs, and Article 517 for Health Care Facilities. *NEC®* 210.52 gives us the guidelines necessary to avoid the use of extension cords in dwellings. 210.52(A) requires outlets to be located so that no point along the floor line is more than 6 feet from an outlet and wall spaces 2 feet wide have an outlet in them. This requirement ensures that it is not necessary to lay cords across door openings to reach an outlet. This section also points out that outlets that are a part of luminaires (light fixtures) and appliances, located within cabinets, or above 5½ feet (1.7 m) cannot be counted as one of those required. 210.52(A) through (H) gives us the requirements for our outlets including the appliance outlets.

You will find as you study Article 210 and throughout the *Code* numerous references to "210.8(A) or 210.8(B). These sections cover requirements for Class "A" Ground-Fault Circuit-Interrupter protection for personnel (GFCI). 210.8(A) covers the GFCI requirements for dwelling units. 210.8(B) covers the GFCI requirements for other than dwelling units. These Class A GFCIs can be found in all supply warehouses and retail stores that carry electrical supplies. They are readily available as circuit breakers or GFCI-type receptacles. They are covered under UL Standard 943 and have a trip level of 4ma to 6ma.

Now let us look at *NEC®* 210.70(A) for the required lighting outlets and wall switches. Extra lighting can be added for the customer, but the required lighting will provide a safe illumination. *NEC®* 210.70(A) generally requires a switch-controlled light in each room. Stairways are required to be lighted with a switch at the top and bottom of the stairs. *NEC®* 210.70(A) now requires a switch-controlled lighting outlet in attics and under floor spaces where used for storage or equipment that requires servicing. This section also requires a switch-controlled light at outside exits or entrances. After we have established the branch-circuit requirements, we can do the general lighting load calculations in accordance with Table 220.12 and footnotes (see pages 59–60).

A contractor recently informed me that because of the number of handheld appliances now being used in bathrooms, it is advisable that two bathrooms not share the same circuit; not a *Code* requirement but good design. The total calculated load cannot be ascertained

220.10 Branch-circuit Load Calculations. Branch-circuit loads shall be calculated as shown in 220.12, 220.14, and 220.16.

220.12 Lighting Load for Specified Occupancies. A unit load of not less than that specified in Table 220.12 for occupancies specified therein shall constitute the minimum lighting load. The floor area for each floor shall be calculated from the outside dimensions of the building, dwelling unit, or other area involved. For dwelling units, the calculated floor area shall not include open porches, garages, or unused or unfinished spaces not adaptable for future use.

FPN: The unit values herein are based on minimum load conditions and 100 percent power factor and may not provide sufficient capacity for the installation contemplated.

220.14 Other Loads — All Occupancies. In all occupancies, the minimum load for each outlet for general-use receptacles and outlets not used for general illumination shall not be less than that calculated in 220.14(A) through (L), the loads shown being based on nominal branch-circuit voltages.

Exception: The loads of outlets serving switchboards and switching frames in telephone exchanges shall be waived from the calculations.

(A) Specific Appliances or Loads. An outlet for a specific appliance or other load not covered in 220.14(B) through (L) shall be calculated based on the ampere rating of the appliance or load served.

(B) Electric Dryers and Household Electric Cooking Appliances. Load calculations shall be permitted as specified in 220.54 for electric dryers and in 220.55 for electric ranges and other cooking appliances.

(C) Motor Loads. Outlets for motor loads shall be calculated in accordance with the requirements in 430.22, 430.24, and 440.6.

(D) Luminaires (Lighting Fixtures). An outlet supplying luminaire(s) [lighting fixture(s)] shall be calculated based on the maximum volt-ampere rating of the equipment and lamps for which the luminaire(s) [fixture(s)] is rated.

(E) Heavy-Duty Lampholders. Outlets for heavy-duty lampholders shall be calculated at a minimum of 600 volt-amperes.

(F) Sign and Outline Lighting. Sign and outline lighting outlets shall be calculated at a minimum of 1200 volt-amperes for each required branch circuit specified in 600.5(A).

Table 220.12 General Lighting Loads by Occupancy

Type of Occupancy	Unit Load	
	Volt-Amperes per Square Meter	Volt-Amperes per Square Foot
Armories and auditoriums	11	1
Banks	39[b]	3½[b]
Barber shops and beauty parlors	33	3
Churches	11	1
Clubs	22	2
Court rooms	22	2
Dwelling units[a]	33	3
Garages — commercial (storage)	6	½
Hospitals	22	2
Hotels and motels, including apartment houses without provision for cooking by tenants[a]	22	2
Industrial commercial (loft) buildings	22	2
Lodge rooms	17	1½
Office buildings	39	3½[b]
Restaurants	22	2
Schools	33	3
Stores	33	3
Warehouses (storage)	3	¼
In any of the preceding occupancies except one-family dwellings and individual dwelling units of two-family and multifamily dwellings:		
Assembly halls and auditoriums	11	1
Halls, corridors, closets, stairways	6	½
Storage spaces	3	¼

[a]See 220.12(J)

[b]See 220.14(K)

(G) Show Windows. Show windows shall be calculated in accordance with either of the following:

(1) The unit load per outlet as required in other provisions of this section
(2) At 200 volt-amperes per 300 mm (1 ft) of show window

(continued on page 60)

(H) Fixed Multioutlet Assemblies. Fixed multioutlet assemblies used in other than dwelling units or the guest rooms or guest suites of hotels or motels shall be calculated in accordance with H(1) or H(2). For the purposes of this section, the calculation shall be permitted to be based on the portion that contains receptacle outlets.

(1) Where appliances are unlikely to be used simultaneously, each 1.5 m (5 ft) or fraction thereof of each separate and continuous length shall be considered as one outlet of not less than 180 volt-amperes.

(2) Where appliances are likely to be used simultaneously, each 300 mm (1 ft) or fraction thereof shall be considered as an outlet of not less than 180 volt-amperes.

(I) Receptacle Outlets. Except as covered in 220.14(J) and (K), receptacle outlets shall be calculated at not less than 180 volt-amperes for each single or for each multiple receptacle on one yoke. A single piece of equipment consisting of a multiple receptacle comprised of four or more receptacles shall be calculated at not less than 90 volt-amperes per receptacle. This provision shall not be applicable to the receptacle outlets specified in 210.11 (C)(1) and (2).

(J) Dwelling Occupancies. In one-family, two-family, and multifamily dwellings and in guest rooms or guest suites of hotels and motels, the outlets specified in (J)(1), (J)(2), and (J)(3) are included in the general lighting load calculations of 220.12. No additional load calculations shall be required for such outlets.

(1) All general-use receptacle outlets of 20-ampere rating or less, including receptacles connected to the circuits in 210.11(C)(3)

(2) The receptacle outlets specified in 210.52(E) and (G)

(3) The lighting outlets specified in 210.70(A) and (B)

(K) Banks and Office Buildings. In banks or office buildings, the receptacle loads shall be calculated to be the larger of (1) or (2):

(1) The calculated load from 220.14

(2) 11 volt-amperes/m^2 or 1 volt-ampere/ft^2

(L) Other Outlets. Other outlets not covered in 220.14 (A) through (K) shall be calculated based on 180 volt-amperes per outlet.

until all of the equipment loads are known and added to the general-purpose branch-circuits. Chapter 2, Articles 215 and 225, will also guide us as we select and size our inside and outside feeders.

Article 225 for outside branch-circuits and feeders has been revised in the past two *NEC®* editions, and Part II provides a mirror image of the service requirements for "more than one building or structure on the same property under single management." The requirements for the feeder supplying that building or structure from the service are almost identical to those for a service. *NEC®* 225.30 limits the building to be supplied by one feeder or branch-circuit except for special conditions such as fire pumps and emergency circuits and for special occupancies. These are requirements similar to those of *NEC®* 230.2. *NEC®* 225.31 through 225.39 cover the disconnect requirements. *NEC®* 225.40 requires ready access to the overcurrent devices. Part III covers the requirements for over-600-volt feeders.

Article 230—Services is the logical next step and also the next article. We are beginning to see some rationale for the chapter and article arrangement and layout. The diagram 230.1 is an excellent aid for using this article as the sections apply to the installation. Although materials are not specified, the methods and requirements are quite specific (see page 63).

Note: It is very important when designing the service, or solving questions related to the service, that you properly define the components of the installation; see Article 100—Definitions. *NEC®* 230.42(A) requires the service-entrance conductors to be of sufficient size to carry the loads as calculated in Article 220.

Article 240 covers the overcurrent protection for conductors requirements. This part of the *Code* has some very specific rules that will apply to all areas in the *Code*. For example, Parts I through VII of Article 240 cover the general requirements for overcurrent protection and overcurrent protective devices 600 volts or less. Part VIII covers overcurrent protection for those portions of supervised industrial installations operating at voltages of not more than 600 volts, nominal. Part IX covers overcurrent protection over 600 volts, nominal.

NEC® 240.2 was new to the 2002 *NEC®*. This section contains special definitions that apply to this article of the *Code*.

Figure 4–1 A typical single family dwelling with an attached garage and covered porch on the front with a covered patio in the back. Neither the porch nor the patio is enclosed.

Example:
In Figure 4-1, what is the general lighting load for this house?
- A. 11,181 VA
- B. 11,180.25 VA
- C. 7152 VA
- D. 9024 VA

Answer: C. *NEC*® 220.12 states that the computed area does not include porches and garages in *NEC*® Table 220.12. The general lighting load for a dwelling unit[a] is 3 volt-amperes per square foot.

The superscript [a] directs us to a note at the bottom of Table 220.12 that references *NEC*® 220.14(J), which states that no additional load calculations are required.

FPN: Overcurrent protection for conductors and equipment is provided to open the circuit if the current reaches a value that will cause an excessive or dangerous temperature in conductors or conductor insulation. See also 110.9 for requirements for interrupting ratings and 110.10 for requirements for protection against fault currents.

Reprinted with permission from NFPA 70-2005.

230.6 Conductors Considered Outside the Building. *Conductors shall be considered outside of a building or other structure under any of the following conditions:*

(1) Where installed under not less than 50 mm (2 in.) of concrete beneath a building or other structure
(2) Where installed within a building or other structure in a raceway that is encased in concrete or brick not less than 50 mm (2 in.) thick
(3) Where installed in any vault that meets the construction requirements of Article 450, Part III
(4) Where installed in conduit and under not less than 450 mm (18 in.) of earth beneath a building or other structure.

Coordination (Selective). *Localization of an overcurrent condition to restrict outages to the circuit or equipment affected, accomplished by the choice of overcurrent protective devices and their ratings or settings.*

Current-Limiting Overcurrent Protective Device. *A device that, when interrupting currents in its current-limiting range, reduces the current flowing*

(continued on page 63)

① 10 feet (3.0 m) – Lowest point of the drip loop of the electric entrance, and above areas or sidewalks accessible only to pedestrians, measured from final grade or other accessible surface where the voltage does not exceed 150volts to ground.

② 12 feet (3.7 m) – Point of attachment should not be farther than 2 feet from the Weatherhead.

③ 15 feet (4.5 m) – Over residential property and driveways, and commercial areas not subject to truck traffic where the voltage exceeds 300 volts to ground.

NEC ® Article 230

Table 310.15(B)(6) Conductor Types and Sizes for 120/240-Volt, 3-Wire, Single-Phase Dwelling Services and Feeders. Conductor Types RHH, RHW, RHW-2, THHN, THHW, THW, THW-2, THWN, THWN-2, XHHW, XHHW-2, SE, USE, USE-2 *

Conductor (AWG or kcmil)		
Copper	Aluminum or Copper-Clad Aluminum	Service or Feeder Rating (Amperes)
4	2	100
3	1	110
2	1/0	125
1	2/0	150
1/0	3/0	175
2/0	4/0	200
3/0	250	225
4/0	300	250
250	350	300
350	500	350
400	600	400

④ 18 feet (5.5 m) – Over public streets, alleys, roads, parking areas subject to truck traffic, driveways on other than residential property, and other land such as cultivated, grazing, forest, and orchard. Grounding electrode conductors shall be unspliced and have 6" exposed for the interconnection of other systems. Electrical nonmetallic or metal conduit shall be used for protection. If metallic, bond both ends (see section 4-1 & 3).

Minimum distance to final grade to attachment

5'-0" min
6'-0" max

12"
24" min

Final grade

Second ground rod

6'-0" min

Required ground rod

NEC ® 250.52

* Reprinted with permission from NFPA 70-2005.

Figure 4-2 An example of a simple electrical service supplying a building. (1) Vertical clearing requirements; (2) Point of attachment should be located to remove stress on the mast; (3) *NEC®* Table 310.15(B)(6) [reprinted with permission from NFPA 70-2005]); (4) Example of grounding electrodes as required by *NEC®* 250.50 and 250.52.

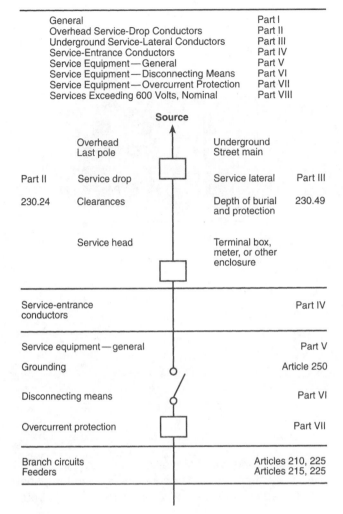

General	Part I
Overhead Service-Drop Conductors	Part II
Underground Service-Lateral Conductors	Part III
Service-Entrance Conductors	Part IV
Service Equipment—General	Part V
Service Equipment—Disconnecting Means	Part VI
Service Equipment—Overcurrent Protection	Part VII
Services Exceeding 600 Volts, Nominal	Part VIII

Figure 230.1 Services.

in the faulted circuit to a magnitude substantially less than that obtainable in the same circuit if the device were replaced with a solid conductor having comparable impedance.

Supervised Industrial Installation. *For the purposes of Part VIII, the industrial portions of a facility where all of the following conditions are met:*

(1) Conditions of maintenance and engineering supervision ensure that only qualified persons will monitor and service the system.
(2) The premises wiring system has 2500 kVA or greater of load used in industrial process(es), manufacturing activities, or both, as calculated in accordance with Article 220.
(3) The premises has at least one service that is more than 150 volts to ground and more than 300 volts phase-to-phase.

This definition excludes installations in buildings used by the industrial facility for offices, warehouses, garages, machine shops, and recreational facilities that are not an integral part of the industrial plant, substation, or control center.

Tap Conductors. *As used in this article, a tap conductor is defined as a conductor, other than a service conductor, that has overcurrent protection ahead of its point of supply that exceeds the value permitted for similar conductors that are protected as described elsewhere in 240.4.*

240.4(E) Tap Conductors. *Tap conductors shall be permitted to be protected against overcurrent in accordance with the following:*

(1) 210.19(A)(3) and (A)(4) Household Ranges and Cooking Appliances and Other Loads
(2) 240.5(B)(2) Fixture Wire
(3) 240.21 Location in Circuit
(4) 368.17(B) Reduction in Ampacity Size of Busway
(5) 368.17(C) Feeder or Branch Circuits (busway taps)
(6) 430.53(D) Single Motor Taps

240.21 Location in Circuit. *Overcurrent protection shall be provided in each ungrounded circuit conductor and shall be located at the point where the conductors receive their supply except as specified in 240.21(A) through (G). No conductor supplied under the provisions of 240.21(A) through (G) shall supply another conductor under those provisions, except through an overcurrent protective device meeting the requirements of 240.4.*

Note: This text in *NEC®* 240.21 clarifies that you can never tap a tap! (See Figure 4-3.)

NEC® 240.3 contains a list of about 40 *NEC®* articles that contains specific overcurrent requirements for systems and equipment such as emergency systems, motors generators, and the like, that are in addition to Article 240.

NEC® 240.4 covers the protection of conductors and states that other than flexible cords and fixture wires, all conductors shall be protected against overcurrent in accordance with their ampacities specified in *NEC®*310.15, unless otherwise permitted or required in 240.4(A) through 240.4(G).

* Reprinted with permission from NFPA 70-2005.

Figure 4–3 This drawing depicts a violation of the 2005 *NEC®*. The *Code* permits smaller conductor taps from a feeder to supply more than one piece of equipment. However, the tap from the tap conductors feeding the panelboard to supply the switch is not allowed. You are not permitted to "tap a tap." (See 240.2 Definition of a Tap, 240.4 Protection of Conductors, and 240.21.)

Example: Conductor overload protection is not required for which of the following where the interruption of the circuit would create a hazard?

A. Material-handling magnet
B. Motors
C. Lighting
D. All of these

Answer: A. *NEC®* covers power loss hazards such as a material-handling magnet. If there was a power loss while the magnet was moving a heavy object, the object would fall and possibly cause a severe injury.

Example: For overcurrent devices 800 amperes or less, the next higher overcurrent device is always permitted. True or False?

Answer: False. Although this is permitted where all of the conditions in *NEC®* 240.4(B) are met. It is not always permitted.

Example: For overcurrent devices 800 amperes or more, the next higher overcurrent device is never permitted. True or False?

Answer: True. *NEC®* 240.4(C) requires the overcurrent rating to be greater than the ampacity of the conductors it protects.

Small Size Conductor Overcurrent Protection Rules

The overcurrent protection ratings (circuit breaker or fuse) cannot exceed 15 amperes for 14 AWG copper conductor, 20 amperes for 12 AWG copper conductor, and 30 amperes for 10 AWG copper conductor; or 15 amperes for 12 AWG and 25 amperes for 10 AWG aluminum and copper-clad aluminum conductor after any correction factors for ambient temperature and number of conductors have been applied except as permitted in *NEC®* 240.4(E) through 240.4 (G).

Example: A 120-volt, single-phase circulating hot water pump in a multifamily dwelling is supplied by a 12-2 W/Grd Type NM cable. What is the maximum size overcurrent circuit breaker or fuse protection for this circuit?

    A. 10 ampere
    B. 20 ampere
    C. 25 ampere
    D. None of these

Answer: D. There is not enough information to answer this question. *NEC®* 240.3 states that motor overcurrent protection is governed by Article 430 Part III.

NEC® 240.4(E) permits tap conductors to be protected in accordance with the applicable *Code* section. For example, the rules for 25-foot tap conductors per 240.21(B)(2) would require the tap conductors to be protected from physical damage and have an ampacity of not less than one third the rating of the overcurrent device protecting the feeder conductors terminating in a single circuit breaker or a single set of fuses that limit the load to the ampacity of the tap conductors. The device can supply any number of additional overcurrent devices on its load side.

Example: A 3 AWG THWN feeder protected by a 100-ampere circuit breaker is tapped in a junction box with _____ AWG THWN conductors 20 feet (6 m) long to feed a 30-ampere single set of fuses. What is the minimum size THWN conductor permitted to be used for making this tap?

    A. 3 AWG
    B. 6 AWG
    C. 8 AWG
    D. 10 AWG

Answer: D. *NEC®* 240.21(B)(2) would require 1/3 of 100 or 33.34-ampere wire.

Table 310.16
• 3 AWG has an allowable ampacity of 100 amperes.
• 10 AWG has an allowable ampacity of 35 amperes.

Section 240.21 covers the tap rules as they are permitted throughout the *Code*. Experts generally agree that in all cases the conductors extending from the secondary of a transformer to the overcurrent device is a tap. They also agree that you can never tap a tap.

(text continues on page 68)

240.21 Location in Circuit. Overcurrent protection shall be provided in each ungrounded circuit conductor and shall be located at the point where the conductors receive their supply except as specified in 240.21(A) through (G). No conductor supplied under the provisions of 240.21(A) through (G) shall supply another conductor under those provisions, except through an overcurrent protective device meeting the requirements of 240.4.

(A) Branch-Circuit Conductors. Branch-circuit tap conductors meeting the requirements specified in 210.19 shall be permitted to have overcurrent protection located as specified in that section.

(B) Feeder Taps. Conductors shall be permitted to be tapped, without overcurrent protection at the tap, to a feeder as specified in 240.21(B)(1) through (B)(5). The provisions of 240.21(B) shall not be permitted for tap conductors.

(1) Taps Not Over 3 m (10 ft) Long. Where the length of the tap conductors does not exceed 3 m (10 ft) and the tap conductors comply with all of the following:

(1) The ampacity of the tap conductors is

    a. Not less than the combined calculated loads on the circuits supplied by the tap conductors, and
    b. Not less than the rating of the device supplied by the tap conductors or not less than the rating of the overcurrent-protective device at the termination of the tap conductors.

(2) The tap conductors do not extend beyond the switchboard, panelboard, disconnecting means, or control devices they supply.

(3) Except at the point of connection to the feeder, the tap conductors are enclosed in a raceway, which shall extend from the tap to the enclosure of an enclosed switchboard, panelboard, or control devices, or to the back of an open switchboard.

(4) For field installations where the tap conductors leave the enclosure or vault in which the tap is made, the rating of the overcurrent device on the line side of the tap conductors shall not exceed 10 times the ampacity of the tap conductor.

FPN: For overcurrent protection requirements for lighting and appliance branch-circuit panelboards and certain power panelboards, see 408.36(A), (B), and (E).

(2) Taps Not Over 7.5 m (25 ft) Long. Where the length of the tap conductors does not exceed 7.5 m (25 ft) and the tap conductors comply with all the following:

(1) The ampacity of the tap conductors is not less than one-third of the rating of the overcurrent device protecting the feeder conductors.
(2) The tap conductors terminate in a single circuit breaker or a single set of fuses that will limit the load to the ampacity of the tap conductors. This device shall be permitted to supply any number of additional overcurrent devices on its load side.
(3) The tap conductors are protected from physical damage by being enclosed in an approved raceway or by other approved means.

(3) Taps Supplying a Transformer [Primary Plus Secondary Not Over 7.5 m (25 ft) Long]. Where the tap conductors supply a transformer and comply with all the following conditions:

(1) The conductors supplying the primary of a transformer have an ampacity at least one-third the rating of the overcurrent device protecting the feeder conductors.
(2) The conductors supplied by the secondary of the transformer shall have an ampacity that is not less than the value of the primary-to-secondary voltage, ratio multiplied by one-third of the rating of the overcurrent device protecting the feeder conductors.
(3) The total length of one primary plus one secondary conductor, excluding any portion of the primary conductor that is protected at its ampacity, is not over 7.5 m (25 ft).
(4) The primary and secondary conductors are protected from physical damage by being enclosed in an approved raceway or by other approved means.
(5) The secondary conductors terminate in a single circuit breaker or set of fuses that limit the load current to not more than the conductor ampacity that is permitted by 310.15.

(4) Taps Over 7.5 m (25 ft) Long. Where the feeder is in a high bay manufacturing building over 11 m (35 ft) high at walls and the installation complies with all the following conditions:

(1) Conditions of maintenance and supervision ensure that only qualified persons service the systems.
(2) The tap conductors are not over 7.5 m (25 ft) long horizontally and not over 30 m (100 ft) total length.
(3) The ampacity of the tap conductors is not less than one-third the rating of the overcurrent device protecting the feeder conductors.
(4) The tap conductors terminate at a single circuit breaker or a single set of fuses that limit the load to the ampacity

of the tap conductors. This single overcurrent device shall be permitted to supply any number of additional overcurrent devices on its load side.

(5) The tap conductors are protected from physical damage by being enclosed in an approved raceway or by other approved means.
(6) The tap conductors are continuous from end-to-end and contain no splices.
(7) The tap conductors are sized 6 AWG copper or 4 AWG aluminum or larger.
(8) The tap conductors do not penetrate walls, floors, or ceilings.
(9) The tap is made no less than 9 m (30 ft) from the floor.

(5) Outside Taps of Unlimited Length. Where the conductors are located outdoors of a building or structure, except at the point of load termination, and comply with all of the following conditions:

(1) The conductors are protected from physical damage in an approved manner.
(2) The conductors terminate at a single circuit breaker or a single set of fuses that limit the load to the ampacity of the conductors. This single overcurrent device shall be permitted to supply any number of additional overcurrent devices on its load side.
(3) The overcurrent device for the conductors is an integral part of a disconnecting means or shall be located immediately adjacent thereto.
(4) The disconnecting means for the conductors is installed at a readily accessible location complying with one of the following:

   a. Outside of a building or structure
   b. Inside, nearest the point of entrance of the conductors
   c. Where installed in accordance with 230.6, nearest the point of entrance of the conductors

(C) Transformer Secondary Conductors. Each set of conductors feeding separate loads shall be permitted to be connected to a transformer secondary, without overcurrent protection at the secondary, as specified in 240.21(C)(1) through (C)(6). The provisions of 240.4(B) shall not be permitted for transformer secondary conductors.

FPN: For overcurrent protection requirements for transformers, see 450.3.

(1) Protection by Primary Overcurrent Device. Conductors supplied by the secondary side of a single-phase transformer having a 2-wire (single-voltage) secondary, or a three-phase, delta-delta connected transformer having a 3-wire (single-voltage) secondary, shall be permitted to be protected by overcurrent protection provided on the primary (supply) side of the transformer, provided this protection is in accordance with 450.3 and does not exceed the value determined by multiplying the secondary conductor ampacity by the secondary to primary transformer voltage ratio.

Single-phase (other than 2-wire) and multiphase (other than delta-delta, 3-wire) transformer secondary conductors are not considered to be protected by the primary overcurrent protective device.

(2) Transformer Secondary Conductors Not Over 3 m (10 ft) Long. Where the length of secondary conductor does not exceed 3 m (10 ft) and complies with all of the following:

(1) The ampacity of the secondary conductors is
    a. Not less than the combined calculated loads on the circuits supplied by the secondary conductors, and
    b. Not less than the rating of the device supplied by the secondary conductors or not less than the rating of the overcurrent-protective device at the termination of the secondary conductors, and
    c. Not less than one-tenth of the rating of the overcurrent device protecting the primary of the transformer, multiplied by the primary to secondary transformer voltage ratio

(2) The secondary conductors do not extend beyond the switchboard, panelboard, disconnecting means, or control devices they supply.
(3) The secondary conductors are enclosed in a raceway, which shall extend from the transformer to the enclosure of an enclosed switchboard, panelboard, or control devices or to the back of an open switchboard.

> FPN: For overcurrent protection requirements for lighting and appliance branch-circuit panelboards and certain power panelboards, see 408.16(A), (B), and (E).

(3) Industrial Installation Secondary Conductors Not Over 7.5 m (25 ft) Long. For industrial installations only, where the length of the secondary conductors does not exceed 7.5 m (25 ft) and complies with all of the following:

(1) The ampacity of the secondary conductors is not less than the secondary current rating of the transformer, and the sum of the ratings of the overcurrent devices does not exceed the ampacity of the secondary conductors.
(2) All overcurrent devices are grouped.
(3) The secondary conductors are protected from physical damage by being enclosed in an approved raceway or by other approved means.

(4) Outside Secondary of Building or Structure Conductors. Where the conductors are located outdoors of a building or structure, except at the point of load termination, and comply with all of the following conditions:

(1) The conductors are protected from physical damage in an approved manner.
(2) The conductors terminate at a single circuit breaker or a single set of fuses that limit the load to the ampacity of the conductors. This single overcurrent device shall be permitted to supply any number of additional overcurrent devices on its load side.
(3) The overcurrent device for the conductors is an integral part of a disconnecting means or shall be located immediately adjacent thereto.
(4) The disconnecting means for the conductors is installed at a readily accessible location complying with one of the following:

    a. Outside of a building or structure
    b. Inside, nearest the point of entrance of the conductors
    c. Where installed in accordance with 230.6, nearest the point of entrance of the conductors

(5) Secondary Conductors from a Feeder Tapped Transformer. Transformer secondary conductors installed in accordance with 240.21(B)(3) shall be permitted to have overcurrent protection as specified in that section.

(6) Secondary Conductors Not Over 7.5 m (25 ft) Long. Where the length of secondary conductor does not exceed 7.5 m (25 ft) and complies with all of the following:

(1) The secondary conductors shall have an ampacity that is not less than the value of the primary-to-secondary voltage ratio multiplied by one-third of the rating of the overcurrent device protecting the primary of the transformer.
(2) The secondary conductors terminate in a single circuit breaker or set of fuses that limit the load current to not more than the conductor ampacity that is permitted by 310.15.
(3) The secondary conductors are protected from physical damage by being enclosed in an approved raceway or by other approved means.

(D) Service Conductors. Service-entrance conductors shall be permitted to be protected by overcurrent devices in accordance with 230.91.

(E) Busway Taps. Busways and busway taps shall be permitted to be protected against overcurrent in accordance with 368.17.

(F) Motor Circuit Taps. Motor-feeder and branch-circuit conductors shall be permitted to be protected against overcurrent in accordance with 430.28 and 430.53, respectively.

(G) Conductors from Generator Terminals. Conductors from generator terminals that meet the size requirement in 445.13 shall be permitted to be protected against overload by the generator overload protective device(s) required by 445.12.

This article also contains guidance for fuses and circuit breakers, their application, and some manufacturing requirements. For instance, 240.60(B) regulates current-limiting fuses. They must be manufactured so that they are different from standard fuses: equipment designed or modified to accept current-limiting fuses will reject standard fuses. Manufacturers accomplish this by placing a notch in the load side (bottom) of the fuse blade, and equipment designed to accept them has a pin in the load side fuse holder. So current-limiting types can be installed, but standard types cannot. *NEC®* 240.83 gives similar specifics for circuit breakers.

Article 240 is a general use article and must be referred to on all installations. Part VIII covers those portions of supervised industrial installations operating at voltages not exceeding 600 volts nominal. *NEC®* 240 Part VIII applies only to large industrial facilities and only to those portions of the electrical system in the supervised industrial installation used exclusively for manufacturing or process control activities.

NEC® 240.2 defines supervised installations as the industrial portions of a facility where all conditions are met. This definition makes it clear that it does not apply to buildings used as offices, warehouses, garages, and so on, that are not part of the industrial plant, substation, or control center.

Part IX covers overcurrent protection over 600 volts nominal. This information was previously located in the 1996 *NEC®* Article 710. Article 710 was deleted, and all requirements for above-600-volt installations were relocated into the appropriate article.

NEC® Article 250—Grounding

The last major article in Chapter 2 is Article 250. (I did not forget Article 280—Surge Arresters and Article 285—Transient Voltage Surge Suppressors: TVSS. If you have a question related to surge arresters or TVSS, do not forget these articles. Grounding is one of the very important parts of any electrical installation

---

250.4 General Requirements for Grounding and Bonding. The following general requirements identify what grounding and bonding of electrical systems are required to accomplish. The prescriptive methods contained in Article 250 shall be followed to comply with the performance requirements of this section.

(A) Grounded Systems.

(1) Electrical System Grounding. Electrical systems that are grounded shall be connected to earth in a manner that will limit the voltage imposed by lightning, line surges, or unintentional contact with higher-voltage lines and that will stabilize the voltage to earth during normal operation.

(2) Grounding of Electrical Equipment. Non–current-carrying conductive materials enclosing electrical conductors or equipment, or forming part of such equipment, shall be connected to earth so as to limit the voltage to ground on these materials.

(3) Bonding of Electrical Equipment. Non–current-carrying conductive materials enclosing electrical conductors or equipment, or forming part of such equipment, shall be connected together and to the electrical supply source in a manner that establishes an effective ground-fault current path.

(4) Bonding of Electrically Conductive Materials and Other Equipment. Electrically conductive materials that are likely to become energized shall be connected together and to the electrical supply source in a manner that establishes an effective ground-fault current path.

(5) Effective Ground-Fault Current Path. Electrical equipment and wiring and other electrically conductive material likely to become energized shall be installed in a manner that creates a permanent, low-impedance circuit facilitating the operation of the overcurrent device or ground detector for high-impedance grounded systems. It shall be capable of safely carrying the maximum ground-fault current likely to be imposed on it from any point on the wiring system where a ground fault may occur to the electrical supply source. The earth shall not be considered as an effective ground-fault current path.

(B) Ungrounded Systems.

(1) Grounding Electrical Equipment. Non-current-carrying conductive materials enclosing electrical conductors or equipment, or forming part of such equipment, shall be connected to earth in a manner that will limit the voltage imposed by lightning or unintentional contact with higher-voltage lines and limit the voltage to ground on these materials.

(2) Bonding of Electrical Equipment. Non–current-

---

carrying conductive materials enclosing electrical conductors or equipment, or forming part of such equipment, shall be connected together and to the supply system grounded equipment in a manner that creates a permanent, low-impedance path for ground-fault current that is capable of carrying the maximum fault current likely to be imposed on it.

(3) Bonding of Electrically Conductive Materials and Other Equipment. Electrically conductive materials that are likely to become energized shall be connected together and to the supply system grounded equipment in a manner that creates a permanent, low-impedance path for ground-fault current that is capable of carrying the maximum fault current likely to be imposed on it.

(4) Path for Fault Current. Electrical equipment, wiring, and other electrically conductive material likely to become energized shall be installed in a manner that creates a permanent, low-impedance circuit from any point on the wiring system to the electrical supply source to facilitate the operation of overcurrent devices should a second fault occur on the wiring system. The earth shall not be considered an effective fault-current path.

FPN No. 1: A second fault that occurs through the equipment enclosures and bonding is considered a ground fault.

FPN No. 2: See Figure 250.4 for information on the organization of Article 250.

Figure 250.4 Grounding.

(Reprinted with permission from NFPA 70-2005, the *National Electrical Code®* Copyright © 2004–, National Fire Protection Association, Quincy, MA 02269. This reprinted material is not the complete and official position of the National Fire Protection Association on the referenced subject, which is represented only by the standard in its entirety.)

project and will be an important part of your studies. Article 250 is very specific. As you design the grounding system for an installation, you will find it to be concise and complete for most installations. However, specific grounding requirements are referenced in several other articles of the *NEC®*. If you are researching a question related to grounding, you may find the answer in the article specifically addressing the question. This is an article that you will become very familiar with during your studies of the *Code*; however, you may find it more efficient to use the index to see all references and where they are located in the *Code*.

NEC® 250.4(A)(5) makes it clear that in all cases the fault-current path must be electrically continuous, permanent, and capable of carrying the maximum fault likely to be imposed on it. It must have sufficiently low impedance to facilitate the operation of the overcurrent device(s) under fault conditions. This

section goes hand in hand with *NEC®* 110.10, which covers all components of a system.

NEC® 110.10 Circuit Impedance and Other Characteristics. The overcurrent protective devices, the total impedance, the component short-circuit current ratings, and other characteristics of the circuit to be protected shall be selected and coordinated to permit the circuit-protective devices used to clear a fault to do so without extensive damage to the electrical components of the circuit. This fault shall be assumed to be either between two or more of the circuit conductors or between any circuit conductor and the grounding conductor or enclosing metal raceway. Listed products applied in accordance with their listing shall be considered to meet the requirements of this section.

Reprinted with permission from NFPA 70-2005.

NEC® 250.24 clearly lays out the requirements for system grounding.

(1) General. *The connection shall be made at any accessible point from the load end of the service drop or service lateral to and including the terminal or bus to which the grounded service conductor is connected at the service disconnecting means.*

(C)(1) Routing and Sizing. *This conductor shall be routed with the phase conductors and shall not be smaller than the required grounding electrode conductor specified in Table 250.66 but shall not be required to be larger than the largest ungrounded service-entrance phase conductor. In addition, for service-entrance phase conductors larger than 1100 kcmil copper or 1750 kcmil aluminum, the grounded conductor shall not be smaller than 12½ percent of the area of the largest service-entrance phase conductor. The grounded conductor of a 3-phase, 3-wire delta service shall have an ampacity not less than that of the ungrounded conductors.**

A premises wiring system supplied by a grounded ac service shall have a grounding electrode conductor connected to the grounded service conductor at each service in accordance with the provisions of 250.24(A)(1) through (A)(5).

NEC® 250.24(D) Grounding Electrode Conductor. *A grounding electrode conductor shall be used to connect the equipment grounding conductors, the service-equipment enclosures, and, where the system is grounded, the grounded service conductor to the grounding electrode(s) required by Part III of this article.*

High-impedance grounded neutral system connections shall be made as covered in 250.36.

Reprinted with permission from NFPA 70-2005.

NEC® 250.32 makes it clear that the neutral shall not be regrounded at the second building or structure if there are any common metallic paths, such as water line, gas line, telephone, CATV, or process piping. This means that, in most cases, an equipment grounding conductor (metal raceway) must be run with the circuit conductors.

NEC® Article 250 Part III has been rewritten with significant changes for the accepted and required grounding electrode system.

NEC® 250.50 now requires that in each building or structure all the electrodes as described in 250.52(A)(1) through 250.52(A)(6) present must be bonded together to form the grounding electrode system. The significance of this is that the concrete-encased (rebar) electrode is now required to be a part of the electrode system unless the footings are poured and the rebar is not accessible.

Where none of the electrodes covered in 250.52(A)(1) through 250.52(6) exists, then one of the electrodes described in 250.52(A)(4) through 250.52(A)(7) must be installed and used.

- *NEC®* 250.52(A) lists the types of electrodes permitted such as underground water pipe in direct contact with earth 10 feet (3.0 m) or more, effectively grounded building steel, concrete encased electrode, and the like.

- *NEC®* 250.52(B) lists the two types of electrodes not permitted:
  (1) Metal underground gas piping system
  (2) Aluminum electrodes

- *NEC®* 250.53(A) through (H) covers the installation requirements for grounding electrodes.

  Note: Unless you test the resistance of the ground rod and it has 25 ohms or less, you must drive two electrodes to supplement the water pipe.

- *NEC®* 250.54 covers supplementary grounding electrodes.

- *NEC®* 250.56 covers resistance requirements for rod, pipe, and plate grounding electrodes.

- *NEC®* 250.58 covers "common" grounding electrodes.

- *NEC®* 250.60 covers the rules for the use of "air terminals."

- *NEC®* 250.62 covers the grounding electrode conductor material. *The material selected shall be resistant to any corrosive condition existing at the installation.*

- *NEC®* 250.64(A) through (F) covers the installation requirements for grounding electrode conductors.

- *NEC®* 250.66(A) through (C) and Table 250.66 cover the size of alternating current (ac) grounding electrode conductors.

- *NEC®* 250.68(A) and (B) covers the grounding electrode conductor and bonding jumper connection to the grounding electrode requirements.

- *NEC®* 250.70 covers the rules for making the connection of the grounding conductor to the electrode.

Now that we have our branch circuitry, the feeders, the service, the overcurrent protection, and the grounding designed into the installation, it is time to select the wiring methods and materials; those can all be found in Chapter 3 of the *NEC®*.

Wiring Methods

Chapter 3 was editorially revised in the 2002 *NEC®*. All articles were renumbered and relocated.

Chapter 3 Article Numbers and Acronyms

Article 300 Wiring Methods
Article 310 Conductors For General Wiring
Article 312 Cabinets, Cutout Boxes and Meter Socket Enclosures
Article 314 Outlet, Device, Pull and Junction Boxes, Conduit Bodies and Fittings
Article 320 Armored Cable **Type AC**
Article 322 Flat Cable Assemblies **Type FC**
Article 324 Flat Conductor Cable **Type FCC**
Article 326 Integrated Gas Spacer Cable **Type IGS**
Article 328 Medium Voltage Cable **Type MV**
Article 330 Metal-Clad Cable **Type MC**
Article 332 Mineral-Insulated Metal-Sheathed Cable **Type MI**
Article 334 Nonmetallic-Sheathed Cable **Types NM, NMC, and NMS**
Article 336 Power and Control Tray Cable **Type TC**
Article 338 Service Entrance Cable **Types SE and USE**
Article 340 Underground Feeder and Branch-Circuit Cable **Type UF**
Article 342 Intermediate Metal Conduit **Type IMC**
Article 344 Rigid Metal Conduit **Type RMC**
Article 348 Flexible Metal Conduit **Type FMC**
Article 350 Liquidtight Flexible Metal Conduit **Type LFMC**
Article 352 Rigid Nonmetallic Conduit **Type RMC**
Article 353 High Density Polyethylene Conduit **Type HDPE**
Article 354 Nonmetallic Underground Conduit with Conductors **Type NUCC**

Article 356 Liquidtight Flexible Nonmetal Conduit **Type LFNC**
Article 358 Electrical Metallic Tubing **Type EMT**
Article 360 Flexible Metallic Tubing **Type FMT**
Article 362 Electrical Nonmetallic Tubing **Type ENT**
Article 366 Auxiliary Gutters
Article 368 Busways
Article 370 Cablebus
Article 372 Cellular Concrete Floor Raceways
Article 374 Cellular Metal Floor Raceways
Article 376 Metal Wireways
Article 378 Nonmetallic Wireways
Article 380 Multioutlet Assembly
Article 382 Nonmetallic Extensions
Article 384 Strut-Type Channel Raceway
Article 386 Surface Metal Raceways
Article 388 Surface Nonmetallic Raceways
Article 390 Underfloor Raceways
Article 392 Cable Trays
Article 394 Concealed Knob and Tube Wiring
Article 396 Messenger Supported Wiring
Article 398 Open Wiring on Insulators

The individual article sections have been renumbered, using a common numbering system. For example, with only a few exceptions, all of the wiring methods in Chapter 3 will start with a "Scope" numbered as 3XX.1. This new common numbering system will make using the *Code* much quicker and easier. Once it is learned that "Securing and Supporting" requirements can be found in each Article 3XX.30 or that "Uses Not Permitted" can be found in 3XX.12, then the entire article will not need to be scanned to find these requirements.

Chapter 3 Wiring Methods Common Numbering

Part I General
3XX.1 Scope
3XX.2 Definitions
3XX.3 Other Articles
3XX.6 Listing Requirements

Part II Installation
3XX.10 Uses Permitted
3XX.12 Uses Not Permitted
3XX.14 Dissimilar Metals
3XX.16 Temperature Limits
3XX.20 Size
3XX.22 Number of Conductors
3XX.24 Bends—How made

(continued)

3XX.26 Bends—Number in one run
3XX.28 Reaming, Threading and Trimming
3XX.30 Securing and Supporting
3XX.40 Boxes and Fittings
3XX.42 Couplings and Connectors
3XX.44 Expansion Fittings
3XX.46 Bushings
3XX.48 Joints
3XX.50 Conductor Terminations
3XX.56 Splices and Taps
3XX.60 Grounding

Part III Construction Specifications
3XX.100 Construction
3XX.110 Corrosion Protection
3XX.120 Marking
3XX.130 Standard Lengths

I hope you have studied Article 300. As I stated earlier, this article covers general requirements that must be applied to all the wiring methods. The wiring method selected must be appropriate for the environment, meet the physical protection requirements, temperature, and all other pertinent conditions that may affect the material selected.

Note: The uses permitted and not permitted appear in each wiring method article. Restrictions of use vary. The answer to your question may be in those *NEC®* sections.

The condition or question may address the wiring method directly or may require you to select an approved method based on these factors. Be sure to study all of the "uses permitted" and the "uses not permitted" in each specific article carefully.

For example, when selecting Rigid Metal Conduit (RMC), you only need to consider the corrosion elements (*NEC®* 300.6). However, when selecting Nonmetallic Sheath Cable (Type NM), many factors must be considered (300.4, and 336.10 and 336.12). Once this has been done, a wiring method may be selected from Chapter 3 based on all the pertinent factors.

Warning: You may have wiring methods that are not permitted or have additional restrictions in your jurisdiction. Check local ordinances that amend the *NEC®*.

If you are not a new student of the *NEC®*, you may wonder what happened to the requirements for temporary wiring. This article was relocated to Chapter 5 Article 590 (formerly Article 305 and Article 527—Temporary Installations). The requirements are essentially still the same as it permits the use of a lower-class wiring for any temporary wiring needed during the construction phase or wiring used for short periods of time such as holiday wiring. These requirements closely parallel OSHA requirements on construction sites. However, OSHA requirements may be much broader, and local experience with OSHA provides guidance in avoiding citations.

The conductor type is selected and sized from Article 310. The wiring method we select may need to be considered at the same time as the conductor type. Most installations require the use of several wiring methods to meet all the conditions, such as a method that is acceptable for dry concealed locations. For this location, most of the methods in Chapter 3 are acceptable. We may also need to select a flexible wiring method for equipment connection and a material that can be used in wet locations and underground. There are methods that are acceptable in all locations: All are found in Chapter 3, including the junction, outlet, cabinets, and cutout boxes from Article 314 and Article 312. (To identify differences in these two types of enclosures, check Article 100—Definitions.) Once again, I remind you that you are required to comply with all the applicable provisions in Article 300 and the article that covers the specific wiring method.

Warning: Some wiring methods are permitted in locations only when specific type conductors are installed. For example, Flexible Metal Conduit (FMC) (Article 348) is permitted in a wet location when conductors suitable for wet locations are enclosed (310.8(C) and 348.12(1)).

Equipment for General Use

Article 400 and Article 402 cover fixture wires and cords. These articles are used most in connection with equipment and, therefore, are placed in Chapter 4 rather than in Chapter 3.

The requirements for switches (Article 404) and switchboards and panelboards (Article 408) are now covered in Chapter 4. Read the scope of these articles.

You see that Article 404 applies to all switches, switching devices, and circuit breakers where used as switches. Article 408's scope is not nearly as clear, and this is where the *NEC®* can be quite difficult to interpret. Motor control centers are not mentioned in Article 408; a fine print note (FPN No. 1) to *NEC®* 430.1 states that motor control centers are covered in 110.26(F). You must study the *Code* carefully and thoroughly to successfully pass your examination the first time you take the test.

Article 404 (formerly Article 380) has been appropriately relocated to Chapter 4 from Chapter 3. It is now easier to find, and it applies to all switches, switching devices, and circuit breakers when they are used as switches.

> **Example:** It is a common practice to use circuit breakers in large commercial installations or warehouses to regularly turn the lights on and off daily. Circuit breakers used as switches must be
>   A. hand operable.
>   B. marked "ON" "OFF."
>   C. marked "SWD" or "HID."
>   D. all of these.
>
> **Answer:** D. 404.11, they must be of a "hand operable" type; 240.81, they must clearly indicate when they are "ON" and when they are "OFF" and where mounted in a vertical position the "ON" position must be up. *NEC®* 240.83(D): Circuit breakers used as switches in 120-volt and 277-volt fluorescent lighting circuits are required to be listed and marked SWD or HID. Circuit breakers used as switches in high-intensity discharge lighting circuits are required to be listed and marked as HID.

Article 406 covers the rating, type, and installation of receptacles, cord connectors, and attachment plugs (cord caps). These requirements formerly appeared as Part L to Article 410. The requirements are now appropriately located into a separate article and will be much easier for a new student of the Code to locate and apply.

Article 408 (formerly Article 384) covers switchboards and panelboards. It has been appropriately relocated to Chapter 4 from Chapter 3. It is now easier to find, and it applies to all switchboards, panelboards, and distribution boards installed for the control of light and power circuits. It also applies to battery-charging panels supplied from light or power circuits.

Article 409—Industrial Control Panels has been added to the *NEC®* for the first time. These industrial control panels are intended for general use and operate from a voltage of 600 volts or less. These control panels are normally manufactured to the UL 508A Standard. They are defined in 409.2 as an assembly of two or more components such as motor controllers, overload relays, fused disconnect switches, and circuit breakers and related control devices but do not include the controlled equipment.

Article 410 covers the requirements for luminaires (lighting fixtures), lampholders, and lamps. The introduction of the term "**luminaire**" was new to the *NEC®* and replaced the term "lighting fixture." The *NEC®* is an international document used in many different countries. Design criteria can be found in this article. *NEC®* 410.4 covers wet locations and gives specific guideline rules for bathrooms where pendant-type fixtures are used. *NEC®* 410.8 covers all rules, allowances, and limitations for fixtures in clothes closets.

> **Note:** High-discharge lighting includes fluorescent fixtures as well as other ballast-type fixtures.

As you look through Chapter 4, you will see articles to cover most utilization equipment found in all types of occupancies—that is, appliances, heating, air-conditioning, refrigeration, motors, transformers, phase converters, and generators. Article 490 was added for equipment over 600 volts in the 1999 *NEC®*. It covers the general requirements for equipment operating at more than 600 volts and defines "high voltage" as more than 600 volts. These requirements were formerly found in Article 710, which was deleted.

Special Equipment, Occupancies, and Systems

These chapters amend or modify the first four chapters. Chapter 8 stands alone, and earlier chapters can be applied only where specifically referenced. Chapter 9 contains tables and examples, which are used as applicable. These chapters are addressed in a later chapter of this book.

QUESTION REVIEW

Answer the following Question Review carefully. You will find it necessary to refer to each article in the *National Electrical Code®*, the Table of Contents, and the Index to solve these questions. You will find that answering these questions will prepare you as you proceed through the latter chapters of this book. The next eight chapters of the manual will help you gain that needed understanding of the *NEC®* and give you a wide variety of sample questions on each chapter of the *NEC®*.

> **Each lesson is designed purposely to require the student to apply the entire *NEC®* text, and not specific chapters or articles. It has been found that when studying for a timed, open-book examination, the student must gain proficiency in the Table of Contents, the Index, and the ability to move quickly from cover to cover to find the correct answer to each question in a timely fashion.**

1. Would a floating restaurant located in a river or along a dock in a harbor be covered by the *National Electrical Code®*?

   Answer: _____

   _____

   _____

   _____

   Reference: _____

2. In which chapter in the *NEC®* would you find the requirements for grounding a community antenna television and radio distribution system?

   Answer: _____

   _____

   _____

   _____

   Reference: _____

3. In which article in the *NEC®* would you find the requirements for receptacles, cord connectors, and attachment plugs?

   Answer: _____

   _____

   _____

   _____

   Reference: _____

4. All requirements for the installation and calculation of branch-circuits are found in Chapter 2 of the *NEC®*. True or False?

   Answer: _____

   _____

   _____

   _____

   Reference: _____

5. In what article would the requirements for installing luminaires (light fixtures) in a paint spray booth be found?

   Answer: _____

   _____

   _____

   _____

   Reference: _____

6. In what article would you find the requirements for the installation of central heating equipment, such as a natural gas furnace?

   Answer: _____

   _____

   _____

   _____

   Reference: _____

7. In what section of the *National Electrical Code*® would you find the allowable number of conductors in a conduit?

   Answer: _____

   _____

   _____

   _____

   Reference: _____

8. Which article of the *NEC*® covers the power and lumination in mine shafts?

   Answer: _____

   _____

   _____

   _____

   Reference: _____

9. Where would you find the cover requirements for underground conductors running from building A to building B? (The conductors are direct burial conductors, the circuit is a 3-phase, 2300-volt, 4-wire system.)

   Answer: _____

   _____

   _____

   _____

   Reference: _____

10. The conductors in Question 9 must terminate in a disconnecting means. What article and section of the *NEC*® cover the requirements for the disconnecting equipment at building B?

    Answer: _____

    _____

    _____

    _____

    Reference: _____

11. A trade size 4 (103) rigid nonmetallic conduit is being installed on the outside of a building. The length of the conduit from the junction box to the point that it enters the building is 300 feet (91 m). Where in the *National Electrical Code®* is the amount of the expansion of the rigid nonmetallic conduit found?

Answer: _____

_____

_____

_____

Reference: _____

12. You are going to wire a nightclub that will seat 300 people. Which article in the *NEC®* covers the wiring requirements for this nightclub?

Answer: _____

_____

_____

_____

Reference: _____

13. Where in the *National Electrical Code®* would the requirements for grounding a portable generator be found?

Answer: _____

_____

_____

_____

Reference: _____

14. In which section in the *NEC®* would the requirements for bonding the forming shell of an inground swimming pool be found?

Answer: _____

_____

_____

_____

Reference: _____

15. In which section in the *NEC®* is material extracted from another NFPA standard explained?

Answer: _____

_____

_____

_____

Reference: _____

16. Where in the *NEC®* are the requirements for the location and installation of smoke detectors found?

Answer: _____

_____

_____

_____

Reference: _____

17. Under what article of the *NEC*® would the wiring requirements for fire alarms be found?

    Answer: _____

    _____

    _____

    _____

    Reference: _____

18. An existing building is being remodeled. Permanent receptacles in the building are being used to supply temporary power for the construction. Must these permanent receptacles be GFCI protected?

    Answer: _____

    _____

    _____

    _____

    Reference: _____

19. You have been given the responsibility for wiring a phosphoric acid fertilizer facility. What section of the *NEC*® covers this type of hazardous installation? (a) What class wiring is required? (b) If the hazardous dusts are not controlled and will be found on the day-to-day work schedule, what division must this area be wired in?

    Answer: _____

    _____

    _____

    _____

    Reference: _____

20. You have been asked to wire a neon sign in the bedroom of a new home you are wiring. To which section of the *NEC*® would you refer to wire this neon sign for the bedroom?

    Answer: _____

    _____

    _____

    _____

    Reference: _____

21. You are asked to make an installation for a lawnmower repair shop where gasoline will be regularly used. The customer also will be repairing chainsaws and boat motors. To which article of the *NEC*® would you refer to wire this facility when gasoline is present in and around the equipment used for cleaning and testing these lawnmowers?

    Answer: _____

    _____

    _____

    _____

    Reference: _____

22. You have been asked to submit a bid for a new restaurant. After calculating the load requirements for the building, you determine that the requirements are too high. Is there an optional calculation you can use for restaurants?

    Answer: _____

    _____

    _____

    _____

    Reference: _____

23. You have been employed to wire a commercial building in electrical nonmetallic tubing. In which section in the *NEC*® are the requirements for running parallel to framing members with this raceway system?

    Answer: _____

    _____

    _____

    _____

    Reference: _____

24. You have been employed to install an optional standby generator system. To which article of the *NEC*® would you refer to make this installation?

    Answer: _____

    _____

    _____

    _____

    Reference: _____

25. A company that you have been working with has decided to burn their garbage, using the heat derived as an energy source to drive a generator and generate power to supplement the power being bought from the utility. To which article of the *NEC*® would you refer to make this installation?

    Answer: _____

    _____

    _____

    _____

    Reference: _____

Chapter Five

OBJECTIVES

Studying this chapter along with the 2005 *NEC®* gives the reader a basic introduction to the general requirements in the *Code* that apply to all installations. The introduction to *NEC®* Chapter 1 covers the general requirements that apply to all installations no matter how small or how large. The reader will study the application of installation rules for above and below 600 volts. The reader will learn how to apply all requirements and understand them.

After completing this chapter, you should know:

- The definitions of words and terms that are unique to the 2005 *NEC®*
- The application of Article 100—Definitions:
    1. Article 100, definitions of words and terms unique to the *NEC®*
    2. Why there are definitions elsewhere in the 2005 *NEC®*
    3. The importance of checking for a term as well as a word
- The application of Article 110—General Requirements
- The definitions of terms such as "listed" or "approved" as they apply to the *NEC®*
- The minimal performance requirements such as "workmanship" and the "application of equipment"
- The application of temperature requirements
- The application of the complex working clearance requirements

GENERAL WIRING METHODS

Article 100

This article contains only definitions that are unique or essential to the application of the *Code*. It does not include commonly defined general terms or commonly defined technical terms from related codes and standards. In general, only those terms used in two or more articles are defined in Article 100. Other definitions are included in the article in which they are used, but may be referenced in Article 100.

Part I of Article 100 contains definitions that apply wherever the terms are used. Part II contains definitions of terms applicable only to those parts of the *Code* specifically covering installations and equipment operating at over 600 volts, nominal.

Electrical terms that are not specific to the *NEC®* may be defined in the *IEEE Dictionary*. Other words are defined in the standard *Webster's Dictionary*. An example of a definition with a specific meaning in the *NEC®* found in Article 100 would be "**accessible**." Accessible is defined both for wiring methods and equipment. The definition found in Article 100 applies specifically to its use throughout the *NEC®*. Other terms or words found in Article 100 have meanings that generally differ from a standard *Webster's Dictionary*.

Terms with unique meanings that have been added to the *NEC®* include such terms as bathroom:

Example:
A bathroom is defined as
    A. a room with a basin, toilet and tub, or a shower.
    B. in a standard dictionary.
    C. an area including a basin, a toilet, and tub, or a shower.
    D. an area including a basin with one or more of the following: a toilet, a tub, or a shower.

Answer: D. This definition, found in Article 100, is unique to the *NEC®*. (See Figure 5-1.)

This meaning is different than one would find in a standard dictionary, which defines "bathroom" as a room equipped for taking a bath or shower and usually includes a sink (basin) and toilet. This definition is needed since handheld grooming appliances have been the source of many serious and sometimes fatal

Figure 5–1 This is one example of a "bathroom area." It includes the separate water closet (toilet) room cut off by a door. Another example would be a typical Hotel-Motel "bathroom area" where the basin(s) are located outside the "bathroom" in the sleeping area.

accidents. These appliances are often used in and around the tub and basin areas. Basins are sometimes located outside the room commonly defined as the bathroom. Thus, the *NEC®* defines the area that includes these items and not the room.

Other words or terms are:

Energized. *Electrically connected to, or is, a source of voltage.*

Live Parts. *Energized conductive components.*

Damp Locations (under **Location, Damp**). *Locations protected from weather and not subject to saturation with water or other liquids but subject to moderate*

degrees of moisture. Examples of such locations include partially protected locations under canopies, marquees, roofed open porches, and like locations, and interior locations subject to moderate degrees of moisture, such as some basements, some barns, and some cold-storage warehouses. (See Figure 5-2.)

Motor Control Center. *An assembly of one or more enclosed sections having a common power bus and principally containing motor control units.*

Nonlinear Load. *A load where the wave shape of the steady-state current does not follow the wave shape of the applied voltage. FPN: Electronic equipment, electronic/electric-discharge lighting, adjustable-speed drive systems, and similar equipment may be nonlinear loads.*

Qualified Person. *One who has the skills and knowledge related to the construction and operation of the equipment and installations and has received safety training on the hazards involved.* (See Figure 5-3.)

Revision of the definition of a **Receptacle**. *A receptacle is a contact device installed at the outlet for the connection of* [a single contact device]. *A single receptacle is a single contact device with no other contact device on the same yoke. A multiple receptacle is two or more contact devices on the same yoke.*

Service Conductors. *The conductors from the service point to the service disconnecting means.* (See Figure 5-4.)

Switchboard. *A large single panel, frame, or assembly of panels on which are mounted on the face, back, or both, switches, overcurrent and other protective devices, buses, and usually instruments. Switchboards are generally accessible from the rear as well as from the front and are not intended to be installed in cabinets.* (See Figure 5-5.)

Transfer Switch (under **Switch, Transfer**). *An automatic or nonautomatic device for transferring one or more load conductor connections from one power source to another.*

Electronically Actuated Fuse. *An overcurrent protective device that generally consists of a control module that provides current sensing, electronically derived time-current characteristics, energy to initiate tripping, and an interrupting module that interrupts current when an overcurrent occurs. Electronically actuated fuses may or may not operate in a current-limiting fashion, depending on the type of control selected.*

There were other modifications to the definitions, but these were the most significant.

Figure 5–2 This is one example of a location that meets the Article 100 definition for "damp location." It is an industrial area protected on one side and open on the other three sides. It is covered with a canopy.

Figure 5–3 An example of "qualified person(s)" as defined in *NEC*® Article 100. Construction electricians are typically licensed based on their technical training and work experience. They typically receive regular safety and OSHA training at their workplace. (Refer to NFPA 70E-2004 Standard for Electrical Safety in the Workplace for Electrical Training Requirements.)

Figure 5–4 Specific system equipment and circuit types. Article 100 defines "service drop," "service conductors," "feeder," "branch-circuit," and so on.

Figure 5–5 An example of an electrical "switchboard." (*Courtesy of Square D Company*)

Article 110

Article 110 contains the general requirements for all electrical installations. *NEC®* 110.2 clarifies the term "approval." The word approved is defined in Article 100; however, "approval" that is required for installations, equipment, and instructions to judge it suitable is the responsibility of the authority having jurisdiction to approve or give approval. Many inspectors require listing and labeling by a recognized testing laboratory before they will give approval.

NEC® 110.3(A) and (B) requires the examination of equipment for identification and use. It also requires that the installation be made in accordance with its listing and or manufacturing instructions and in accordance with the provisions of the *Code*. *NEC®* 110.4 states that the voltages considered are those at which the circuit operates, and *NEC®* 110.6 points out that the *Code* sizes conductors in AWG (American Wire Gauge) and circular mils.

Workmanship

NEC® 110.7, 110.8, 110.11, 110.12, and 110.13 remind us that only wiring methods recognized as suitable are included in the *Code*, and that only recognized methods of wiring are permitted to be installed in any type of building or occupancy as provided in the *NEC®*. This is important, because often materials are misused and wiring methods are devised in the field that are not in accordance with the manufacturer's instructions or design recommendations of the equipment being used.

All equipment and materials must be designed for the environment they will be subjected to over the life of the installation—that is, wet and damp locations, excessive temperatures, exposure to fumes, gases, vapors, liquids, or other elements that may have a deteriorating effect on them. It reminds us that all work must be installed in a neat and professional manner, and that unused openings must be closed. Conductors and cables must be racked and supported neatly in manholes. All equipment must be installed carefully and protected throughout construction phases from contamination from paint, plaster, solvents, and cleaners that might harm contacts, bus bars, terminations, and insulators. All equipment must be firmly and securely mounted. Adequate clearances for all equipment must be mounted and located so that heat can dissipate. Equipment provided with ventilated openings must be mounted so that the openings are not obstructed.

Although the *NEC®* is a minimal requirement, it still contains some very specific rules for good workmanship and to ensure a safe installation.

NEC® 110.9 and 110.10 require that all circuitry and equipment intended to break current must be properly sized and installed so that it will operate correctly and have an interrupting rating sufficient for the nominal circuit voltage and current available at the terminals of the equipment. *The overcurrent protective devices, the total impedance, the component short-circuit current ratings, and other characteristics of the circuit to be protected shall be selected and coordinated to permit the circuit-protective devices used to clear a fault to do so without extensive damage to the electrical components of the circuit.*

Wire Temperature Termination Requirements

Lack of knowledge of wire temperature ratings and temperature termination requirements for equipment results in rejected installations or installations that fail long before they should need replacing. Information about this topic can be found in testing agency directories, product testing standards, and manufacturers' literature, but most do not consult these sources or read the information provided with the equipment until it is too late.

NEC® 110.14 warns that dissimilar metals, such as copper and aluminum, should not be mixed. Only terminals identified for this purpose are permitted to splice these dissimilar metals together. Although 14 AWG and 12 AWG aluminum building wire are no longer manufactured, there are still some installations that were originally wired with aluminum conductors. Until recently it has been relatively difficult to splice these aluminum conductors to copper conductors. However, there are now several listed products on the market to accomplish this without difficulty. These methods vary from some requiring special tools to a simple twist-on connector. (Note that the twist-on device is only listed for copper-to-aluminum and not aluminum-to-aluminum terminations.)

This section also requires that materials, such as solder, flux, and inhibitors, be suitable for the use and not damage the insulation or conductors. Solder connections must be made so as not to depend on the solder for making a good connection.

* Reprinted with permission from NFPA 70-2005.

NEC® 110.14(C) was added in the 1993 edition, and it has now been clarified. The new wording makes it clear that terminations must be considered as a part of every circuit when calculating the allowable ampacity the circuit conductors can carry.

110.14(C) Temperature Limitations. *The temperature rating associated with the ampacity of a conductor shall be selected and coordinated so as not to exceed the lowest temperature rating of any connected termination, conductor, or device. Conductors with temperature ratings higher than specified for terminations shall be permitted to be used for ampacity adjustment, correction, or both.*

(1) Equipment Provisions. The determination of termination provisions of equipment shall be based on 110.14(C)(1)(a) or (C)(1)(b). Unless the equipment is listed and marked otherwise, conductor ampacities used in determining equipment termination provisions shall be based on Table 310.16 as appropriately modified by 310.15(B)(6).

(a) Termination provisions of equipment for circuits rated 100 amperes or less, or marked for 14 AWG through 1 AWG conductors, shall be used only for one of the following:
(1) Conductors rated 60°C (140°F).
(2) Conductors with higher temperature ratings, provided the ampacity of such conductors is determined based on the 60°C (140°F) ampacity of the conductor size used.
(3) Conductors with higher temperature ratings if the equipment is listed and identified for use with such conductors.
(4) For motors marked with design letters B, C, or D, conductors having an insulation rating of 75°C (167°F) or higher shall be permitted to be used, provided the ampacity of such conductors does not exceed the 75°C (167°F) ampacity.

(b) Termination provisions of equipment for circuits rated over 100 amperes, or marked for conductors larger than 1 AWG, shall be used only for one of the following:
(1) Conductors rated 75°C (167°F)
(2) Conductors with higher temperature ratings, provided the ampacity of such conductors does not exceed the 75°C (167°F) ampacity of the conductor size used, or up to their ampacity if the equipment is listed and identified for use with such conductors

(2) Separate Connector Provisions. Separately installed pressure connectors shall be used with conductors at the ampacities not exceeding the ampacity at the listed and identified temperature rating of the connector.

> *FPN: With respect to 110.14(C)(1) and (C)(2), equipment markings or listing information may additionally restrict the sizing and temperature ratings of connected conductors.*

Reprinted with permission from NFPA 70-2005.

Why Are Temperature Ratings Important?

Conductors carry a specific temperature rating based on the type of insulation employed on the conductor. Common insulation types can be found in Table 310.13 of the *NEC®*, and corresponding ampacities can be found in Table 310.16.

Example: A 1/0 AWG copper conductor ampacity based on different conductor insulation types:

Insulation Type	Temperature Rating	Ampacity
TW	60°C	125 amperes
THW	75°C	150 amperes
THHN	90°C	170 amperes

Even though the wire size has not changed (1/0 AWG Cu), the ampacity *has* changed due to the temperature rating of the insulation on the conductor. Higher-rated insulation allows a smaller conductor to be used at the same ampacity as a larger conductor with lower-rated insulation, and, as a result, the amount of copper and even the number of conduit runs needed for the job may be reduced.

One of the most common misapplications of conductor temperature ratings occurs when the rating of the equipment termination is ignored. Conductors must be sized by giving consideration to where they will terminate and how that termination is rated. If a termination is rated for 75°C, this means that the temperature at that termination may rise up to 75°C when the equipment is loaded to its ampacity. If 60°C insulated conductors were employed in this example, the additional heat at the connection above the 60°C conductor insulation rating could result in failure of the conductor insulation.

When a conductor is selected to carry a specific load, the user/installer or designer must know termination ratings for the equipment involved in the circuit.

Figure 5–6 (*Courtesy of Square D Company*)

Example: Using a circuit breaker with 75°C termination and a 150-ampere load, if a THHN (90°C) conductor is selected for the job, from Table 310.16 select a conductor that will carry the 150 amperes. Although Type THHN has a 90°C ampacity rating, the ampacity from the 75°C column must be selected because the circuit breaker termination is rated at 75°C. Looking at the table, a 1/0 AWG copper conductor is acceptable. The installation would be as shown in Figure 5–6, with proper heat dissipation at the termination as well as along the conductor length. Had the temperature rating of the termination not been considered, a 1 AWG conductor, based on the 90°C ampacity, may have been selected, which may have led to overheating at the termination or premature opening of the overcurrent device due to the smaller conductor size. (See Figure 5–7.)

In this example, a conductor with a 75°C insulation type (THW, RHW, USE, etc.) would also be acceptable because the termination is rated at 75°C. A 60°C insulation type (TW or perhaps UF) is not acceptable because the temperature at the termination could rise to a value greater than the insulation rating.

General Rules for Application

When applying equipment with conductor terminations, the following basic rules apply:

- The termination and splicing provisions of equipment shall be based on 110.14(C)(1)(a) or (C)(1)(b).
- Conductor ampacities used in determining equipment termination provisions shall be based on Table 310.16 as appropriately modified by 310.15 (B)(6).
- **110.14(C)(1)(a).** *Termination provisions of equipment for circuits rated 100 amperes or less, or marked for 14 AWG through 1 AWG, shall be used only for one of the following:*
  (1) Conductors rated 60°C (140°F).
  (2) Conductors with higher temperature ratings, provided the ampacity of such conductors is determined based on the 60°C (140°F) ampacity of the conductor size used.
  (3) Conductors with higher temperature ratings if the equipment is listed and identified for use with such conductors.
  (4) For motors marked with design letters B, C, or D, conductors having an insulation rating

Figure 5–7 (*Courtesy of Square D Company*)

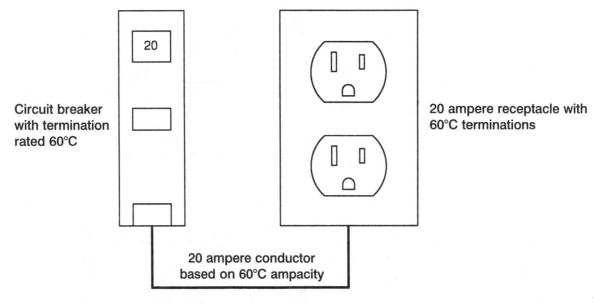

Figure 5–8 (*Courtesy of Square D Company*)

of 75°C (167°F) or higher shall be permitted to be used, provided the ampacity of such conductors does not exceed the 75°C (167°F) ampacity. (See Figure 5-8.)

- **110.14(C)(1)(b).** *Termination provisions of equipment for circuits rated over 100 amperes, or marked for conductors larger than 1 AWG, shall be used only for one of the following:*
  (1) Conductors rated 75°C (167°F)
  (2) Conductors with higher temperature ratings, provided the ampacity of such conductors does not exceed the 75°C (167°F) ampacity of the conductor size used, or up to their ampac-
*ity if the equipment is listed and identified for use with such conductors.** (See Figure 5-9.)

There are exceptions to the rules. Two of the most important exceptions are as follows:

- Conductors with higher temperature insulation may be terminated on lower temperature-rated terminations provided the ampacity of the conductor is based on the lower rating. This is illustrated in the example, where the THHN (90°C) conductor ampacity is based on the 75°C rating to terminate in a 75°C termination. The following table provides a quick reference of how this exception would apply to common terminations.

* Reprinted with permission from NFPA 70-2005.

Figure 5–9 (*Courtesy of Square D Company*)

Conductor Insulation versus Equipment Termination Ratings

Termination Rating	Conductor Insulation Rating		
	60°C	75°C	90°C
60°C	OK	OK (at 60°C ampacity)	OK (at 60°C ampacity)
75°C	No	OK	OK (at 75°C ampacity)
60/75°C	OK	OK (at 60°C or 75°C ampacity)	OK (at 60°C or 75°C ampacity)
90°C	No	No	OK*

* The equipment must have a 90°C rating to terminate 90°C wire at its 90°C ampacity.

- The termination may be rated for a value higher than the value permitted in the general rules if the equipment is listed and marked for the higher temperature rating.

  Example: A 30-ampere safety switch could have 75°C rated terminations if the equipment was listed and identified for use at this rating.

A Word of Caution

When terminations are inside equipment, such as panelboards, motor control centers, switchboards, enclosed circuit breakers, safety switches, and the like, it is important to note that the temperature rating identified on the equipment labeling should be followed—not the rating of the lug itself. It is common to use 90°C rated lugs (that is, marked AL9CU), but the equipment rating may be only 60°C or 75°C. The use of the 90°C rated lugs in this type of equipment does not give permission for the installer to use 90°C wire at the 90°C ampacity.

The labeling of all devices and equipment should be reviewed for installation guidelines and possible restrictions.

Equipment Terminations Available Today

Remember, a conductor has two ends, and the termination on each end must be considered when applying the sizing rules.

Example: Consider a conductor that will terminate in a 75°C rated termination on a circuit breaker at one end, and a 60°C rated termination on a receptacle at the other end. This circuit must

be wired with a conductor that has an insulation rating of at least 75°C (due to the circuit breaker), and sized based on the ampacity of 60°C (due to the receptacle).

In the realm of electrical equipment, terminations are typically rated at 60°C, 75°C, or 60/75°C. There is no listed distribution or utilization equipment that is listed and identified for the use of 90°C wire at its 90°C ampacity. This includes distribution equipment, wiring devices, transformers, motor control devices, and even utilization equipment, such as HVAC, motors, and luminaires (light fixtures). Installers and designers who have not realized this fact have been faced with jobs that do not comply with the *National Electrical Code®* and jobs that have been turned down by the electrical inspectors.

Example: 90°C wire may be used at its 90°C ampacity (see Figure 5–10). Note that the 2 AWG 90°C rated conductor does not terminate directly in the distribution equipment, but in a terminal or tap box with 90°C rated terminations.

Frequently, manufacturers are asked when distribution equipment will be available with terminations that will permit 90°C conductors at the 90°C ampacity. The answer is complex and requires not only significant equipment redesign (to handle the additional heat), but also coordination of the downstream equipment where the other end of the conductor will terminate. Significant changes in the product testing/listing standards would also have to occur.

A final note about equipment: Generally, equipment requiring the conductors to be terminated in the equipment has an insulation rating of 90°C but has an ampacity based on 75°C or 60°C. This type of equipment might include 100 percent rated circuit breakers, fluorescent luminaires (fixtures), and so on, and will include a marking to indicate such a requirement. Check with the manufacturer of the equipment to see whether any special considerations need to be taken into account.

Higher-Rated Conductors and Derating Factors

One advantage to conductors with higher insulation ratings is noted when derating factors are applied. Derating factors may be required due to the number of conductors in a conduit, higher ambient temperatures, or possibly internal design requirements for a facility.

Safety switch with 75°C terminations

Panelboard with 75°C terminations

1 AWG

90°C Insulated conductors sized based on 75°C ampacity

3 AWG

2 AWG

90°C Insulated conductors sized based on 75°C ampacity

Tap box enclosure with termination lugs or splices rated for 90°C terminations

Figure 5–10 *(Courtesy of Square D Company)*

By beginning the derating process at the ampacity of the conductor based on the higher insulation value, upsizing the conductors to compensate for the derating may not be required.

For the following example of this derating process, the following two points must be considered:

1. The ampacity value determined after applying the derating factors must be equal to or less than the ampacity of the conductor, based on the temperature limitations at its terminations.

2. The derated ampacity becomes the allowable ampacity of the conductor, and the conductor must be protected against overcurrent in accordance with this allowable ampacity.

Example (of the derating process): Assume a 480Y/277 Vac, 3Ø4W feeder circuit to a panelboard supplying 200 amperes of fluorescent luminaire (lighting) load, and assume that the conductors will be in a 40°C ambient temperature. Also, assume that the conductors originate and terminate in equipment with 75°C terminations.

1. Because the phase and neutral conductors will all be in the same conduit, the issue of conduit fill must be considered. Table 310.15(B)(4)

states that the neutral must be considered to be a current-carrying conductor because it is supplying electric discharge lighting.

2. With four current-carrying conductors in the raceway, apply Table 310.15(B)(2). This requires an 80 percent reduction in the conductor ampacity based on four to six current-carrying conductors in the raceway.

3. The correction factors at the bottom of Table 310.16 must also be applied. An adjustment of .88 for 75°C and .91 for 90°C is required where applicable.

Now for the calculations:

Using a 75°C conductor such as THWN:

300 kcmil copper has a 75°C ampacity of 285 amperes. Using the factors from the example, the calculations are as follows:

$$285 \times .80 \times .88 = 201 \text{ amperes}$$

Thus, 201 amperes is now the allowable ampacity of the 300 kcmil copper conductor for this circuit. Had the derating factors for conduit fill and ambient temperatures not been required, a 3/0 AWG copper conductor would have met these requirements.

Using a 90°C conductor such as THHN:

250 kcmil copper has a 90°C ampacity of 290 amperes. Using the factors from the example, the calculations are as follows:

$$290 \times .8 \times .91 = 211 \text{ amperes}$$

211 amperes is less than the 75°C ampacity of a 250 kcmil copper conductor (255 amperes), so the 211 amperes would now be the allowable ampacity of the 250 kcmil conductor. Had the calculation resulted in a number larger than the 75°C ampacity, the actual 75°C ampacity would have been required to be used as the allowable ampacity of the conductor. This is critical because the terminations are rated at 75°C.

> *Note:* The primary advantage to using 90°C conductors is exemplified here. The conductor is permitted to be reduced by one size (300 kcmil to 250 kcmil) and still accommodate all the required derating factors for the circuit.

In summary, when using 90°C wire for derating purposes, begin by derating at the 90°C ampacity. Compare the result of the calculation to the ampacity of the conductor based on the termination rating (60°C or 75°C). The smaller of the two numbers then becomes the allowable ampacity of the conductor. Note that if the load dictates the size of the required overcurrent device (that is, continuous load × 125%), then the conductor allowable ampacity must be protected by the required overcurrent device. (See 240.4, which permits the conductor to be protected by the next higher standard size overcurrent device as per 240.6) This may require moving up to a larger conductor size and beginning the derating. In the example, if the 200 amperes of load were continuous, a 250-ampere overcurrent device would be required. The 201 or 211 amperes from our calculations would not be protected by the 250-ampere overcurrent device and, as such, we would be required to move to a larger conductor meeting these requirements to begin the derating process.

Summary

There are a number of factors that affect how the allowable ampacity of a conductor is determined. The key is not to treat the wire as a system in itself, but as a component of the total electrical system. The terminations, the equipment ratings, and the environment all affect the ampacity that can be assigned to the conductor. If the designer and the installer keep each of the rules in mind, the installation will go much more smoothly from the beginning.

NEC® 110.14 is especially important in that the requirements for all electrical connections must be made in the proper manner. The 2005 *Code* has added text clarifying the termination requirements to meet the listing laboratories recommendations.

High-Leg Marking

NEC® 110.15 High-Leg Marking. The general requirements for marking the high leg has been appropriately located in Article 110. The rules for services remain in 230.56.

> ***110.15 High-Leg Marking.*** *On a 4-wire, delta-connected system where the midpoint of one phase winding is grounded, only the conductor or busbar having the higher phase voltage to ground shall be durably and permanently marked by an outer finish that is orange in color or by other effective means. Such identification shall be placed at each point on the system where a connection is made if the grounded conductor is also present.*
>
> Reprinted with permission from NFPA 70-2005.

New requirements for worker protection against electrical burns from "arc-flash" has been introduced into *NEC®* 110.16. These new requirements only require field marking warnings to persons performing maintenance or servicing certain electrical equipment in other than dwellings.

> ***110.16 Flash Protection.*** *Switchboards, panelboards, industrial control panels, meter socket enclosures, and motor control centers that are in other than dwelling occupancies and are likely to require examination, adjustment, servicing, or maintenance while energized shall be field marked to warn qualified persons of potential electric arc flash hazards. The marking shall be located so as to be clearly visible to qualified persons before examination, adjustment, servicing, or maintenance of the equipment.*
>
> *FPN No. 1: NFPA 70E-2004,* Standard for Electrical Safety in the Workplace, *provides assistance in determining severity of potential exposure, planning safe work practices, and selecting personal protective equipment.*
>
> *FPN No. 2: See ANSI Z535.4-1998,* Product Safety Signs and Labels, *provides guidelines for the design of safety signs and labels for application to products.*
>
> Reprinted with permission from NFPA 70-2005.

110.22 Identification of Disconnecting Means.
Each disconnecting means shall be legibly marked to indicate its purpose unless located and arranged so the purpose is evident. The marking shall be of sufficient durability to withstand the environment involved.

Where circuit breakers or fuses are applied in compliance with the series combination ratings marked on the equipment by the manufacturer, the equipment enclosure(s) shall be legibly marked in the field to indicate the equipment has been applied with a series combination rating. The marking shall be readily visible and state the following:

*CAUTION — SERIES COMBINATION SYSTEM
RATED ____ AMPERES. IDENTIFIED
REPLACEMENT COMPONENTS REQUIRED.**

Working Clearances

NEC® 110.26 reminds us that we must provide adequate working clearances around electrical equipment of 600 volts nominal or less. This section contains requirements for providing sufficient access and working space about all electric equipment for the ready and safe operation and maintenance of this equipment (see Figure 5-5 and Figure 5-11).

Lockable enclosures housing electrical apparatus that are controlled by lock and key shall be considered accessible to qualified persons.

Working space for equipment operating at 600 volts, nominal, or less to ground and likely to require examination, adjustment, servicing, or maintenance while energized shall comply with the dimensions of 110.26(A)(1), (A)(2), and (A)(3) or as required or permitted elsewhere in this Code.* 110.26(B) reminds us that this space is not to be used for storage but must be kept clear for servicing and maintenance purposes, and shall be suitably guarded when located in passage ways or clear spaces.

NEC® 110.26(C), retitled "Entrance to Working Space," now covers the minimum entrance space and the requirements for access to electrical rooms or spaces containing large electrical equipment.

Large equipment is considered as equipment rated 1200 amperes or more and over 6 ft (1.8 m) wide that contains overcurrent devices, switching devices, or control devices.

* Reprinted with permission from NFPA 70-2005.

110.26(C)(2) Large Equipment. For equipment rated 1200 amperes or more that contains overcurrent devices, switching devices, or control devices, there shall be one entrance to the required working space not less than 610 mm (24 in.) wide and 2.0 m (6½ ft) high at each end of the working space. Where the entrance has a personnel door(s), the door(s) shall open in the direction of egress and be equipped with panic bars, pressure plates, or other devices that are normally latched but open under simple pressure. [See Figure 5-11.]

Reprinted with permission from NFPA 70-2005.

NEC® 110.26(E) states that a minimum of 6½ feet (2 m) head room or to the top of the equipment, whichever is greater, must be maintained in the required working spaces to work on the equipment. This means that the area 30 inches (750 mm) wide or the width of the equipment, whichever is greater, by the depth required in Table 110.26(A)(1) must have a minimum headroom clearance of the height of the equipment or 6½ feet (2 m), whichever is greater, to provide a safe place for the worker to work on the equipment. Proper illumination must be provided in these working spaces and around the equipment. If an adequate lighting source is provided in these rooms or areas, additional specified lighting is not required.

Working Clearances Over 600 Volts

Over-600-volt requirements are found in Part III, which supplements or modifies the preceding sections. In no case shall the provisions of this part apply to equipment on the supply-side of the service conductors. These requirements are very specific regarding entrance and access to working space, guarding barriers, enclosures for electrical installations, and separation from other circuitry (see Figure 5–11 and Figure 5–12).

Over-600-volt installations must be located in locked rooms or enclosures, except where under the observation of qualified persons at all times. Table 110.34(E) covers the elevation of unguarded live parts above working space for nominal voltages 601 through 35 kV.

Many of the requirements for over-600-volt installations are similar to those for 600 volts or less; however, when preparing for an examination or applying these sections, it is very important that they be studied carefully.

Doors shall open

Panic bars, pressure
plates or hardware that
open under pressure

Figure 5–11 *NEC*® 110.26(C) (up to 600 volts) and 110.33(A)(1) (over 600 volts) defines large equipment and the required entrance and access to large electrical equipment. Where the qualified worker is required to exit through a door to get out of the working space, the door shall open out and be equipped with panic-type hardware.

Top view of electrical equipment

Wall line

Front work clearances
as per Table 110.26(A)

30" min
(750 mm)

In all cases the door
must be able to open 90°.

Figure 5–12 *NEC*® 110.26(A) requires that a minimum of clearance of 30 inches (750 mm) wide be maintained in front of electrical equipment or the width of the equipment, whichever is greater; in addition, all doors must have clearance to open at least 90°. The distance in front of the equipment must be in accordance with Table 110.26(A). See Part III to Article 110 for required clearances over 600 volts and Table 110.34(A) for clearances.

QUESTION REVIEW

Each lesson is designed purposely to require the student to apply the entire *NEC*® text, and not specific chapters or articles. It has been found that when studying for a timed, open-book examination, the student must gain proficiency in the Table of Contents, the Index, and the ability to move quickly from cover to cover to find the correct answer to each question in a timely fashion.

The following drawing applies to Questions 1–5.

This proposed electrical equipment room is to be constructed of concrete floor and concrete. Block walls will be painted.

1. What is the space required between the switchboard and the 112½ kVA transformer?

   Answer: _____

   _____

   _____

   _____

   Reference: _____

2. How many doors are required for entering and leaving this room?

   Answer: _____

   _____

   _____

   _____

   Reference: _____

3. Is this electrical equipment room required to be illuminated?

   Answer: _____

   _____

   _____

   _____

   Reference: _____

4. What is the required headroom in front of the electrical equipment needed for those performing tests or maintenance on this electrical equipment?

Answer: _____

_____

_____

_____

Reference: _____

5. If the proposed electrical room size was increased from 10 × 18 to 18 × 18, how many doors would be required, provided that the room had no additional electrical equipment added?

Answer: _____

_____

_____

_____

Reference: _____

6. Does the heat pump installation shown here comply with the *NEC*®? This is a single-family dwelling. The supplementary overcurrent device and required disconnecting means are located on the dwelling wall directly behind the heat pump for convenience.

Answer: _____

_____

_____

_____

Reference: _____

7. The working space in front of enclosed electrical equipment operating at 480 volts must have a width when facing the equipment of at least _____ inches (_____ mm).

Answer: _____

_____

_____

_____

Reference: _____

8. In an electrical system, an overcurrent device shall be placed in series with each _____ conductor.

Answer: _____

_____

_____

_____

Reference: _____

9. A conduit nipple of 18 inches (450 mm) in length is installed between two boxes. This nipple can be filled to a maximum of _____ percent without derating the conductors.

Answer: _____

_____

_____

_____

Reference: _____

10. A distribution panelboard must *not* be installed in a _____.

Answer: _____

_____

_____

_____

Reference: _____

11. An unprotected cable is installed through bored holes in wood studs. The holes must be bored so that the edge of the hole is at least _____ inch(es) (_____ mm) from the nearest edge of the stud.

Answer: _____

_____

_____

_____

Reference: _____

12. Circuit conductors run in electrical nonmetallic tubing (ENT) cannot exceed _____ volts.

    Answer: _____

    _____

    _____

    _____

    Reference: _____

13. Can liquidtight flexible nonmetallic conduit be used as a service raceway?

    Answer: _____

    _____

    _____

    _____

    Reference: _____

14. When a metal raceway is used as physical protection for a grounding electrode conductor, must this metallic raceway be bonded even though the grounding electrode conductor is bare copper? If the answer is yes, how must this metal raceway be bonded?

    Answer: _____

    _____

    _____

    _____

    Reference: _____

15. The size of the equipment grounding conductor routed with the feeder or branch-circuit conductors is determined by what?

    Answer: _____

    _____

    _____

    _____

    Reference: _____

16. How many outdoor receptacle outlets are required for a one-family dwelling?

    Answer: _____

    _____

    _____

    _____

    Reference: _____

17. Are the metal parts of electrical equipment associated with a hydromassage tub required to be bonded?

    Answer: _____

    _____

    _____

    _____

    Reference: _____

18. Where rooms within dwellings are separated by railings, planters, and so forth, are receptacles required to be installed in or along these items?

    Answer: _____

    _____

    _____

    _____

    Reference: _____

19. Are the receptacle outlet spacings for room dividers, bar-type counters, and fixed room dividers the same for site-built dwellings as they are for mobile homes?

    Answer: _____

    _____

    _____

    _____

    Reference: _____

20. You have been asked to install a receptacle on a rooftop for the air-conditioning repair people to plug their instruments and portable tools into. In which section of the *NEC*® would this requirement be found?

    Answer: _____

    _____

    _____

    _____

    Reference: _____

21. You have encountered a problem in making a new service installation and need to know the requirements of the meter enclosure on the outside of the building. Which article of the *NEC*® covers meter enclosures?

    Answer: _____

    _____

    _____

    _____

    Reference: _____

22. In making an installation in a home, the homeowners ask you to install a luminaire (light fixture) over their hydromassage bathtub. Which section of the *NEC®* governs luminaires (light fixtures) over hydromassage tubs?

    Answer: _____

    _____

    _____

    _____

    Reference: _____

23. As you are getting ready to install the conductors for the service drop on a large residence, you discover that all you have on the truck is a roll of 1 AWG THW copper. The service is 200 amperes. Realizing that this is not large enough, you parallel two 1 AWGs for each phase and two 1 AWGs for the neutral. Is this method acceptable for making this installation in accordance with the *NEC®*?

    Answer: _____

    _____

    _____

    _____

    Reference: _____

24. In Question 23, what size single conductors would be required for a 200 ampere service?

    Answer: _____

    _____

    _____

    _____

    Reference: _____

25. In making an installation of over 200 feet (61 m), a concern is voltage drop. Where does one obtain the resistance for conductors to be installed in a raceway or in direct burial for the purposes of calculating voltage drop?

    Answer: _____

    _____

    _____

    _____

    Reference: _____

Chapter Six

BRANCH CIRCUITS

Branch circuits are covered in *NEC®* Article 210. Part I covers the general provisions for branch-circuits; Part II covers the branch-circuit ratings; and Part III gives the required outlets. Outdoor branch-circuits and feeders are found in *NEC®* Article 225. Article 220 covers the calculations for branch-circuits, feeders, and services, as well as the optional calculations for calculating feeder and service loads. However, branch-circuits are found throughout the *Code*. Air-conditioning branch-circuits are found in *NEC®* Article 440, appliance branch-circuits in *NEC®* Article 422, heating

Branch Circuit

The circuit conductors between the final overcurrent device protecting the circuit and the outlet(s).

Branch Circuit, Appliance

A branch circuit that supplies energy to one or more outlets to which appliances are to be connected and that has no permanently connected luminaires (lighting fixtures) that are not a part of an appliance.

Branch Circuit, General-Purpose

A branch circuit that supplies two or more receptacles or outlets for lighting and appliances.

Branch Circuit, Individual

A branch circuit that supplies only one utilization equipment.

Branch Circuit, Multiwire

A branch circuit that consists of two or more ungrounded conductors that have a voltage between them, and a grounded conductor that has equal voltage between it and each ungrounded conductor of the circuit and that is connected to the neutral or grounded conductor of the system.

Reprinted with permission from NFPA 70-2005.

branch-circuits in *NEC®* Article 424, and motor branch-circuits in *NEC®* Article 430, and so on. Chapters 5, 6, and 7 supplement or augment the first four chapters of the *Code*. Therefore, for branch-circuit studies, it is necessary that you be familiar with the entire *NEC®* and utilize the index to quickly find the specific type branch-circuits with which you are dealing.

Branch-circuits are defined in *NEC®* Article 100 and are defined as a branch-circuit, an appliance branch circuit, a general-purpose branch-circuit, an individual branch circuit, a multiwire branch-circuit, and the branch-circuit selection current. It is very important that you read these definitions before applying the requirements in the *NEC®*, so that you are properly applying the requirements. (See Figure 6-1.)

Figure 6–1 Specific conductor types as defined in *NEC®* Article 100.

General-Purpose Branch-Circuits

The simplest forms of branch-circuits are general-purpose branch-circuits, which supply a number of outlets for lighting and appliances, and these are in 15-, 20-, 30-, 40-, and 50-ampere sizes. Where conductors or higher ampacity are used for any reason, the ampere rating or the setting of the specified overcurrent device determines the circuit classification. Multiwire branch-circuits greater than 50 amperes may be permitted for nonlighting outlet loads on industrial premises where maintenance and supervision indicate that a qualified person will service the equipment. A common multioutlet branch-circuit greater than 50 amperes is often found in industrial buildings utilizing welding receptacles so that the maintenance personnel can move their welders throughout the plant on an as-needed basis. In these instances, you may find many receptacles on one 60-ampere or above rated circuit to supply these receptacles. There is generally no danger of overloading this circuit because the plant has a limited number of welding machines and personnel capable of utilizing those machines. There are no voltage limitations on branch-circuits, because branch-circuits

can supply the equipment and motors of many varying different voltages. However, branch-circuits exceeding 600 volts nominal between conductors are only permitted to supply utilization equipment in establishments where the conditions of maintenance and supervision ensure that only qualified persons service the installation.

> **Qualified Person.** *One who has the skills and knowledge related to the construction and operation of the electrical equipment and installations and has received safety training on the hazards involved.*
> *(FPN): Refer to NFPA 70E-2004 Standard for Electrical Safety in the Workplace for Electrical Safety Training Requirements.*
> Reprinted with permission from NFPA 70-2005.

NEC® 210.6 lists the limitations for branch-circuit voltages. In occupancies such as dwelling units and guest rooms of hotels, motels, and similar occupancies, the voltage is not permitted to exceed 120 volts between conductors that supply terminals of luminaires (lighting fixtures) and cord-and-plug-connected loads 1440 VA nominal or less, or less than ¼ horse-power, and permitted to supply terminals of medium-base, screw-shell lampholders, or lampholders of other types applied within their voltage ratings, auxiliary equipment of electric discharge lamps, and cord-and-plug-connected or permanently connected utilization equipment. Circuits exceeding 120 volts, nominal, and not exceeding 277 volts, nominal, to ground are permitted to supply listed electric-discharge luminaires (lighting fixtures) equipped with medium-base screw-shell lampholder, luminaires (lighting fixtures) with mogul-base screw-shell lampholders, and lampholders other than the screw-shell type applied within their voltage ratings, auxiliary equipment of electric discharge lamps, and cord-and-plug connected or permanently connected utilization equipment. Circuits exceeding 277 volts, nominal, to ground and not exceeding 600 volts, nominal, between conductors are permitted to supply auxiliary equipment of electric discharge lamps mounted in permanently installed luminaires (fixtures) where the luminaires (fixtures) are mounted in accordance with *NEC*® 210.6(D) (see Figure 6–2). Branch circuits supplying equipment over 600 volts are required to comply with the appropriate *Code* sections and articles related to that equipment.

Example:
The circuit conductors supplying a 600-horse-power, 2300-volt, 3-phase induction motor are described in the 2002 *NEC*® as a
A. feeder.
B. branch circuit.
C. service conductors.
D. none of the above.

Answer: B. **Branch circuit.** *The circuit conductors between the final overcurrent device protecting the circuit and the outlet(s).**

* Reprinted with permission from NFPA 70-2005.

Example: What is the maximum circuit rating for a cord-and-plug-connected load that can be connected to a circuit that contains two or more 15- ampere receptacles? What is the maximum load that may be connected?

Ranges. For the computation of the range loads in these examples, Column C of Table 220.55 has been used. For optional methods, see Columns A and B of Table 220.55. Except where the computations result in a major fraction of a kilowatt (0.5 or larger), such fractions are permitted to be dropped.

SI Units. For metric conversions, 0.093 m² = 1 f² and 0.3048 m = 1 ft.

Example D6 Maximum Demand for Range Loads

Table 220.55, Column C applies to ranges not over 12 kW. The application of Note 1 to ranges over 12 kW (and not over 27 kW) and Note 2 to ranges over 8¾ kW (and not over 27 kW) is illustrated in the following two examples.

A. Ranges All the Same Rating *(see Table 220.55, Note 1)*
Assume 24 ranges, each rated 16 kW.

From Table 220.55 Column C, the maximum demand for 24 ranges of 12-kW rating is 39 kW. 16 kW exceeds 12 kW by 4.
5% × 4 = 20% (5% increase for each kW in excess of 12)
39 kW × 20% = 7.8 kW increase

Solution: 39 kW + 7.8 kW = 46.8 kW (value to be used in selection of feeders)

B. Ranges of Unequal Rating *(see Table 220.55, Note 2)*
Assume 5 ranges, each rated 11 kW; 2 ranges, each rated 12 kW; 20 ranges, each rated 13.5 kW; 3 ranges, each rated 18 kW.

5 ranges × 12 kW	=	60 kW (use 12 kW for range rated less than 12)
2 ranges × 12 kW	=	24 kW
20 ranges × 13.5 kW	=	270 kW
3 ranges × 18 kW	=	54 kW

30 ranges   Total kW = 408 kW

408 kW ÷ 30 ranges = 13.6 kW (average to be used for calculation)

From Table 220.55 Column C, the demand for 30 ranges of 12-kW rating is 15 kW + 30 (1 kW × 30 ranges) = 45 kW. 13.6 kW exceeds 12 kW by 1.6 kW (use 2 kW).

5% × 2 = 10% (5% increase for each kW in excess of 12 kW)

45 kW × 10% = 4.5 kW increase

Solution: 45 kW + 4.5 kW = 49.5 kW (value to be used in selection of feeders)

Figure 6–2 Branch circuit, feeder, and service-load calculations for a single-family dwelling. (Reprinted with permission from NFPA 70-2005, the *National Electrical Code*®, Copyright © 2004–, National Fire Protection Association, Quincy, MA 02269. This reprinted material is not the complete and official position of the National Fire Protection Association on the referenced subject, which is represented only by the standard in its entirety.)

Step 1. Refer to the Index "Branch Circuits, Overcurrent Protection," which references 210.20.

Step 2. *NEC®* 210.20 gives the rules for 15- and 20-ampere branch-circuits and references Table 210.21(B)(2).

Step 3. Table 210.21(B)(2).

Answer: A 15- or 20-ampere circuit with a maximum connected load of 12 amperes.

Example: How many general-use, duplex 20-ampere receptacles can be connected to one 120-volt, 20-ampere circuit breaker?

Step 1. This question is not complete enough to reach a single answer; therefore, we must look at both residential dwelling units and at other type installations.

Step 2. Table of Contents; branch-circuit calculations are found in Article 220.

Step 3. Calculation of branch-circuits, 220.10.

Step 4. Table 220-12; from the table we see "Dwelling Units."* See 220.14(J).

Step 5. **Dwelling Occupancies.** *In one-family, two-family, and multifamily dwellings and in guest rooms or guest suites of hotels and motels, the outlets specified in (J)(1), (J)(2), and (J)(3) are included in the general lighting load calculations of 220.12. No additional load calculations shall be required for such outlets.*

(1) All general-use receptacle outlets of 20-ampere rating or less, including receptacles connected to the circuits in 210.11(C)(3)

(2) The receptacle outlets specified in 210.52(E) and (G)

*(3) The lighting outlets specified in 210.70(A) and (B)**

Answer: For dwelling units, the number is not limited.

Step 6. For all other occupancies we must go to 220.16(B) because we have no special instructions under Table 220.12.

Step 7. **Other Outlets.** Other outlets not covered in (1) through (10) shall be computed based on 180 volt-amperes per outlet.*

* Reprinted with permission from NFPA 70-2005.

Step 8. I = VA ÷ E or I = 180 ÷ 120 or I = 1.5 amperes.

Step 9. Table 210.24 references 210.23(A), which limits cord-and-plug-connected loads to 80 percent of the branch-circuit rating.

Step 10. 20-ampere circuit breaker multiplied by 80 percent equals 16 amperes.

Step 11. 16 ÷ 1.5 = 10.667.

Answer: For other occupancies, ten duplex receptacles are permitted.

Ground-Fault Protection

The *Code* requires 120-volt 15- and 20-ampere receptacles protected by a ground-fault circuit-interrupter protection (GFCI) device in many locations where safety of the person using that circuit is of greater concern than in normal locations. In all but a few applications this protection can be achieved by installing a GFCI circuit breaker or a GFCI receptacle. However, 620.85 requires that receptacles installed in hoistways, elevator pits, on elevator car tops, and in escalator and moving walk wellways shall be of the GFCI type. But either type can protect receptacles installed in machine rooms and machinery spaces.

The GFCI rules for dwellings are found in 210.8(A), and 210.8(B) covers locations in other than dwelling units.

***210.8(A) Dwelling Units.** All 125-volt, single-phase, 15- and 20-ampere receptacles installed in the locations specified in (1) through (8) shall have ground-fault circuit-interrupter protection for personnel.*

(1) Bathrooms
(2) Garages, and also accessory buildings that have a floor located at or below grade level not intended as habitable rooms and limited to storage areas, work areas, and areas of similar use

Exception No. 1 to (2): Receptacles that are not readily accessible.
Exception No. 2 to (2): A single receptacle or a duplex receptacle for two appliances located within dedicated space for each appliance that, in normal use, is not easily moved from one place to another and that is cord-and-plug connected in accordance with 400.7(A)(6), (A)(7), or (A)(8).

> *Receptacles installed under the exceptions to 210.8(A)(2) shall not be considered as meeting the requirements of 210.52(G).*
>
> *(3) Outdoors*
> *Exception to (3): Receptacles that are not readily accessible and are supplied by a dedicated branch circuit for electric snow-melting or deicing equipment shall be permitted to be installed in accordance with 426.28.*
>
> *Receptacles installed under the exceptions to 210.8(A)(5) shall not be considered as meeting the requirements of 210.52(G).*
>
> *(4) Crawl spaces—at or below grade level*
> *(5) Unfinished basements—for purposes of this section, unfinished basements are defined as portions or areas of the basement not intended as habitable rooms and limited to storage areas, work areas, and the like*
>
> *Exception No. 1 to (5): Receptacles that are not readily accessible.*
> *Exception No. 2 to (5): A single receptacle or a duplex receptacle for two appliances located within dedicated space for each appliance that, in normal use, is not easily moved from one place to another and that is cord-and-plug connected in accordance with 400.7(A)(6), (A)(7), or (A)(8).*
> *Exception No. 3 to (5): A receptacle supplying only a permanently installed fire alarm or burglar alarm system shall not be required to have ground-fault circuit-interrupter protection.*
>
> *(6) Kitchens—where the receptacles are installed to serve the countertop surfaces*
> *(7) Laundry, utility, and wet bar sinks—where the receptacles are installed within 1.8 m (6 ft) of the outside edge of the sink.*
> *(8) Boathouses*
>
> Reprinted with permission from NFPA 70-2005.

New to the 2002 *NEC®* was the requirement that all 15- and 20-ampere receptacles installed in kitchens must be GFCI protected. The term "kitchen" is not defined, and therefore the AHJ will approve these locations and make the final determination which receptacles are covered.

> *(B) Other Than Dwelling Units. All 125-volt, single-phase, 15- and 20-ampere receptacles installed in the locations specified in (1) through (5) shall have ground-fault circuit-interrupter protection for personnel:*
> *(1) Bathrooms*
> *(2) Commercial and institutional kitchens—for the purposes of this section, a kitchen is an area with a sink and permanent facilities for food preparation and cooking*
> *(3) Rooftops*
> *(4) Outdoors in public spaces—for the purpose of this section a public space is defined as any space that is for use by, or is accessible to, the public*
>
> *Exception to (3) and (4): Receptacles that are not readily accessible and are supplied from a dedicated branch circuit for electric snow-melting or deicing equipment shall be permitted to be installed in accordance with the applicable provisions of Article 426*
>
> *(5) Outdoors, where installed to comply with 210.63*
>
> Reprinted with permission from NFPA 70-2005.

The *Code* is not generally retroactive; however, the requirements in *NEC®* 406.3(D)(3) make the GFCI safety requirements retroactive.

> *406.3(D)(2) Ground-fault circuit-interrupter protected receptacles shall be provided where replacements are made at receptacle outlets that are required to be so protected elsewhere in this* Code.
>
> Reprinted with permission from NFPA 70-2002.

NEC® 210.8(B) applies to locations other than in dwelling units. This would include common areas in multifamily dwellings, commercial, and industrial installations. The important thing to remember about these requirements is that normally in other than dwelling units a receptacle is not required to be installed. If they are installed, then the requirement must be followed.

The requirement for placing a receptacle on a rooftop is found in 210.63. Where equipment is located on the rooftop, then this requirement must be followed.

Other requirements for ground-fault circuit interrupters can be found in specific articles, such as Article 511 for commercial garages, Article 517 for health care facilities, and Article 680 for swimming pools, and so on. 210.52 covers the requirements and general provisions for installing receptacle outlets in dwellings and does not cover installations other than dwelling installations generally; there are some exceptions. There are provisions in the *Code*, in Article 220, that do cover these receptacle outlets in other than dwelling installations when installed. Lighting outlets for dwellings are covered in 210.70. Just as for receptacle outlets, the lighting requirements are generally not specified in the *NEC*® for other than dwellings. However, the installation procedures are specified when they are installed in these installations. When calculating the loads for branch-circuit conductors, one must verify whether these loads are rated as continuous duty or as noncontinuous duty. Continuous duty is defined in Article 100. Most loads in dwellings are considered to be noncontinuous loads, because they would be on for less than three hours at any given time. An example of a continuous load would be the lighting in a commercial or industrial establishment, which would generally be energized for more than three hours at a time. Most branch-circuit calculations are made in Article 220. However, specific branch-circuit requirements for branch-circuits supplying specific utilization equipment are also found in the appropriate article for that equipment, such as Article 440 for air-conditioning equipment and Article 430 for motors. There are also many other specific requirements throughout the latter chapters in the *NEC*®. Several examples for calculating branch-circuits are found in the latter sections of this chapter. Article 215 covers the installation requirements and minimum size and ampacity for conductors for feeders supplying branch-circuit loads, as calculated in Article 220.

Example: Is it permissible for the required single 20-ampere laundry circuit in a dwelling to supply both an automatic washer with a nameplate of 9.8 amperes and a gas dryer with a nameplate of 7.2 amperes?

Step 1. *NEC*® 210.52(F) requires at least one receptacle outlet.

Step 2. *NEC*® 210.11(C)(2) requires at least one 20-ampere circuit for laundry outlet(s) to supply the receptacle(s) required by 210.52(F); 422.10(B); also see references in 210.23.

Note: This means all outlets for laundry purposes, that is, washer, 120-volt dryer, and ironing closet.

Step 3. *NEC*® 210.24 and Table 210.24 reference 210.23(A).

Step 4. *NEC*® 210.23(A); the rating of any one utilization equipment supplied from a cord and plug shall not exceed 80 percent of the branch-circuit rating.

Step 5. 20×80 percent = 16 amperes. Therefore, that requirement is acceptable.

Step 6. $9.8 + 7.2 = 17$ amperes. The rating of the circuit is not exceeded.

Answer: Yes. Reference: 422.10(B), 210.24, 210.23, 210.23(A)(1), and 210.11(C)(2), also see definition of "**continuous load**." These would not be considered a continuous load.

The requirements for providing arc-fault circuit-interrupter protection (AFCI) for dwellings was first introduced into the 1999 *NEC*®. The initial rule was to require that all bedroom receptacle circuits be protected by AFCI. In the 2005 *NEC*® the requirement is that all 15- or 20-ampere circuits supplying bedrooms must be AFCI protected. This means that only an AFCI-type circuit breaker can be used to supply the bedroom outlets, including the lighting outlets and the receptacle outlets.

210.12. Arc-Fault Circuit-Interrupter Protection.
(A) Definition. *An arc-fault circuit interrupter is a device intended to provide protection from the effects of arc faults by recognizing characteristics unique to arcing and by functioning to de-energize the circuit when an arc fault is detected.*
(B) Dwelling Unit Bedrooms. *All 120-volt, single phase, 15- and 20-ampere branch circuits supplying outlets installed in dwelling unit bedrooms shall be protected by a listed arc-fault circuit interrupter, combination type installed to provide protection of the branch circuit.*

Reprinted with permission from NFPA 70-2005.

Outside Feeders and Branch Circuits

Article 225 covers outside branch-circuits and feeders. Article 225 covers the electrical equipment and wire for the supply of utilization equipment located on or attached to the outside of public and private buildings, or run between buildings, other structures, or poles on premises served. For clearances, see Figure 6–4 and Figure 6–5. An example of a feeder would be those conductors being fed from an overcurrent device in the service equipment to a subpanel in another part of the building or feeding a separate building on the premises, terminating in a panel or group of overcurrent devices within the second building or in another location. A feeder also serves separately derived systems in many cases, such as a transformer where the voltage is reduced to serve lighting and receptacle branch-circuit loads.

Arc-fault Circuit-interrupter Protection

All branch-circuits that supply
125-volt, 15- and 20-ampere
outlets in dwelling unit bedrooms
required to have AFCI protection.

Typical dwelling unit bedroom

Figure 6–3 The present ruling requires that branch-circuits supplying all 125-volt, 15- and 20-ampere outlets be AFCI protected. (See definition of "dwelling unit" in Article 100.)

NEC® 225.32. This section provides that where more than one building or structure is on the same property and under single management, each building or other structure served shall be provided with a means of disconnecting all ungrounded conductors, which must be installed either inside or outside the building or structure in a readily accessible location nearest the point of entrance of the supply feeder conductors. The disconnect disconnecting these feeder conductors must be installed in accordance with 230.70 and 230.72 and must be suitable as service equipment. However, it is required to have overcurrent protection in accordance with Article 220 for branch-circuits and Article 215 for feeders. The reference to Article 220 is necessary because residential buildings and structures, such as garages and other outbuildings, are permitted to be supplied with branch-circuits where the loads permit. For examples of branch-circuit and feeder calculations, see Figure 6–6.

A new tap rule was added to 240.21: 240.21(B)(5). This addition was necessary to cover the conductors from the secondary of an outside transformer to the building, which are now clearly defined as feeders.

FEEDERS

A feeder is defined in Article 100 as the circuit conductor between the service equipment or the source of a separately derived system and the final branch-circuit overcurrent device.

Caution: A feeder generally terminates in more than one overcurrent device.

For instance, many luminaires (lighting fixtures) and other equipment contain supplementary overcurrent devices at the luminaire or equipment. The conductors feeding that equipment should still be considered as the branch-circuit conductors and not feeder conduc-

Branch-circuits supplying lampholders—Exceptions to 150 volt limit

Incandescent lamps

480Y/277 volt 3-phase circuit

277 volt switch not part of fixture

Mogul base screw shell lampholder

Not over 300 volts to ground

Not less than 8 feet (2.5 m) from floor

Floor line

Electric discharge lamps

480Y/277 volt 3-phase 4-wire circuit

277 volt switch not part of fixture

Fluorescent fixture

Screw shell lamp holder

No minimum height

Not over 300 volts to ground

Not less than 8 feet (2.5 m)

Floor line

Figure 6–4 Examples of incandescent and electric discharge luminaires (lighting fixtures). (*Courtesy of American Iron & Steel Institute*)

tors. Feeders are generally found in *NEC*® Articles 215 and 225. Calculations for sizing feeders are found in *NEC*® Article 220 (see Figure 6–6). Other references to feeders may be found in *NEC*® Article 368 for busways, *NEC*® Article 430 for motors (see Figure 6–7), *NEC*® Article 550 for mobile homes, and *NEC*®

Article 530 for motion picture studios. See the index for a complete listing of where feeders may be found in the *NEC*®.

Examples for calculating feeders and outside branch-circuits are shown in this chapter.

(*Text continues on page 112*)

Figure 6–5 Required clearances for service and feeder conductors.

Example D1(a). One-Family Dwelling

The dwelling has a floor area of 1500 ft², exclusive of an unfinished cellar not adaptable for future use, unfinished attic, and open porches. Appliances are a 12-kW range and a 5.5-kW, 240-V dryer. Assume range and dryer kW ratings equivalent to kVA ratings in accordance with 220.54 and 220.55.

Calculated Load *[see 220.40]*
General Lighting Load: 1500 ft² at 3 VA per ft² = 4500 VA

Minimum Number of Branch Circuits Required *[see 210.11(A)]*
General Lighting Load: 4500 VA ÷ 120 V = 37.5 A

 This requires three 15-A, 2-wire or two 20-A, 2-wire circuits
Small Appliance Load: Two 2-wire, 20-A circuits *[see 210.11(C)(1)]*
Laundry Load: One 2-wire, 20-A circuit *[see 210.11(C)(2)]*
Bathroom Branch Circuit: One 2-wire, 20-A circuit (no additional load calculation is required for this circuit) *[see 210.11(C)(3)]*

Minimum Size Feeder Required *[see 220.40]*

General Lighting		4500 VA
Small Appliance		3000 VA
Laundry		1500 VA
	Total	9000 VA
3000 VA at 100%		3000 VA
9000 VA – 3000 VA =		
6000 VA at 35%		2100 VA
	Net Load	5100 VA
Range *(see Table 220.19)*		8000 VA
Dryer *(see Table 220.54)*		5500 VA
	Net Calculated Load	18,600 VA

Net Calculated Load for 120/240-V, 3-wire, single-phase service or feeder

$$18,600 \text{ VA} \div 240 \text{ V} = 77.5 \text{ A}$$

Sections 230.42(B) and 230.79 require service conductors and disconnecting means rated not less than 100 amperes.

Calculation for Neutral for Feeder and Service

Lighting and Small Appliance Load	5100 VA
Range: 8000 VA at 70% *(see 220.61)*	5600 VA
Dryer: 5500 VA at 70% *(see 220.61)*	3850 VA
Total	14,550 VA

Calculated Load for Neutral

$$14,550 \text{ VA} \div 240 \text{ V} = 60.6 \text{ A}$$

Example D1(b). One-Family Dwelling

Assume same conditions as Example No. D1(a), plus addition of one 6-A, 230-V, room air-conditioning unit and one 12-A, 115-V, room air-conditioning unit,* one 8-A, 115-V rated waste disposer, and one 10-A, 120-V rated dishwasher.* See Article 430 for general motors and Article 440, Part VII, for air-conditioning equipment. Motors have nameplate ratings of 115 V and 230 V for use on 120-V and 240-V nominal voltage systems.

* (For feeder neutral, use larger of the two appliances for unbalance.)

From Example No. D1(a), feeder current is 78 A (3-wire, 240 V).

	Line A	Neutral	Line B
Amperes from Example D1(a)	78	61	78
One 230-V air conditioner	6	—	6
One 115-V air conditioner and 120-V dishwasher	12	12	10
One 115-V disposer	—	8	8
25% of largest motor *(see 430.24)*	3	3	2
Total amperes per line	99	84	104

Therefore, the service would be rated 110 A.

Example D2(a). Optional Calculation for One-Family Dwelling, Heating Larger than Air-Conditioning *[see 220.82]*

The dwelling has a floor area of 1500 ft², exclusive of an unfinished cellar not adaptable for future use, unfinished attic, and open porches. It has a 12-kW range, a 2.5-kW water heater, a 1.2-kW dishwasher, 9 kW of electric space heating installed in five rooms, a 5-kW clothes dryer, and a 6-A, 230-V, room air-conditioning unit. Assume range, water heater, dishwasher, space heating, and clothes dryer kW ratings equivalent to kVA.

Air Conditioner kVA Calculation

$$6 \text{ A} \times 230 \text{ V} \div 1000 = 1.38 \text{ kVA}$$

This 1.38 kVA [Item 1 from 220.82(C)] is less than 40% of

Figure 6–6 Examples of branch-circuit, feeder, and service-load calculations for different types of occupancies. (Reprinted with permission from NFPA 70-2005, the *National Electrical Code*®, Copyright © 2004–, National Fire Protection Association, Quincy, MA 02269. This reprinted material is not the complete and official position of the National Fire Protection Association on the referenced subject, which is represented only by the standard in its entirety.)

9 kVA of separately controlled electric heat [Item 6 from 220.82(C)], so the 1.38 kVA need not be included in the service calculation.

General Load

1500 ft² at 3 VA	4500 VA
Two 20-A appliance outlet circuits at 1500 VA each	3000 VA
Laundry circuit	1500 VA
Range (at nameplate rating)	12,000 VA
Water heater	2500 VA
Dishwasher	1200 VA
Clothes dryer	5000 VA
Total	29,700 VA

Application of Demand Factor [See 220.82(B)]

First 10 kVA of general load at 100%	10,000 VA
Remainder of general load at 40% (19.7 kVA × 0.4)	7880 VA
Total of general load	17,880 VA
9 kVA of heat at 40% (9000 VA × 0.4) =	3600 VA
Total load	21,480 VA

Calculated Load for Service Size

21.48 kVA = 21,480 VA

21,480 VA ÷ 240 V = 89.5 A

Therefore, the minimum service size would be 100 A in accordance with 230.42 and 230.79.

Feeder Neutral Load, per 220.61

1500 ft² at 3 VA	4500 VA
Three 20-A circuits at 1500 VA	4500 VA
Total	9000 VA
3000 VA at 100%	3000 VA
9000 VA – 3000 VA = 6000 VA at 35%	2100 VA
Subtotal	5100 VA
Range: 8 kVA at 70%	5600 VA
Clothes dryer: 5 kVA at 70%	3500 VA
Dishwasher	1200 VA
Total	15,400 VA

Calculated Load for Neutral

15,400 VA ÷ 240 V = 64.2 A

Example D2(b). Optional Calculation for One-Family Dwelling, Air Conditioning Larger than Heating [see 220.82(A) and 220.82.(C)]

The dwelling has a floor area of 1500 ft², exclusive of an unfinished cellar not adaptable for future use, unfinished attic, and open porches. It has two 20-A small appliance circuits, one 20-A laundry circuit, two 4-kW wall-mounted ovens, one 5.1-kW counter-mounted cooking unit, a 4.5-kW water heater, a 1.2-kW dishwasher, a 5-kW combination clothes washer and dryer, six 7-A, 230-V room air-conditioning units, and a 1.5-kW permanently installed bathroom space heater. Assume wall-mounted ovens, counter-mounted cooking unit, water heater, dishwasher, and combination clothes washer and dryer kW ratings equivalent to kVA.

Air Conditioning kVA Calculation

Total amperes = 6 units × 7 A = 42 A

42 A × 240 V ÷ 1000 = 10.08 kVA (assume PF = 1.0)

Load Included at 100%
Air-Conditioning: Included below [see item 1 in 220.82(C)]
Space Heater: Omit [see item 5 in 220.82(C)]

General Load

1500 ft² at 3 VA	4500 VA
Two 20-A small appliance circuits a at 1500 VA each	3000 VA
Laundry circuit	1500 VA
Two ovens	8000 VA
One cooking unit	5100 VA
Water heater	4500 VA
Dishwasher	1200 VA
Washer/dryer	5000 VA
Total general load	32,800 VA
First 10 kVA at 100%	10,000 VA
Remainder at 40% (22.8 kVA × 0.4 × 1000)	9120 VA
Subtotal general load	19,120 VA
Air conditioning	10,080 VA
Total	29,200 VA

Calculated Load for Service

29,200 VA ÷ 240 V = 122 A (service rating)

Feeder Neutral Load, per 220.61

Assume that the two 4-kVA wall-mounted ovens are supplied

Figure 6–6 *(continued)* (Reprinted with permission from NFPA 70-2005, the *National Electrical Code®*, Copyright © 2004–, National Fire Protection Association, Quincy, MA 02269. This reprinted material is not the complete and official position of the National Fire Protection Association on the referenced subject, which is represented only by the standard in its entirety.)

by one branch circuit, the 5.1-kVA counter-mounted cooking unit by a separate circuit.

1500 ft² at 3 VA		4500 VA
Three 20-A circuits at 1500 VA		4500 VA
	Subtotal	9000 VA
3000 VA at 100%		3000 VA
9000 VA – 3000 VA = 6000 VA at 35%		2100 VA
	Subtotal	5100 VA

Two 4-kVA ovens plus one 5.1-kVA cooking unit = 13.1 kVA. Table 220.55 permits 55% demand factor or, 13.1 kVA × 0.55 = 7.2 kVA feeder capacity.

Subtotal from above	5100 VA
Ovens and cooking unit: 7200 VA × 70% for neutral load	5040 VA
Clothes washer/dryer: 5 kVA × 70% for neutral load	3500 VA
Dishwasher	1200 VA
Total	14,840 VA

Calculated Load for Neutral

14,840 VA ÷ 240 V = 61.83 A (use 62 A)

Example D2(c). Optional Calculation for One-Family Dwelling with Heat Pump (Single-Phase, 240/120-Volt Service) *(see 220.82)*

The dwelling has a floor area of 2000 ft², exclusive of an unfinished cellar not adaptable for future use, unfinished attic, and open porches. It has a 12-kW range, a 4.5-kW water heater, a 1.2-kW dishwasher, a 5-kW clothes dryer, and a 2½-ton (24-A) heat pump with 15 kW of backup heat.

Heat Pump kVA Calculation

24 A × 240 V ÷ 1000 = 5.76 kVA

This 5.76 kVA is less than 15 kVA of the backup heat; therefore, the heat pump load need not be included in the service calculation *[see 220.82(C)]*.

General Load

2000 ft² at 3 VA	6000 VA
Two 20-A appliance outlet circuits at 1500 VA each	3000 VA
Laundry circuit	1500 VA
Range (at nameplate rating)	12,000 VA

Water heater	4500 VA
Dishwasher	1200 VA
Clothes dryer	5000 VA
Subtotal general load	33,200 VA
First 10 kVA of general load at 100%	10,000 VA
Remainder of general load at 40% (23,200 VA × 0.4)	9280 VA
Total net general load	19,280 VA

Heat Pump and Supplementary Heat*

240 V × 24 A = 5760 VA

15-kW Electric Heat:

5760 VA + (15,000 VA × 65%) = 5.76 kVA + 9.75 kVA = 15.51 kVA

*If supplementary heat is not on at same time as heat pump, heat pump kVA need not be added to total.

Totals

Net general load	19,280 VA
Heat pump and supplementary heat	15,510 VA
Total	34,790 VA

Calculated Load for Service

34.79 kVA × 1000 ÷ 240 V = 144.96 A

This dwelling unit would be permitted to be served by a 150-A service.

Example D3. Store Building

A store 50 ft by 60 ft, or 3000 ft², has 30 ft of show window. There are a total of 80 duplex receptacles. The service is a 120/240 V, single-phase 3-wire service. Actual connected lighting load is 8500 VA.

Computed Load *(see 220.40)*

Noncontinuous Loads
Receptacle Load *(see 220.44)*

80 receptacles at 180 VA		14,400 VA
10,000 VA at 100%		10,000 VA
14,400 VA – 10,000 VA = 4400 VA at 50%		2,200 VA
	Subtotal	12,200 VA

Figure 6–6 *(continued)* (Reprinted with permission from NFPA 70-2005, the *National Electrical Code*®, Copyright © 2004–, National Fire Protection Association, Quincy, MA 02269. This reprinted material is not the complete and official position of the National Fire Protection Association on the referenced subject, which is represented only by the standard in its entirety.)

Continuous Loads

General Lighting*	
3000 ft² at 3 VA per ft²	9000 VA
Show Window Lighting Load	
30 ft at 200 VA per ft	6000 VA
Outside Sign Circuit *[see 220.14(F)]*	1200 VA
Subtotal	16,200 VA
Subtotal from noncontinuous	12,200 VA
Total noncontinuous +	
continuous loads =	28,400 VA

*In the example, 125% of the actual connected lighting load (8500 VA × 1.25 = 10,625 VA) is less than 125% of the load from Table 220.12, so the minimum lighting load from Table 220.12 is used in the calculation. Had the actual lighting load been greater than the value calculated from Table 220.12, 125% of the actual connected lighting load would have been used.

Minimum Number of Branch Circuits Required

General Lighting: Branch circuits need only be installed to supply the actual connected load *[see 210.11(B)]*.

$$8500 \text{ VA} \times 1.25 = 10{,}625 \text{ VA}$$

$$10{,}625 \text{ VA} \div 240 \text{ V} = 44 \text{ A for 3-wire, 120/240 V}$$

The lighting load would be permitted to be served by 2-wire or 3-wire, 15- or 20-A circuits with combined capacity equal to 44 A or greater for 3-wire circuits or 88 A or greater for 2-wire circuits. The feeder capacity as well as the number of branch-circuit positions available for lighting circuits in the panelboard must reflect the full calculated load of 9000 VA × 1.25 = 11,250 VA.

Show Window

$$6000 \text{ VA} \times 1.25 = 7500 \text{ VA}$$

$$7500 \text{ VA} \div 240 \text{ V} = 31 \text{ A for 3-wire, 120/240 V}$$

The show window lighting is permitted to be served by 2-wire or 3-wire circuits with a capacity equal to 31 A or greater for 3-wire circuits or 62 A or greater for 2-wire circuits.

Receptacles required by 210.62 are assumed to be included in the receptacle load above if these receptacles do not supply the show window lighting load.

Receptacles

Receptacle Load: 14,400 VA ÷ 240 V = 60 A for 3-wire, 120/240 V

The receptacle load would be permitted to be served by 2-wire or 3-wire circuits with a capacity equal to 60 A or greater for 3-wire circuits or 120 A or greater for 2-wire circuits.

Minimum Size Feeder (or Service) Overcurrent Protection *[see 215.3 or 230.90]*

Subtotal noncontinuous loads	12,200 VA
Subtotal continuous load at 125%	
(16,200 VA × 1.25)	20,250 VA
Total	32,450 VA

$$32{,}450 \text{ VA} \div 240 \text{ V} = 135 \text{ A}$$

The next higher standard size is 150 A *(see 240.6)*.

Minimum Size Feeders (or Service Conductors) Required *[see 215.2 and 230.42(A)]*

For 120/240-V, 3-wire system,

$$32{,}450 \text{ VA} \div 240 \text{ V} = 135 \text{ A}$$

Service or feeder conductor is 1/0 Cu per 215.3 and Table 310.16 (with 75°C terminations).

Figure 6–6 *(concluded)* (Reprinted with permission from NFPA 70-2005, the *National Electrical Code*®, Copyright © 2004–, National Fire Protection Association, Quincy, MA 02269. This reprinted material is not the complete and official position of the National Fire Protection Association on the referenced subject, which is represented only by the standard in its entirety.)

Example D8. Motor Circuit Conductors, Overload Protection, and Short-Circuit and Ground-Fault Protection *(see 240.6, 430.6, 430.22, 430.23, 430.24, 430.32, 430.52, and 430.62, Table 430.52 and Table 430.250)*

Determine the minimum required conductor ampacity, the motor overload protection, the branch-circuit short-circuit and ground-fault protection, and the feeder protection, for three induction-type motors on a 480-V, 3-phase feeder, as follows:

(a) One 25-hp, 460-V, 3-phase, squirrel-cage motor, nameplate full-load current 32 A, Design B, Service Factor 1.15

(b) Two 30-hp, 460-V, 3-phase, wound-rotor motors, nameplate primary full-load current 38 A, nameplate secondary full-load current 65 A, 40°C rise.

Conductor Ampacity

The full-load current value used to determine the minimum required conductor ampacity is obtained from Table 430.250 *[see 430.6(A)]* for the squirrel-cage motor and the primary of the wound-rotor motors. To obtain the minimum required conductor ampacity, the full-load current is multiplied by 1.25 *[see 430.22 and 430.23(A)]*.

For the 25-hp motor,

$$34 \text{ A} \times 1.25 = 42.5 \text{ A}$$

For the 30-hp motors,

$$40 \text{ A} \times 1.25 = 50 \text{ A}$$

$$65 \text{ A} \times 1.25 = 81.25 \text{ A}$$

Motor Overload Protection

Where protected by a separate overload device, the motors are required to have overload protection rated or set to trip at not more than 125% of the nameplate full-load current *[see 430.6(A) and 430.32(A)(1)]*.

For the 25-hp motor,

$$32 \text{ A} \times 1.25 = 40.0 \text{ A}$$

For the 30-hp motors,

$$38 \text{ A} \times 1.25 = 47.5 \text{ A}$$

Where the separate overload device is an overload relay (not a fuse or circuit breaker), and the overload device selected at 125% is not sufficient to start the motor or carry the load, the trip setting is permitted to be increased in accordance with 430.32(C).

Branch-Circuit Short-Circuit and Ground-Fault Protection

The selection of the rating of the protective device depends on the type of protective device selected, in accordance with 430.52 and Table 430.52. The following is for the 25-hp motor.

(a) Nontime-Delay Fuse: The fuse rating is 300% × 34 A = 102 A. The next larger standard fuse is 110 A *[see 240.6 and 430.52(C)(1), Exception No. 1]*. If the motor will not start with a 110-A nontime-delay fuse, the fuse rating is permitted to be increased to 125 A because this rating does not exceed 400% *[see 430.52(C)(1), Exception No. 2(a)]*.

(b) Time-Delay Fuse: The fuse rating is 175% × 34 A = 59.5 A. The next larger standard fuse is 60 A *[see 240.6 and 430.52(C)(1), Exception No. 1]*. If the motor will not start with a 60-A time-delay fuse, the fuse rating is permitted to be increased to 70 A because this rating does not exceed 225% *[see 430.52(C)(1), Exception No. 2(b)]*.

Feeder Short-Circuit and Ground-Fault Protection

The rating of the feeder protective device is based on the sum of the largest branch-circuit protective device (example is 110 A) plus the sum of the full-load currents of the other motors, or 110 A + 40 A + 40 A = 190 A. The nearest standard fuse that does not exceed this value is 175 A *[see 240.6 and 430.62(A)]*.

Figure 6–7 Examples of motor calculations. (Reprinted with permission from NFPA 70-2005, the *National Electrical Code*®, Copyright © 2004–, National Fire Protection Association, Quincy, MA 02269. This reprinted material is not the complete and official position of the National Fire Protection Association on the referenced subject, which is represented only by the standard in its entirety.)

VOLTAGE DROP

The resistance or impedance of conductors may cause a substantial difference between the voltage measured at the service-entrance of a facility and the point of use at the equipment utilizing the said voltage. It is necessary to minimize the voltage drop in conductors to ensure good service. Excessive voltage drop impairs the operation of electrical equipment. Lower than rated voltage will result not only in decreased light output from fluorescent lighting, but also may cause starting and operating difficulties. The total lumen output for a 2-lamp rapid-stop fluorescent fixture in general will vary about 1 percent for each percent change in line voltage. A similar voltage drop of 1 percent causes about 3½ percent loss in lumen output for filament-type lamps.

Although motors generally are designed for a 10 percent overall voltage variation, they operate more efficiently at nameplate voltages. Excessive voltage drop (more than 5 percent of the line voltage) can cause serious motor overheating. If the overheating is serious enough, it could cause a fire, should the overcurrent or thermal protective devices not operate properly.

Although the *NEC®* does not make mandatory requirements to govern voltage drop generally, it would be a violation of *NEC®* 110.3(B) to supply voltage to equipment at less than the nameplate rating. *NEC®* 695.7 for Fire Pumps contains specific requirements for limiting the voltage drop to 5 percent of the voltage rating of the motor. *NEC®* 210.19 includes a fine print note (FPN), that includes guides to the allowable voltage drop in branch-circuit conductors; *NEC®* includes a similar fine print note (FPN) for feeder conductors. Adherence to these fine print notes provides a reasonable efficiency of operation. As a matter of information, the *National Electrical Code®* states in fine print notes the following:

Conductors for branch circuits as defined in Article 100 size to prevent voltage drop exceeding 3 percent at the farthest outlet of power, heating, and lighting loads, or combinations of such loads, and where the maximum total voltage drop on both feeders and branch circuits to the farthest outlet does not exceed 5 percent provide reasonable efficiency of operation. *

This means that the voltage drop must be fairly equally divided between the feeder and the branch-circuit conductors. If the branch-circuit voltage drop is 3 percent of the rated voltage, then the voltage drop in the feeder should not exceed more than 2 percent. A

voltage drop greater than 5 percent of line voltage can result in a variation greater than 10 percent from designed operating voltage, as in the case of a 240-volt motor connected to a 208 branch-circuit having an excessive voltage drop.

Available heat from an electric heating appliance is proportional to the square of the applied voltage. If a heating unit is rated at 240 volts and the terminal voltage is 5 percent less, or 228 volts, the heat generated at the appliance is reduced by 10 percent. At a 10 percent reduction, or 216 volts, the heat generated at the appliance is reduced by nearly 20 percent.

It is important, therefore, that proper precautions be taken to avoid excessive voltage drop in all types of circuits, whether for lighting, power, or heating. The basic formula for determining voltage drop in a 2-wire, direct current circuit, a 2-wire, single-phase, or a 3-wire, single-phase alternating current circuit with a balanced load at 100 percent power factor (where reactance is negligible) is:

$$\text{Voltage Drop} = \frac{2 \times R \times L \times I}{\text{Circular Mils}}$$

where: R = Resistivity of conductor material (12 ohms per circular mil foot for copper, 19 ohms per circular mil foot for aluminum); L = one way length of circuit (feet); I = current in one conductor (amperes).

For 3-phase, 4-wire, alternating current circuits, the voltage drop between 1-phase wire and the neutral is ½ of the value derived from this formula. For 3-phase circuits at 100 percent power factor, the voltage drop between any 2-phase wires is 0.866 × the voltage drop calculated by the formula.

In circuits where each phase and neutral conductor consists of two or more wires, the voltage drop is determined by using the current in only one wire of a phase conductor and the neutral. For loads of other than 100 percent (unity) power factor, the formula above does not apply. For example, where motors are a major part of a circuit load, the power factor may be considerably less than one. Determining voltage drop in circuits having a load at less than unity power factor involves the use of a more complicated formula.

For 3-phase circuits, the voltage drop may be computed by multiplying the single-phase voltage drop by .86. Should the voltage drop so determined be greater than the suggested desirable percentage of the circuit voltage, a larger size conductor should be used, the circuit length shortened, or the circuit load reduced until the voltage drop is within the required or desired limits.

QUESTION REVIEW

> Each lesson is designed purposely to require the student to apply the entire *NEC®* text, and not specific chapters or articles. It has been found that when studying for a timed, open-book examination, the student must gain proficiency in the Table of Contents, the Index, and the ability to move quickly from cover to cover to find the correct answer to each question in a timely fashion.

1. A 240/480 3-phase power panelboard supplies only one 15,000 VA, 480-volt, 3-phase balanced resistive load. Each ungrounded conductor in the subfeeder to this power panel has a total net computed load of _____ amperes.

   Answer: _____

   _____

   _____

   _____

   Reference: _____

2. Fluorescent luminaires (lighting fixtures), each containing two ballasts rated 0.8 ampere each at 120 volts, are to be installed for general lighting in a store. The overcurrent protection devices are not listed for continuous operation at 100 percent of its rating. What is the maximum number of these luminaires (lighting fixtures) that may be permanently wired to a 20-ampere, 120-volt branch-circuit?

   Answer: _____

   _____

   _____

   _____

   Reference: _____

The following drawing and the drawing at the top of the next page apply to Questions 3 through 20.

Nonaccessible attic

Accessible crawl space

Example for Questions 3 through 21: You are going to build an investment single-family dwelling 28 feet by 40 feet (8.5 m by 12 m). There will be three bedrooms and one bathroom with central natural gas heat located under the floor in the crawl space. There will be no air conditioning. There will be an electric water heater (40-gallon, quick recovery) with two 4500-watt elements, a gas cooking range, and electric dryer. There will be a detached single-car garage. The electrical installation will be made in accordance with the 2005 *NEC®*. The attic in this dwelling is not suitable for the location of equipment or for storage.

3. How many square feet (square meters) will this house contain?

   Answer: _____

   _____

   _____

   _____

   Reference: _____

4. How many 15-ampere general lighting and receptacle branch-circuits are required?

   Answer: _____

   _____

   _____

   _____

   Reference: _____

5. How many 20-ampere lighting and receptacle branch-circuits are required?

   Answer: _____

   _____

   _____

   _____

   Reference: _____

6. How many appliance branch-circuits are required?

Answer: _____

_____

_____

_____

Reference: _____

7. Is a special circuit required for the laundry?

Answer: _____

_____

_____

_____

Reference: _____

8. What is the maximum standard size overcurrent device permitted to protect the electric water heater?

Answer: _____

_____

_____

_____

Reference: _____

9. How many branch-circuits are required to supply the detached garage?

Answer: _____

_____

_____

_____

Reference: _____

10. How many outdoor receptacles are required?

Answer: _____

_____

_____

_____

Reference: _____

11. What is the minimum number of ground-fault circuit interrupter (GFCI) devices required?

    Answer: _____

    _____

    _____

    _____

    Reference: _____

12. Are lighting outlets required at each outside door?

    Answer: _____

    _____

    _____

    _____

    Reference: _____

13. Are lighting outlets required in the attic not used for storage?

    Answer: _____

    _____

    _____

    _____

    Reference: _____

14. Is a receptacle required under the floor containing equipment requiring servicing?

    Answer: _____

    _____

    _____

    _____

    Reference: _____

15. If the answer to Question 7 is yes, what size branch-circuit would be required?

    Answer: _____

    _____

    _____

    _____

    Reference: _____

16. Would it be permissible to connect the furnace to a general lighting and receptacle branch-circuit provided it does not overload the circuit?

    Answer: _____

    _____

    _____

    _____

    Reference: _____

17. Are lighting outlets required in each closet?

    Answer: _____

    _____

    _____

    _____

    Reference: _____

18. Is it permissible to connect the electric ignition system for gas ranges to one of the required small-appliance branch-circuits?

    Answer: _____

    _____

    _____

    _____

    Reference: _____

19. If an outdoor receptacle is located 6 feet (1.8 m) above the ground level, is it required to be GFCI protected?

    Answer: _____

    _____

    _____

    _____

    Reference: _____

20. Could the outdoor receptacle described in Question 19 serve as one of the required outdoor receptacle?

    Answer: _____

    _____

    _____

    _____

    Reference: _____

21. What is the minimum size service required for this dwelling? (Calculate answers where required.)
    (a) General purpose lighting and receptacle load
    (b) Appliance branch-circuit load
    (c) Water heater
    (d) Laundry
    (e) Dryer
    (f) Detached garage
    (g) Furnace
    (h) Outdoor receptacle
    (i) Air conditioning
    (j) Range
    (k) Bathroom

    (a) _____

    (b) _____

    (c) _____

    (d) _____

    (e) _____

    (f) _____

    (g) _____

    (h) _____

    (i) _____

    (j) _____

    (k) _____

22. A branch-circuit supplies a single, continuous-duty, pump motor for a residential water supply. The motor-circuit conductors must have an ampacity of at least _____ percent of the motor full-load current rating.

    Answer: _____

    _____

    _____

    _____

    Reference: _____

23. The minimum lighting load required for the general lighting (only) of a church building having outside dimensions of 100 feet × 200 feet (30.5 m × 61 m) is _____ volt-amps.

    Answer: _____

    _____

    _____

    _____

    Reference: _____

24. For the kitchen small-appliance load, a dwelling unit requires at least _____ circuit(s).

    Answer: _____

    _____

    _____

    _____

    Reference: _____

25. To lower the voltage drop in a wire, reduce the conductors' length or increase their _____ .

    Answer: _____

    _____

    _____

    _____

    Reference: _____

Chapter Seven

OBJECTIVES

Studying this chapter along with the 2005 *NEC®* gives the reader a basic introduction to the service requirements in the *Code* that apply to all installations for above and below 600 volts.

After studying this chapter, you should know:

- The definitions and the application of terms that are applicable and are unique to the 2005 *NEC®*
- How to find the many *Code* requirements in which services are addressed as they apply to the *NEC®*
- Service circuit sizing
- Service conductor sizing
- Service overcurrent sizing
- Service raceway sizing
- How to find the *Code* requirements in which high-voltage services are addressed as they apply to the *NEC®*

SERVICES 600 VOLTS OR LESS

NEC® Article 230 covers the requirements for services, service conductors, and equipment for the control and protection of services and their installation requirements. Parts I through VII cover services, 600 volts, nominal, or less; Part VIII has additional requirements for services exceeding 600 volts, nominal. It should be remembered that all of Article 230 applies to these services, and Part VIII is additional provisions that supplement or modify the rest of Article 230. Part VIII shall not apply to the equipment on the supply-side of the service point. Clearances for conductors over 600 volts are found in ANSI C2, the *National Electrical Safety Code. NEC®* 230.2 generally limits a building or structure to one service. There are several special conditions that permit more than one service per building or structure. It is necessary to comply with one of these special conditions any time two or more services are being installed on a building or structure. Con-

fusing to many is the permissiveness for more than one set of service conductors permitted by the Exceptions of 230.40. These exceptions frequently are used on strip shopping centers, condominium complexes, and apartment buildings. The rule satisfies the needs for individual supply, control, and metering of occupancies in these multi-occupancy buildings. However, it does not relieve the requirement of 230.2 for one service for each building or structure (see Figure 7–1 and Figure 7–2). The service conductors, as defined in Article 100, are required to go directly to the disconnecting means of that building or structure, because the disconnecting means is required to be installed at a readily accessible location either outside of the building or structure, or inside nearest the point of the entrance of the service conductors. *NEC®* 230.6 defines that conductors shall be considered outside of a building or other structure where installed under not less than 2 inches (50 mm) of concrete beneath a building or structure, where installed within a building or structure in a raceway that is encased in concrete or brick not less than 2 inches (50 mm) thick, or where installed in a vault that complies with the requirements of *NEC®* Article 450, Part III. A typical service supplying a building or structure contains several carefully defined parts. These parts are defined in Article 100 as service, service cable, service conductors, service drop, service-entrance conductors, overhead system, underground system, service lateral, service equipment, and service point. Each service will contain several of these parts—however, not all of these parts. It is important that the components being installed in accordance with the *NEC®* be correctly defined before sizing, calculating, or installing. For instance, a typical residential service can be supplied by the serving utility either overhead or underground. When supplied overhead, it shall be supplied as a service drop. When supplied underground, it shall be supplied as a service lateral. These terms should be studied carefully in *NEC®* Article 100 so that calculations and design can be made correctly. The diagram in 230.1 should help the reader identify and use these terms correctly. (See Figure 7–3.)

Electrical service components

Figure 7–1 *NEC®* 230.2 permits only one service to each building. There are several conditions to this rule where more than one service is permitted. The service may be supplied with aerial spans overhead or by underground "service laterals" from the "service point." (See Article 100—Definitions.) **Note:** A vault is not required for underground runs but may be installed where necessary.

Figure 7–2 Illustration of two common types of residential services, overhead service drop and underground service lateral.

Article 230 – Services

Source

Part II	Overhead last pole		Underground street main	Part III
	Service drop		Service lateral	
230.24	Clearances		Depth of burial and protection	230.49
	Service head		Terminal box, meter, or other enclosure	

Service-entrance conductors — Part IV

Service equipment—general — Part V

Grounding — Article 250

Disconnecting means — Part VI

Overcurrent protection — Part VII

Branch circuits — Articles 210, 225
Feeders — Articles 215, 225

Figure 7–3 Article 230 is divided into several parts. Each covers a specific part of service installation. (Reprinted with permission from NFPA 70-2005, the *National Electrical Code®*, Copyright © 2004–, National Fire Protection Association, Quincy, MA 02269. This reprinted material is not the complete and official position of the National Fire Protection Association on the referenced subject, which is represented only by the standard in its entirety.)

Part II of Article 230 covers the overhead service-drop conductors, size and rating, and clearances required for making this installation. These conductors are generally installed by the serving utility; however, it is the responsibility of the installer to locate the service mast so that these clearances will be met. Careful and close coordination between the serving utility, the customer to be served, and the installer is necessary to ensure compliance with this section of the *NEC*®. (Refer to Figure 6–5 for clearance examples.)

Part III of Article 230 covers underground service-laterals (see Figure 7–4). Care should be taken so that the service-lateral is installed correctly and so that where it emerges from the ground it is amply protected by rigid or IMC steel conduit or Schedule 80 PVC to provide these conductors with physical protection against damage. Article 230, Part IV covers the requirements for the service-entrance conductors. These service-entrance conductors, as defined in *NEC*® Article 100, must be installed in one of the wiring methods listed in 230.43, and sized and rated in accordance with 230.42. Specific requirements for these service-entrance conductors are covered in Article 230 Part IV and must be complied with regardless of how they are routed. The general requirements for service equipment can be found in Article 230, Part V, and the disconnecting means is found in Part VI.

Example: Where a trade size 3 (78) steel intermediate metal conduit (IMC) contains 120/208 3-phase service conductors consisting of eight size 3/0 AWG THWN copper conductors, each phase consists of two conductors per phase, and the full size neutral (grounded conductor) consists of two conductors per phase terminating in a 400-ampere service-disconnecting means consisting of an inverse time circuit breaker. The data processing, fluorescent lighting, and air-conditioning load is evenly divided. Does this installation meet *Code*?

Step 1. Chapter 9 Table 1 permits a 40-percent fill of this conduit. 3/0 AWG THWN conductors; Annex C Table C4A would allow eight 3/0 AWG THWN conductors in a trade size 2½ (63) conduit. Therefore, the trade size 3 (78) IMC conduit is acceptable.

Note: Due to the change in the tables in Chapter 9 and the addition of Annex C in the *NEC*®, each type of raceway must be calculated individually. For example, in this problem if nonmetallic conduit Schedule 40 were used, a trade size 3 (78) would be required. If more than one size conductor was contained within the raceway, calculations would be necessary

Electrical service components

Figure 7–4 *NEC*® 230.2 permits only one service per building. It is permissible to supply overhead or underground.

using Chapter 9 Tables 1, 4, and 5 for the number of conductors, the specific type and size, and the specific raceway type.

Step 2. The ampacity of the 3/0 AWG THWN conductors is 200 amperes each based on Table 310.16. But Table 310.15(B)(2) requires a deration to 70 percent where eight conductors are in one raceway. Therefore, 200 amperes × 2 = 400 × 70% = 280-ampere ampacity. *NEC*® 230.90(A) Exception 2 references 240.4(B) and (C) and 240.6, which permits overload protection of the conductors at the next higher standard size overcurrent device in accordance with 240.6. Thus, the largest size overcurrent device permitted would be a 300-ampere device.

Step 3. Assuming the calculated load required the 400-ampere device, you would be required to either increase the conductor size, which would indubitably require a larger conduit or two conduits in parallel with conductors sized to carry the calculated load after derating.

Step 4. A 400-ampere calculated load would require a trade size 3½ (91) IMC and a trade size 4 (103) PVC Schedule 40 non-metallic raceway to contain the required eight 350-kcmil type THWN conductors.

Answer: Table 310.16 and Table 310.15 (B)(2): 310 × 2 = 620 × 70% = 434.

Each service disconnecting means is required to be suitable for the prevailing equipment conditions, and when installed in hazardous locations must comply with Chapter 5 of the *National Electrical Code*®. Each service disconnecting means permitted by 230.2 or for each set of service-entrance conductors permitted by 230.40, Exception 1 (see Figure 7–5) shall consist of not more than six switches or six circuit breakers mounted in a single enclosure, in a group of separate enclosures, or on a switchboard. (See Figure 7–6.) There shall be no more than six disconnects per service grouped in any one location. These requirements are critical and must be followed. The disconnecting means must be grouped; each disconnect must be marked to indicate the load served. An exception permits one of the six disconnecting means permitted where used for a water pump intended to provide fire protection to be located

remotely from the other disconnecting means. In multiple-occupancy buildings, each occupant shall have access unless local building management, which supplies continuous supervision, is present. In such a case, the service disconnecting means shall be permitted to be accessible to the authorized management personnel only. Each service disconnecting means shall simultaneously disconnect the ungrounded service conductors from the premise wiring. Where the service disconnecting means does not disconnect the grounded conductor from the premise wiring, other means shall be provided for this purpose in the service equipment. A terminal or bus to which all grounded conductors can be attached by means of pressure connections shall be permitted for this purpose.

Example: A 277/480-volt, 3-phase service to a building is rated at 1600 amperes. The required main disconnecting means consists of two 800-ampere fused disconnects. Is this service required to have ground-fault protection (GFP)?

Index "Ground-Fault Protection," *Ground-fault protection for service disconnects, 230.95.*

Note: To answer this question, some basic electrical knowledge is required. First, the voltage tells the story. 277/480 indicates it is first a solidly grounded system. It also indicates it is a wye system because 480 equals 1.73 of 277. This information can be obtained from additional reference books listed as study material.

Answer: No. References: 230.71, the maximum number of service disconnects six (6); 230.95 states that solidly grounded wye services of more than 150 volts to ground not exceeding 600 volts phase-to-phase and over 1000 amperes requires ground-fault protection.

Part VII covers the overcurrent protection and the location of the overcurrent protection requirements for services. All solidly grounded wye services of more than 150 volts to ground but not exceeding 600 volts phase-to-phase rated 1000 amperes or more must be provided with ground-fault protection of equipment. Two exceptions to this: Exception 1 states that ground-fault protection does not apply to the service disconnecting means of a continuous industrial process where a nonorderly shutdown will introduce additional or increased hazard. Exception 2 states that it does not apply to fire pumps. Specific settings and testing

Figure 7–5 Typical riser sketch of a multiple occupancy building. *NEC*® 230.40 Exception 1 permits the main disconnecting means for that occupancy to be located on the load end of the feeder within the occupancy.

Figure 7–6 Typical sketch of six individual disconnecting means on one service as permitted by *NEC*® 230.71(A).

procedures for this ground-fault protection of equipment are covered in 230.95. Additional ground-fault protection of feeders where ground-fault protection has not been provided on the service is covered in both Article 215 and Article 240.

Example: Where a steel conduit contains eight (8) 3/0 AWG THWN insulated copper conductors with two of the 3/0 AWG conductors being in parallel for each phase, and two 3/0 AWG conductors being in parallel for the neutral, (a) would this installation be permitted to be terminated into a 400-ampere main service breaker where the system voltage is 120/208? (The expected loads are evenly divided between fluorescent lighting, data processing equipment, and air-conditioning equipment.) (b) If not, what would be the maximum ampere rating for this service?

Answer: Eight (8) 3/0 AWG THWN conductors in a single conduit, Annex C, Table C8, would require a trade size 3 (78) rigid metal conduit or tubing.

Table 310.16—3/0 AWG THWN has an ampacity of 200 amperes (from 75°C column). Where the number of current-carrying conductors exceeds three, the allowable ampacity shall be reduced as shown in the table with seven to nine conductors, a derating of 70 percent shall be applied. See Table 310.15(B)(2).

Two conductors per phase:

$$2 \times 200 \times 70\% = 280 \text{ amperes}$$

The next higher overcurrent device per 240.3 would be 300 amperes. The service would be a 300-ampere service.

The 400-ampere overcurrent device would require 2-350 kcmil per phase and with eight 350 kcmil THWN would require the rigid metal conduit to be increased to trade size 4 (103). For conductor termination, see 110.14(C).

For services exceeding 600 volts, nominal, Part VIII must be read and adhered to carefully. The definition of service point formerly appeared in this section. The examples given in this section help clarify the application of *NEC*® Article 230.

SERVICES OVER 600 VOLTS

Services over 600 volts nominal introduce an additional level of hazard. Article 230 Part VIII covers the requirements for services exceeding 600 volts nominal, which modify or amend the rest of Article 230. All of Article 230 is applicable and, in addition, Part VIII must be followed. Clearance requirements for over-600-volt services can be governed by ANSI C2, the *National Electrical Safety Code*, as noted in the FPN in 230.200. The service-entrance conductors to buildings or enclosures shall not be smaller than 6 AWG unless in cable, and in cable not smaller than 8 AWG, and installed by the specific methods listed in 230.202(B). It should be noted that Article 300, Part II and Article 110, Part II also will need to be regularly referenced in making installations over 600 volts. The requirements for support guarding and draining cables are covered in this section. Warning signs with the words "Danger. High Voltage. Keep Out." shall be posted where unauthorized people come in contact with energized parts. The disconnecting means must comply with 230.70 or 230.208(B) and shall simultaneously disconnect all underground conductors and shall have fault closing rating of not less than the maximum short circuit available in supply terminals, except where used switches or separate mining fuses are installed, the fuse characteristic shall be permitted to contribute to the fault closing rating of the disconnecting means. These requirements for over-600-volt services create the need for your studies to take you into new areas of the *Code*, such as Part II of Article 100, which covers the general requirements for over 600 volts nominal; Part II of Article 100, which gives you the unique definitions for those circuits; Part IX of Article 240; Part XI of Article 250 for grounding of system; and Part II of Article 300; and the requirements of Article 490. You must be aware that these additional requirements amend or augment the primary requirements in each article. In most instances, those requirements also apply to over-600-volt nominal services. Therefore, as you study this section, familiarize yourself with these new parts as they apply to your studies.

QUESTION REVIEW

> **Each lesson is designed purposely to require the student to apply the entire *NEC*® text, and not specific chapters or articles. It has been found that when studying for a timed, open-book examination, the student must gain proficiency in the Table of Contents, the Index, and the ability to move quickly from cover to cover to find the correct answer to each question in a timely fashion.**

1. Where a circuit breaker is provided as the short-circuit protective device for service-entrance conductors exceeding 600 volts, the breaker shall have a trip setting of not more than _____ times the ampacity of the conductors.

   Answer: _____

   _____

   _____

   _____

   Reference: _____

2. An electrical service transmission line drops the voltage from 255 volts to 240 volts. The service transmission line efficiency is _____ percent.

   Answer: _____

   _____

   _____

   _____

   Reference: _____

3. Where a group of buildings are served by electrical feeders from a single service drop or electrical service point in another building, is a disconnecting means required at each of the other buildings where the feeders terminate or can the feeders terminate in main lug-only panels?

   Answer: _____

   _____

   _____

   _____

   Reference: _____

4. Is there a limit to the number of overcurrent devices that can be used on the secondary side of a transformer if the total of the device ratings does not exceed the allowed value for a single secondary overcurrent device?

   Answer: _____

   _____

   _____

   _____

   Reference: _____

5. Does it make a difference which opening in the service head the service-entrance conductors are brought out of? If so, why?

   Answer: _____

   _____

   _____

   _____

   Reference: _____

6. If the clearances required by *NEC*® 110.26 can be maintained in front of a panelboard or a service panel to be installed in a bathroom in a dwelling, would this installation be permitted by the *NEC*®?

   Answer: _____

   _____

   _____

   _____

   Reference: _____

7. When service-entrance cable is used within a dwelling for branch-circuit wiring, is the Type SE cable considered the same as Type NM nonmetallic-sheathed cable?

   Answer: _____

   _____

   _____

   _____

   Reference: _____

8. Can a 120/240-volt, single-phase, 3-wire service be located closer than 3 feet (900 mm) to a window of a dwelling?

   Answer: _____

   _____

   _____

   _____

   Reference: _____

9. The *Code* generally prohibits splicing a grounding-electrode conductor.
   (a) Is this conductor permitted to be tapped?
   (b) If so, how are these taps sized?

   Answer: _____

   _____

   _____

   _____

   Reference: _____

10. Circuit breakers are required to open all ungrounded conductors of a circuit simultaneously. Obviously, fuses cannot be required to open all ungrounded conductors of a circuit simultaneously. Why the difference in the requirements?

    Answer: _____

    _____

    _____

    _____

    Reference: _____

11. What is a separately derived system? Where is it mentioned in the Code?

    Answer: _____

    _____

    _____

    _____

    Reference: _____

12. How many overcurrent devices are permitted to be used on the secondary side of a transformer if the total of the device ratings does not exceed the allowed value for a single overcurrent device?

    Answer: _____

    _____

    _____

    _____

    Reference: _____

13. Is it necessary to identify the higher voltage-to-ground phase (hi-leg) at the disconnect of a 3-phase motor where the service available is 120/240-volt, 3-phase, 4-wire?

    Answer: _____

    _____

    _____

    _____

    Reference: _____

14. A service panel is securely bolted to the metal frame of a building that is effectively grounded, and the panelboard's main bonding jumper is properly installed. Is this sufficient to ground the panelboard to the metal building grounding electrode as required in *NEC*® 250.64?

    Answer: _____

    _____

    _____

    _____

    Reference: _____

15. What is the maximum permitted contraction or expansion for a run of Schedule 40 PVC conduit before the installation of an expansion coupling is required?

Answer: _____

_____

_____

_____

Reference: _____

16. How can one distinguish between the mandatory rules of the *NEC®* and the explanatory rules of the *NEC®*?

Answer: _____

_____

_____

_____

Reference: _____

17. On a recent installation, the plans called for the building steel to be used as a grounding electrode if effectively grounded. What does this term mean, and where is it used in the *NEC®*?

Answer: _____

_____

_____

_____

Reference: _____

18. Which article of the *NEC®* covers Type nonmetallic-sheathed cable?

Answer: _____

_____

_____

_____

Reference: _____

19. What is the maximum size flexible metallic tubing available manufactured today?

Answer: _____

_____

_____

_____

Reference: _____

20. Which article of the *National Electrical Code*® contains the requirements for fusible safety switches and other type snap switches?

    Answer: _____

    _____

    _____

    _____

    Reference: _____

21. In making an installation in a residential garage, you have determined that physical protection is needed for the nonmetallic-sheathed cable and that it must be installed in steel conduit. How do you size the conduit for this nonmetallic-sheathed cable?

    Answer: _____

    _____

    _____

    _____

    Reference: _____

22. In making an installation for a large parking garage adjacent to a hotel building, which article of the *NEC*® covers these wiring methods?

    Answer: _____

    _____

    _____

    _____

    Reference: _____

23. You have encountered a nonmetallic wireway to be installed in a wet location. Which article of the *NEC*® covers this nonmetallic wireway?

    Answer: _____

    _____

    _____

    _____

    Reference: _____

24. In attempting to calculate the load on a small office building, the receptacles are to be calculated at how many volt-amperes each?

    Answer: _____

    _____

    _____

    _____

    Reference: _____

25. In wiring a new residence, the most convenient location for the lighting and branch-circuit panelboard is on a bathroom wall. Is this an acceptable wiring method? Which section governs this requirement?

    Answer: _____

    _____

    _____

    _____

    Reference: _____

Chapter Eight

OBJECTIVES

Studying this chapter along with the 2005 *NEC®* gives the reader a basic introduction to the overcurrent requirements in the *Code* that apply to all installations for above and below 600 volts.

After studying this chapter, you should know:

- The definitions and the application of terms that are applicable and are unique to the 2005 *NEC®*
- The many *Code* articles in which overcurrent is addressed as it applies to the *NEC®*
- Overcurrent sizing
- Fuse applications and types
- Circuit breaker types and applications
- Short circuit/fault current calculations

Figure 8–1 Panelboards. *(Courtesy of Square D Company)*

OVERCURRENT PROTECTION

NEC® Article 240 provides the general requirements for overcurrent protection and overcurrent protective devices not more than 600 volts, nominal. Part IX of *NEC®* Article 240 covers the overcurrent protection for over 600 volts, nominal. The overcurrent protection for conductors and equipment is provided to open the circuit if the current reaches a value that will cause an excessive or dangerous temperature in conductors or conductor insulation. *NEC®* 110.9 and 110.10 cover the requirements for the interrupting capacity and protection against fault currents (see Figure 8–1). Article 240 covers the general requirements. The specific overcurrent requirements for equipment can be found in each individual article as it pertains to that specific equipment or circuitry; for example, the overcurrent protection for air-conditioning and refrigeration equipment is covered in Article 440 and appliances in Article 422. A listing of the specific overcurrent requirements can be found in 240.3. *NEC®* 240.4 outlines the protection of conductors. This section provides the rules for the many different applications throughout the *Code*. *NEC®* 240.5 covers the requirements for protecting flexible cords and fixture wires,

and standard overcurrent devices are listed in *NEC®* 240.6 for fuses and fixed-trip circuit breakers. See Figure 8–2 and Figure 8–3 for examples. Adjustable-trip circuit breakers are described and covered in 240.6(B). Part II of Article 240 specifies the location that the overcurrent device must be placed in the circuit. Part III specifies the enclosure required for enclosing the overcurrent devices. Part IV covers the disconnecting requirements for disconnecting the overcurrent devices so that they are accessible to the maintenance personnel performing work on those devices. Parts V and VI cover fuses, fuseholders, and adapters for both plug- and cartridge-type fuses; circuit breakers are covered in Part VII. Article 240, Part VIII covers supervised industrial installations. Part IX

Figure 8–2 Types IFL, IKL, and ILL "1 LIMITER" molded case current-limiting circuit breakers. *(Courtesy of Square D Company)*

Figure 8–3 Class T current-limiting, fast-acting fuse; 200,000-ampere interrupting rating, 300 and 600 volts. Has little time delay. Generally used for protection of circuit-breaker panels and for circuits that do not have high-inrush loads, such as motors. When used on high-inrush loads, generally size at 300 percent so as to be able to override the momentary inrush current. These fuses have different dimensions compared with ordinary fuses. They will not fit into switches made for other classes of fuses, nor will other classes of fuses fit into a Class T disconnect switch. *(Courtesy of Cooper Bussmann)*

covers the overcurrent protection for over 600 volts, nominal, and states that feeders shall have short-circuit protection devices in each ungrounded conductor or comply with 230.208(A) and (B). The protective device(s) shall be capable of detecting and interrupting all values of current that can occur at their location in excess of their trip setting or melting point. In no case shall the fuse rating exceed three times the long-time

trip element setting of the breaker, or six times the ampacity of the conductor. Branch-circuit requirements for over-600-volt circuits are covered in 240.100.

CAPABLE OF CARRYING THE AVAILABLE FAULT CURRENT

Short-Circuit Calculations

Throughout the *National Electrical Code®*, as well as in industry standards, you will find references to fault currents, short-circuit currents, ground-fault currents, interrupting ratings, impedances of transformers, and the like.

NEC® 110.9 states that *equipment intended to interrupt current at fault levels shall have an interrupting rating sufficient for the nominal circuit voltage and the current that is available at the line terminals of the equipment.**

NEC® 110.10 states that *the overcurrent protective devices, the total impedance, the component short-circuit ratings, and other characteristics of the circuit to be protected shall be so selected and coordinated as to permit the circuit-protective devices used to clear a fault to do so without extensive damage to the electrical components of the circuit.**

These tough code requirements mean that we need to know the available fault currents at the main service equipment as well as at all points in the system. Too often, engineers, electricians, and contractors look for adequate interrupting ratings of the fuses or breakers in the main service equipment only. This is wrong. The *Code* tells us to observe available fault currents throughout the electrical system. This means panelboards, motor controllers, motor control centers, distribution centers, and so on.

Determining Short-Circuit Current at Various Distances from Transformers, Switchboards, Panelboards, and Load Centers Using the Point-to-Point Method

A simple method of determining the available short-circuit current (also called fault current) at various distances from a given location is the "**point-to-point method**." Reasonable accuracy is obtained when this method is used with 3-phase and single-phase systems.

The following procedure demonstrates the use of the "**point-to-point method**."

* Reprinted with permission from NFPA 70-2005.

Step 1. Determine the full-load rating of the transformer—in amperes—from the transformer nameplate, tables, or the following formulas:

a) For 3-phase transformers:

$$I_{FLA} = \frac{kVA \times 1000}{E_{L\text{-}L} \times 1.73}$$

where $E_{L\text{-}L}$ = Line-to-line voltage

b) For single-phase transformers:

$$I_{FLA} = \frac{kVA \times 1000}{E_{L\text{-}L}}$$

Step 2. Find the percent impedance (Z) on nameplate of transformer.

Step 3. Find the transformer multiplier "M_1":

$$M1 = \frac{100}{\text{transformer \% impedance (Z)} \times 0.9}$$

(*Note:* Because the marked transformer impedance can vary ±10 percent per the UL standard, the 0.9 factor takes this into consideration to show "worst case" conditions.)

Step 4. Determine the transformer let-through short-circuit current at the secondary terminals of transformer. Use tables or the following formula:

a) For 3-phase transformers (L-L-L):

$$I_{SCA} = \text{transformer}_{FLA} \times \text{multiplier "}M_1\text{"}$$

b) For 1-phase transformers (L-L):

$$I_{SCA} = \text{transformer}_{FLA} \times \text{multiplier "}M_1\text{"}$$

c) For 1-phase transformers (L-N):

$$I_{SCA} = \text{transformer}_{FLA} \times \text{multiplier "}M_1\text{"} \times 1.5$$

(*Note:* The 1.5 factor is explained in Step 5 in the paragraph marked "X.")

Step 5. Determine the "f" factor:

a) For 3-phase faults:

$$f = \frac{1.73 \times L \times I}{N \times C \times E_{L\text{-}L}}$$

b) For 1-phase, line-to-line (L-L) faults on 1-phase, center-tapped transformers:

$$f = \frac{2 \times L \times I}{N \times C \times E_{L\text{-}L}}$$

c) For 1-phase, line-to-neutral (L-N) faults on 1-phase, center-tapped transformers:

$$f = \frac{2 \times L \times I}{N \times C \times E_{L\text{-}N}}$$

where

L = the length of the circuit to the fault, in feet.

I = the available fault-current—in amperes— at the beginning of the circuit.

C = the constant derived from the "C" tables (see tables on pages 137–138) for the specific type of conductors and wiring method.

E = the voltage, line-to-line or line-to-neutral. See Step 4a, b, and c to decide which voltage to use.

N = the number of conductors in parallel.

X = the fault current in amperes at the transformer terminals.

The L-N fault current is higher than the L-L fault current at the secondary terminals of a single-phase, center-tapped transformer. At some distance from the terminals, depending on the wire size, the L-N fault current is lower than the L-L fault current. This can vary from 1.33 to 1.67 times. These figures are based on the change in the turns ratio between primary and secondary, infinite-source impedance, a distance of zero feet from the terminals of the transformer, and 1.2% resistance (X) and 1.5% resistance (R) for the L-N versus L-L resistance and reactance values.

For simplicity, in Step 4c we used an approximate multiplier of 1.5. First, do the L-N calculation at the transformer secondary terminals, Step 4c, then proceed with the point-to-point method.

Step 6. After finding the "f" factor, refer to Chart M (see page 138) the appropriate value of the multiplier "M2" for the specific "f" value. Or, calculate as follows:

$$M_2 = \frac{1}{1 + f}$$

"C" VALUES—Copper Conductors

| | Three Single Conductors | | | | | | Three Conductor Cable | | | | | |
| | Steel Conduit | | | Nonmagnetic Conduit | | | Steel Conduit | | | Nonmagnetic Conduit | | |
AWG or kcmil	600V	5KV	15KV	600V	5KV	15KV	600V	5KV	15KV	600V	5KV	15KV
14	389	—	—	389	—	—	389	—	—	389	—	—
12	617	—	—	617	—	—	617	—	—	617	—	—
10	981	—	—	981	—	—	981	—	—	981	—	—
8	1557	1551	1557	1556	1555	1558	1559	1557	1559	1559	1558	1559
6	2425	2406	2389	2430	2417	2406	2431	2424	2414	2433	2428	2420
4	3806	3750	3695	3825	3789	3752	3830	3811	3778	3837	3823	3798
3	4760	4760	4760	4802	4802	4802	4760	4790	4760	4802	4802	4802
2	5906	5736	5574	6044	5926	5809	5989	5929	5827	6087	6022	5957
1	7292	7029	6758	7493	7306	7108	7454	7364	7188	7579	7507	7364
1/0	8924	8543	7973	9317	9033	8590	9209	9086	8707	9472	9372	9052
2/0	10755	10061	9389	11423	10877	10318	11244	11045	10500	11703	11528	11052
3/0	12843	11804	11021	13923	13048	12360	13656	13333	12613	14410	14118	13461
4/0	15082	13605	12542	16673	15351	14347	16391	15890	14813	17482	17019	16012
250	16483	14924	13643	18593	17120	15865	18310	17850	16465	19779	19352	18001
300	18176	16292	14768	20867	18975	17408	20617	20051	18318	22524	21938	20163
350	19703	17385	15678	22736	20526	18672	22646	21914	19821	24904	24126	21982
400	20565	18235	16365	24296	21786	19731	24253	23371	21042	26915	26044	23517
500	22185	19172	17492	26706	23277	21329	26980	25449	23125	30028	28712	25916
600	22965	20567	17962	28033	25203	22097	28752	27974	24896	32236	31258	27766
750	24136	21386	18888	28303	25430	22690	31050	30024	26932	32404	31338	28303
1000	25278	22539	19923	31490	28083	24887	33864	32688	29320	37197	35748	31959

"C" VALUES—Aluminum Conductors

| | Three Single Conductors | | | | | | Three Conductor Cable | | | | | |
| | Steel Conduit | | | Nonmagnetic Conduit | | | Steel Conduit | | | Nonmagnetic Conduit | | |
AWG or kcmil	600V	5KV	15KV	600V	5KV	15KV	600V	5KV	15KV	600V	5KV	15KV
14	236	—	—	236	—	—	236	—	—	236	—	—
12	375	—	—	375	—	—	375	—	—	375	—	—
10	598	—	—	598	—	—	598	—	—	598	—	—
8	951	950	951	951	950	951	951	951	951	951	951	951
6	1480	1476	1472	1481	1478	1476	1481	1480	1478	1482	1481	1479
4	2345	2332	2319	2350	2341	2333	2351	2347	2339	2353	2349	2344
3	2948	2948	2948	2958	2958	2958	2948	2956	2948	2958	2958	2958
2	3713	3669	3626	3729	3701	3767	3733	3719	3693	3739	3724	3709
1	4645	4574	4497	4678	4631	4580	4686	4663	4617	4699	4681	4646
1/0	5777	5669	5493	5838	5766	5645	5852	5820	5717	5875	5851	5771
2/0	7186	6968	6733	7301	7152	6986	7327	7270	7109	7372	7328	7201
3/0	8826	8466	8163	9110	8851	8627	9077	8980	8750	9242	9164	8977
4/0	10740	10167	9700	11174	10749	10386	11184	11021	10642	11408	11277	10968
250	12122	11460	10848	12862	12343	11847	12796	12636	12115	13236	13105	12661
300	13909	13009	12192	14922	14182	13491	14916	14698	13973	15494	15299	14658
400	16670	15355	14188	18505	17321	16233	18461	18063	16921	19587	19243	18154
500	18755	16827	15657	21390	19503	18314	21394	20606	19314	22987	22381	20978
600	20093	18427	16484	23451	21718	19635	23633	23195	21348	25750	25243	23294
750	21766	19685	17686	25976	23701	20934	26431	25789	23750	29036	28262	25976
1000	23477	21235	19005	28778	26109	23482	29864	29049	26608	32938	31919	29135

"C" VALUES—Busways

Ampacity	Plug-In Busway		Feeder Busway		High Imped. Busway
	Copper	Aluminum	Copper	Aluminum	Copper
225	28700	23000	18700	12000	—
400	38900	34700	23900	21300	—
600	41000	38300	36500	31300	—
800	46100	57500	49300	44100	—
1000	69400	89300	62900	56200	15600
1200	94300	97100	76900	69900	16100
1350	119000	104200	90100	84000	17500
1600	129900	120500	101000	90900	19200
2000	142900	135100	134200	125000	20400
2500	143800	156300	180500	166700	21700
3000	144900	175400	204100	188700	23800
4000	—	—	277800	256400	—

CHART M (multiplier)

f	M	f	M
0.01	0.99	1.20	0.45
0.02	0.98	1.50	0.40
0.03	0.97	2.00	0.33
0.04	0.96	3.00	0.25
0.05	0.95	4.00	0.20
0.06	0.94	5.00	0.17
0.07	0.93	6.00	0.14
0.08	0.93	7.00	0.13
0.09	0.92	8.00	0.11
0.10	0.91	9.00	0.10
0.15	0.87	10.00	0.09
0.20	0.83	15.00	0.06
0.30	0.77	20.00	0.05
0.40	0.71	30.00	0.03
0.50	0.67	40.00	0.02
0.60	0.63	50.00	0.02
0.70	0.59	60.00	0.02
0.80	0.55	70.00	0.01
0.90	0.53	80.00	0.01
1.00	0.50	90.00	0.01
		100.00	0.01

$$M = \frac{1}{1 + f}$$

Step 7. Multiply the available fault current at the beginning of the circuit by the multiplier "M_2" to determine the available symmetrical fault current at the fault.

I_{SCA} at fault = I_{SCA} at beginning of circuit × "M_2"

Example: Provide the fuse sizes for a 10-horsepower, 230-volt, 3-phase motor with a code letter G and a service factor of 1.15. Using *Table 430.250*, the full load current is 28-amperes. [See *430.6(A)(1)*.]

(A) Using *Table 430.52*, the maximum rating of a time-delay fuse is 175 percent of the motor full load current. The fuse size, using time-delay fuses for the branch-circuit, short-circuit, and ground-fault protection is calculated as follows: 28 × 1.75 = 49 amperes. *NEC® 430.52, Exception No. 1* permits where the values for branch-circuit, short-circuit, and ground-fault protective devices determined by *Table 430.52* do not correspond to the standard sizes of fuses, a higher size that does not exceed the next higher standard ampere rating shall be permitted. The next higher fuse rating as shown in *240.6* is 50 amperes. *NEC® 430.52(C)(1), Exception No. 2(b)* permits where this rating is not sufficient for the starting current of the motor, the rating of the time-delay fuse shall be permitted to be increased but shall in no case exceed 225 percent of the full load current.

(B) Motor overload protection based on the motor service factor of 1.15% is calculated in accordance with *430.32*, as follows: 28 × 1.25 = 35 amperes.

(C) Using non–time-delay fuses, sizes for branch-circuit, short-circuit, and ground-fault protective devices are calculated as follows: 28 × 3 =

84 amperes. The next higher rating as permitted by *430.52(C)(1), Exception No. 2(a)* is 90 amperes. Where this rating is not sufficient for the starting current of the motor, the rating of the non–time-delay fuse shall be permitted to be increased but shall in no case exceed 400 percent of the full load current.

Example: To calculate the available short-circuit current at a panelboard located 20 feet (6 m) away from a transformer when the transformers are 500 kcmil copper in steel conduit. The "C" value for the conductors is 22,185. The transformer is marked 300 kVA, 208/120 volts, 3-phase, 4-wire. The transformer impedance is 2 percent. Consider the source to have an infinite amount of fault current available (often referred to as "infinite primary").

(1) To find the transformer's full-load secondary current:

$$I_{fla} = \frac{kVA \times 1000}{E \times 1.73} = \frac{300 \times 1000}{208 \times 1.73} = 834 \text{ amperes}$$

(2) To find the transformer SCA: (multiplier = 100/2 = 50)

Transformer SCA =

$I_{fla} \times$ multiplier = $834 \times 50 = 41,700$ amperes

(3) To find the (F) factor:

$$F = \frac{1.73 \times L \times I}{C \times E_{L-L}} = \frac{1.73 \times 20 \times 41,700}{22,185 \times 208} = 0.3127$$

(4) To find the (M) multiplier:

$$M = \frac{1}{1+F} = \frac{1}{1+0.3127} = 0.76$$

(5) To find the short-circuit current of the panelboard:

$$I_{sca} = \text{transformer}_{sca} \times (M) = 41,700 \times 0.76 = 31,692 \text{ amperes}$$

The fault current available at the panelboard where it is located 20 feet (6 m) away from the transformer is equal to 31,692 amperes.

QUESTION REVIEW

> **Each lesson is designed purposely to require the student to apply the entire *NEC®* text, and not specific chapters or articles. It has been found that when studying for a timed, open-book examination, the student must gain proficiency in the Table of Contents, the Index, and the ability to move quickly from cover to cover to find the correct answer to each question in a timely fashion.**

1. Does the *Code* permit the use of two single-pole circuit breakers in a panelboard to serve a line-to-line connected load such as a household electrical range rated at 120/240 volts or a hot water heater rated at 240 volts?

   Answer: _____

   _____

   _____

   _____

   Reference: _____

2. Where outdoor conductors are tapped and these tapped conductors terminate into a single overcurrent device designed to limit the load so as not to exceed the ampacity of the tap conductors, how long can these tap conductors be at the maximum length?

   Answer: _____

   _____

   _____

   _____

   Reference: _____

3. Is it permissible to use 300-volt cartridge-type fuses and fuseholders to protect a multiwire, 4-wire circuit such as an emergency circuit in a food store where the electrical service is 277/480-volt, 3-phase, 4-wire?

Answer: _____

_____

_____

_____

Reference: _____

4. Service-entrance conductors are required to be protected at their rated ampacity with exception, the exception being the next standard size overcurrent device when the service conductors are protected by a single device. When are multiple overcurrent devices supplied by a service allowed to exceed the ampacity rating of the service-entrance conductors?

Answer: _____

_____

_____

_____

Reference: _____

5. Fixed electric space heating loads shall be computed at _____ percent of the total connected load, generally.

Answer: _____

_____

_____

_____

Reference: _____

6. A motor is to be installed in a Class I, Division 2 location. It is to be connected with 3 feet (900 mm) of trade size ¾ (21) liquidtight flexible metal conduit. Is it permissible to use the liquidtight flexible metal conduit as the grounding path for this installation?

Answer: _____

_____

_____

_____

Reference: _____

7. Are feeder conductors required to be rated for 125 percent of the continuous load plus the noncontinuous load?

Answer: _____

_____

_____

_____

Reference: _____

8. How many duplex receptacles can be installed on a 20-ampere branch-circuit in a dwelling?

Answer: _____

_____

_____

_____

Reference: _____

9. Table 310.16 lists the allowable ampacity of 12 AWG THWN conductors as 25 amperes. Can 16 general-use receptacles in an office building be connected to a 20-ampere circuit?

Answer: _____

_____

_____

_____

Reference: _____

10. Should you use Article 310 or Annex B to calculate underground conductor ampacities?

Answer: _____

_____

_____

_____

Reference: _____

11. When an electric heat pump is installed with backup resistance heat in a building, (a) could one of the loads ever be considered dissimilar for the purposes of calculating the service loads? (b) Would the heat pump load have to be calculated at 125 percent for the service sizing?

Answer: _____

_____

_____

_____

Reference: _____

12. Are all 20-ampere residential underground circuits required to have GFCI protection?

Answer: _____

_____

_____

_____

Reference: _____

13. Can two circuit breakers mounted adjacent to each other in a panelboard with handle ties be used in place of a double-pole breaker to supply a 240-volt electric baseboard heater?

    Answer: _____

    _____

    _____

    _____

    Reference: _____

14. Are water heaters considered to be a continuous load so that a 125 percent load is required for calculating feeder and service on an installation?

    Answer: _____

    _____

    _____

    _____

    Reference: _____

15. Do the tap conductors from the secondary of a transformer always require secondary overcurrent protection? How about a 10-foot (3 m) tap? a 25-foot (7.5 m) tap?

    Answer: _____

    _____

    _____

    _____

    Reference: _____

16. Is it permitted to plug a microwave oven with a nameplate rating of 13 amperes into a 15-ampere receptacle protected on a 20-ampere circuit?

    Answer: _____

    _____

    _____

    _____

    Reference: _____

17. Can a 1400 VA load, a cord-and-plug-connected load, be supplied by a 240-volt circuit in a residence?

    Answer: _____

    _____

    _____

    _____

    Reference: _____

18. Three receptacles on a single yoke or strap are to be installed in a commercial or industrial facility. Are these outlets to be calculated at 180 VA or 540 VA?

Answer: _____

_____

_____

_____

Reference: _____

Use the following diagram for Questions 19–21.

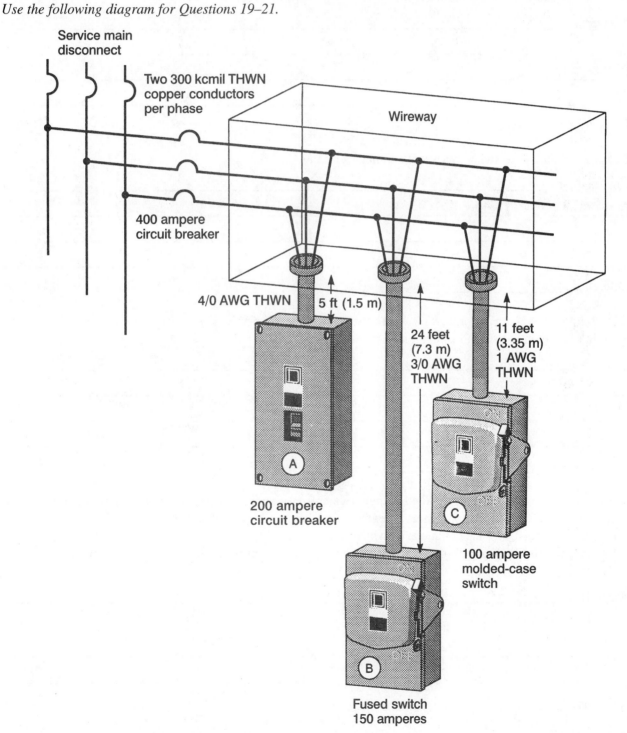

19. Does the circuit breaker marked Ⓐ comply with the *NEC*®? If so, which section?

    Answer: _____

    _____

    _____

    _____

    Reference: _____

20. Does the switch marked Ⓑ comply with the *NEC*®? If so, which section?

    Answer: _____

    _____

    _____

    _____

    Reference: _____

21. Does the molded-case switch marked Ⓒ comply with the *NEC*®?

    Answer: _____

    _____

    _____

    _____

    Reference: _____

22. *NEC*® requires overcurrent protection where the busway is reduced in size for the last 50 feet (15.24 m) or more in other than industrial installations. True or False?

    Answer: _____

    _____

    _____

    _____

    Reference: _____

23. A 10-foot (3 m) tap is made in the field without overcurrent protection where the smaller conductor is tapped to the larger conductor. The conductors extend beyond the switchboard. The tap conductor must be
    (a) at least $\frac{1}{10}$ the size of the conductor to which it is tapped.
    (b) at least $\frac{1}{10}$ the ampere rating of the overcurrent device that protects the conductors from which the tap is made.

    Answer: _____

    _____

    _____

    _____

    Reference: _____

24. In an extremely long feeder circuit, the circuit conductors have been increased in size to compensate for voltage drop. Is the equipment grounding conductor also required to be increased?

    Answer: _____

    _____

    _____

    _____

    Reference: _____

25. The nameplate on an air-conditioning unit is marked "maximum size fuse 40-amperes." Is it permissible to install a 40-ampere circuit breaker, or must a 40-ampere fuse protection be installed?

    Answer: _____

    _____

    _____

    _____

    Reference: _____

Chapter Nine

OBJECTIVES

Studying this chapter along with the 2005 *NEC®* gives the reader a basic introduction to the grounding requirements in the *Code* that apply to all installations for above and below 600 volts.

After studying this chapter, you should know:

- A brief history on grounding and why the system used in the United States was chosen over other systems

- The definitions and the application of terms that are applicable and that are unique to the 2005 *NEC®*

- The many *Code* requirements in which grounding is addressed as it applies to the *NEC®*

- The differences between grounding and bonding

- The differences between the grounded conductor and neutral

- The importance of equipment grounding and the proper sizing of the equipment grounding conductor

- The charts and software available for proper sizing of the equipment grounding conductor

GROUNDING ELECTRICAL SYSTEMS

Important reasons for achieving a properly grounded installation are the safety and property protection afforded by a grounded system.

The key to a reliable electrical system is a well-designed grounded electrical system with all materials properly selected for the application and installed in accordance with the manufacturer's recommended installation guidelines, product "listing," and a well-recognized electrical code.

Note: There are exceptions for unique installations in which an ungrounded system with reliable ground-fault detectors and assured maintenance are acceptable.

It is important to note that although the ground element of all electrical systems is the most important element, it still remains one of the most misunderstood and most errant part of any electrical system. Many engineers, teachers, students, and skilled workers regularly design, teach, and install grounding systems incorrectly, which often negates or exacerbates the systems' operating characteristics. (See Figure 9–1.)

History of Grounding

Since before the turn of the century when electricity began being used for commercial purposes, the topic of grounding and the mysteries surrounding this subject have occupied the minds of those considered to be experts in the electrical field. The mysteries of grounding become a major topic at nearly every meeting of any kind involving the electrical community, regardless of whether it is on the agenda or not. The discussion varies from the mythology to many real-life problem scenarios encountered by those involved in the use of day-to-day electricity.

It was decided many years ago that a grounded system provides the best system, both economically and from a safety point of view. A grounded system provides for a safer system and also stabilizes the voltage levels so as to protect equipment. Even the earliest leaders in this industry seemed to generally agree that with an ac system, all low-voltage secondary should be grounded; this is necessary to abolish the many risks of fatal shock and damage to equipment, and to reduce electrical fire hazards. Many heated discussions took place in the first 20 years of the twentieth century regarding this subject. Although the experts did not agree initially, as time went by it was finally resolved that the grounded system provided the safest economical system for general use. Early discussions on electrical circuit grounding brought about such comments as those from Killingsworth Hedges, an English scientist, who stated, "One precaution is to earth the secondary circuit, another is to connect one or both leads to a safety appliance which would automatically divert any excess current to earth and at the

Requirements for grounding electrode conductor, grounded system

*Grounds both the identified grounded conductor and the equipment grounding conductor. It is connected to the grounded metalic cold water piping on premises, or other available grounding electrode, such as metal building frame if effectively grounded, or a made electrode.

Figure 9–1 Grounding principles. *(Courtesy of American Iron & Steel Institute)*

same time shut off the supply." Another comment from Professor Elihu Thomson suggested grounding the transformer secondary winding or surrounding it with a grounded sheath; as alternatives, he suggested cutting off the secondary by automatic means or grounding it automatically by fuse cutouts when excessive voltage appeared.

AIEE, the predecessor to IEEE, resolved early on that the recommendation of grounding low-tension secondary ac systems was strongly recommended in all cases where a reliable grounding connection could be secured, and in 3-wire systems with a grounded neutral and that solid connections without fuse be permitted on the neutral wire. They further stated, "The grounding of the secondary ac systems inside of buildings to water pipes provided such connection is made nearest the point to cellar wall on water pipes outside of meter is not only safe, but places no burden on the water pipes."

In 1911, the National Fire Protection Association met and resolved that grounding of transformer secondary circuits was highly recommended. In the 1913 *National Electrical Code®*, a mandatory grounding rule, *Rule 15*, was adopted, which stated:

Grounding Low Potential Circuits-Alternating current secondary systems, transformer secondary of distribution systems except when supplied from private, industrial power or lighting plants, when the primary voltage does not exceed 550 volts, must be grounded provided the maximum difference of potential between the grounded point of any other point in the circuit does not exceed 150 volts and may be grounded when the maximum difference of potential between the grounded point and any other point in the circuit exceeds 150 volts. In either case, the following rules must be complied with: (1) The grounding must be made at the neutral point or wire whenever the neutral point or wire is accessible; (2) When no neutral point or wire is accessible, one side of the secondary circuit must be grounded; (3) The ground connection must be at the transformer or on the individual service as provided in Parts III–VII and when transformers feed systems with a neutral wire, the neutral must also be grounded at least every 500 feet (152.4 m).

This rule has not been changed in substance since that time. The main difference is that in the 1923 and 1940 *National Electrical Codes®*, grounding for voltages above 150 to 300 volts was recommended instead of merely permitted.

Since the 1913 resolution requiring mandatory grounding, experts have had many discussions on the merits and methods of grounding. There is no reason to believe that many more discussions on this great subject will not occur. However, the 1996 *National Electrical Code®* was published in the fall of 1995, and it continues to contain mandatory rules for safe grounding practices. These rules are essentially the same as the original rules that went in the 1913 *National Electrical Code®*. Although experts have continued to discuss and debate the methods and merits, the science has not changed and the general agreement is consistent with those of the early fathers of the electrical community—

Grounding provides the safest, most economical electrical system for general use known to humankind

Most Use the Solidly Grounded System

It is enlightening to note that those major countries on the North American continent—Canada, the United States, and Mexico—all supply predominantly grounded systems to residential and commercial facilities; therefore, most electrical equipment manufactured in these three countries can be readily transported between the countries without alteration of the equipment. However, it is equally important to note that each country—Canada, the United States, and Mexico—has quality installation problems. In many areas across the North American continent, good grounding practices are not followed, and grounding installations are not made in accordance with all good engineering guidelines; nor is the *National Electrical Code®* closely followed where adopted. Even though it is considered a minimum standard for an electrical installation, many installers are negligent in making

Types of AC Utilization Systems 600 Volts or Less

a. Single-phase, 3-wire, 120, 240 volts

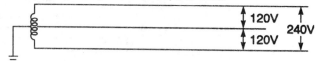

b. Three-phase, 4-wire wye 208Y/120 volts

c. Three-phase, 4-wire delta, 240/120 volts

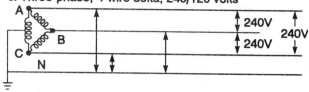

d. Three-phase, 4-wire wye, 480/277 volts

e. Three-phase, 3-wire, 120 delta, 240 or 480 volts

Insulating transformers balanced across phases; secondaries must be grounded

Figure 9–2 Examples of grounded systems. *(Courtesy of American Iron & Steel Institute)*

the grounding installations in accordance with those minimum standards or better. (See Figure 9–1 and Figure 9–2.)

General Requirements

The requirements for systems that must be grounded are covered in *NEC®* Article 250, 250.20 and 250.21. (See Figure 9–2.)

Alternating-Current Systems to Be Grounded. Alternating-current circuits and systems shall be grounded as provided for in *250.20(A), (B), (C), or (D)*. Other systems shall be permitted to be grounded.

(A) Alternating-Current Systems of Less Than 50 Volts. Alternating-current systems of less than 50 volts shall be grounded under any of the following conditions: (1) Where supplied by transformers, if the transformer supply system exceeds 150 volts to ground; (2) Where supplied by transformers, if the transformer supply system is ungrounded; (3) Where installed as overhead conductors outside of buildings.

(B) Alternating-Current Systems of 50 Volts to 1000 Volts. (1) Where the system can be grounded so that the maximum voltage to ground on the ungrounded conductors does not exceed 150 volts; (2) Where the system is 3-phase, 4-wire, wye connected in which the neutral is used as a circuit conductor; (3) Where the system is 3-phase, 4-wire, delta connected in which the midpoint of one phase winding is used as a circuit conductor.

(C) Alternating-Current Systems of 1 kV and Over. Alternating current systems supplying mobile or portable equipment shall be grounded as specified in 250.188. Where supplying other than mobile or portable equipment, such systems shall be permitted to be grounded.

(D) Separately Derived Systems. Separately derived systems, as covered in 250.20(A) or (B) shall be grounded as specified in 250.30.*

Only four system circuits are not permitted to be grounded by the *NEC®*:

1. cranes that operate over combustible materials in a hazardous Class III location
2. isolated power systems in hospitals or other health care facilities in locations of hazardous inhalation, anesthetizing, and in wet locations
3. circuits with electrolytic cells as provided in Article 668
4. lighting systems as provided in 411.5(A)

Ungrounded Permitted

There are delta systems that are permitted by the *NEC®* to be ungrounded. These are generally systems used for industrial or agriculture applications, and where the ungrounded system is installed, usually an additional degree of service continuity is needed. It is important, however, to note that a lesser degree of safety is ensured with this system, and an occurrence of a ground fault on a system will not cause an overcurrent protective device to open. It does, however,

ground the system. A second fault on the same feeder or circuit will cause one or more overcurrent protective devices to open. If the second fault is in another feeder or circuit on the same system, one or more of the overcurrent devices will open. For safety reasons, as well as for system continuity, it is necessary that maintenance personnel locate and correct these ground faults as soon as practical. Systems permitted by the *NEC®* to be installed as ungrounded systems are delta systems, such as 480-volt, 3-phase, 3-wire; 2300-volt, 3-phase, 3-wire; 4600-volt, 3-phase, 3 wire; and 13.8 kV, 3-phase, 3-wire systems. Although not required by the *National Electrical Code®*, ground detector lights that indicate that a ground fault has occurred on an ungrounded system are highly advisable. The light will burn until the phase goes to ground, thus identifying the faulted phase. This indicator light is to alert maintenance personnel so that the ground fault can be corrected, preferably immediately or during the hours while production is not in operation, and the plant can continue operations with the single fault (one phase grounded) thus preventing a costly shutdown. New state-of-the-art equipment is available, which will help identify the exact location of the fault.

Corner Grounded Delta

Many power companies in the United States occasionally furnish a 480-volt corner grounded delta service for secondary use in agriculture and oil field installations. Many questions arise as to the use of this system, which employs the 3-wire delta secondary with one phase intentionally grounded. (See Figure 9–3.) It is not the most common service used today; however, it has provided a grounded system in locations formerly supplied by a 3-wire ungrounded delta system. It should be emphasized, however, that all metal parts that are not intended to carry current must be permanently bonded to an unfused, unswitched conductor, which extends back to the transformer secondary ground. It must be run with the circuit conductors and

Figure 9–3 3-phase, 3-wire delta corner grounded transformer. (See *NEC®* 250.20 [FPN] and 250.26(4).)

sized to the overcurrent protection protecting the circuit. It is not acceptable to bond the equipment to a driven grounding rod at the equipment location. This means that the user's wiring, beyond the service main disconnecting means must consist of four wires, the three current-carrying phase conductors and a fourth wire called an equipment grounding conductor. The equipment grounding conductor is the conductor that is not intended to carry load current, but instead provides an interconnection between the transformer bank ground and all grounds on the user's equipment. All grounding of the electrical circuit conductors, surge arrestors, and conductive noncurrent-carrying materials and equipment are required to be installed and arranged in a manner that will prevent objectionable flow of current over the grounding conductor or grounding paths under normal operating conditions. If these objectionable currents are present, one or more alterations must be made. One must discontinue one or more such grounding connections, change the location of the grounding connections, interrupt the continuity of one or more of the conducting paths interconnecting the grounding connections, or take other suitable remedial action satisfactory to the authority having jurisdiction.

The advantages for this 480-volt corner-grounded system are that two element meters can be used and are less expensive than three element meters and require less maintenance. Triplex (three-conductor) wire is less expensive than quadplex (four-conductor) wire. The meter base is also less expensive. However, dangerous step-and-touch potentials can exist, such as nonautomatic isolation and location of phase-to-ground shorts and ground currents objectionable to residents living in the area. Ground potentials have been known to cause solid-state starters to start in the "off" position. Lightning arrestor applications may be unsatisfactory. There may be interference to communication equipment nearby. This system places a higher dielectric stress on the two ungrounded phases and higher magnitudes of transient overvoltages. In conclusion, for the corner-grounded 480-volt service, the most significant reason for use is that it is less expensive. However, it may not provide a safe working environment, and it is advisable to request service from a 4-wire system from the utility company if it is available. This will reduce normal current flows in the grounds due to operational problems. It is advisable to consider other types of 480-volt delta grounding or solidly grounded wye systems where available.

As you can see, the ungrounded system for some specialty systems is still permitted, and the corner-grounded delta provides an additional level of safety where the traditional grounded system is not available. But, generally, it is not nearly as desirable as a solidly grounded system and should only be selected under engineering guidelines.

Grounded Conductor (Neutral)

The grounded conductor of the system is one of the most important elements. It is that conductor derived from the supplying power source, transformer, generator, or other derived source, that is required by the *NEC*® to be grounded at an accessible location from the load end of the service drop or service lateral to and including the terminal bus at the service disconnecting means. These locations would include current transformer enclosures, meter enclosures, pull boxes, junction boxes, busways, auxiliary gutters, or in the panel box acceptable to the authority having jurisdiction. This grounded service conductor provides a vital good low impedance path for the fault current to return to the service. It must be run to and installed in each service disconnecting means regardless of whether it is needed for the service or not. If the grounded conductor is not run to the service as required, it is possible for ground fault to have a high impedance path back to the source. It is a violation of the *NEC*® 250.4(A), and becomes virtually impossible to clear a ground fault, thus introducing a hazard in the system. However, if we comply with the *NEC*® 250.24 for a single building installation and 250.32 for multiple buildings supplied by one service, we have a low-impedance ground fault path and have complied with 250.4 of the *NEC*®. The safety of the service is improved immeasurably, because the grounded conductor is run to provide a ground fault path. The size of the conductor depends on the size of the phase conductors and must never be smaller than the grounding-electrode conductor, the size of which is given in Table 250.66 of the *NEC*®. (See Figure 9–4.)

Grounding-Electrode System (Conductor)

Earth is considered to be at zero potential. The grounding electrode—see *NEC*® 250.50 and 250.52(A)(1) through (6)—consisting of all effectively grounded (see definition of "**grounded effectively**") components available and where one or more effectively grounded electrodes are not available, a made electrode must be used; see *NEC*® 250.52(A)(4) through

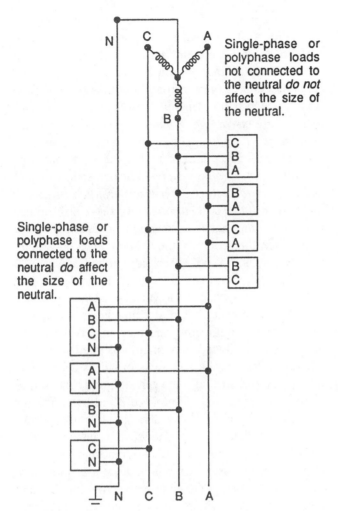

Figure 9–4 Size of a neutral conductor as in *NEC®* 220.61. *(Courtesy of American Iron & Steel Institute)*

NEC® Article 250, Part V

Figure 9–5 Bonding jumpers as in *NEC®* Article 250, Part V. *(Courtesy of American Iron & Steel Institute)*

(7). Where metal water pipe is the only electrode available, it must also be supplemented as per *NEC®* 250.53(D)(2). The electrodes are bonded together to form one common electrode system, as stated in *NEC®* 250.58. The grounding electrode must be large enough to provide a low-impedance path to the electrode system for lightning strikes, voltage surges, or in the event of a large fault. (See Figure 9-5.)

It is important to note that under normal conditions no current potential exists between any conductor or equipment where the system is properly designed, installed, and maintained. Only in abnormal conditions such as lightning, line surges, and high-fault conditions does current flow through this system to earth.

Though it is desirable to have a grounding-electrode system have an impedance as low as practicable, it is imperative that we design and install systems that provide a low-impedance path back to the source that will clear a fault when it occurs without

resulting in loss of property or life. In a properly designed grounded system where a hazard can exist for only the time it takes to clear the fault, it is extremely unlikely that any loss of life would occur and damage to property and equipment would be kept to a minimum.

Equipment Grounding System (Conductor)

Equipment grounding is somewhat different than the system grounded conductor. It relates to the methods for grounding nonelectrical conductive materials, which either enclose energized conductors or are adjacent thereto. These conductive materials are required to be interconnected and grounded for several reasons to achieve the end result of a safe installation. The requirement is to bond all metal enclosures and objects adjacent to conductive energized conductors; to provide current-carrying capabilities large enough to adequately accept any ground-fault current permitted by the overcurrent protective system without creating a fire or explosive hazard to life or property; to ensure freedom from dangerous electrical shock exposure to persons in the area; and to contribute to a superior performance of the system. The equipment ground plays no part in the lighting or power circuit; however, it plays a very important part in ensuring the safety and equipment protection needed for every good reliable system (see the definition of "**Grounding conductor,**

Equipment"). Some examples of components of the equipment-grounding system are metallic conduit, motors, equipment enclosures, and grounding conductor sized as per 250.122. *NEC®* 250.122 and Table 250.122 provide for a minimum size only. The size of the conductor may have to be increased due to the length of the circuit or the amount of fault current that is available and may be imposed upon it. This is clarified in the general requirements of the *National Electrical Code®* Article 110, and the requirements of 110.10 must be complied with.

> *Note:* A grounding conductor is part of the equipment-grounding system as distinguished from a grounded conductor, which is part of the power distribution system. The requirements for this grounding system are covered in Article 250 of the *NEC®*. See 250.4(A).

NEC® **110.10, Circuit Impedance and Other Characteristics.** *The overcurrent protective devices, the total impedance, the component short-circuit current ratings, and other characteristics of the circuit to be protected shall be selected and coordinated to permit the circuit-protective devices used to clear the fault to do so without extensive damage to the electrical components of the circuit. This fault shall be assumed to be either between two or more of the circuit conductors, or between any circuit conductor and the grounding conductor or enclosing metal raceway. Listed products applied in accordance with their listing shall be considered to meet the requirements of this section.*

Reprinted with permission from NFPA 70-2005.

NEC® 250.4 states that the path to ground from circuits, equipment, and metal enclosures for conductors shall

- be permanent and continuous.

- have the capacity to safely conduct any fault current likely to be imposed on it.

- have sufficiently low impedance to limit the voltage to ground and to facilitate the operation of the circuit protective devices in the circuit.

NEC® 250.90 and 250.96 require bonding to ensure the continuity and the capacity to safely carry the fault currents likely to be imposed. *NEC®* 250.122 lists the types of equipment grounding conductors with exceptions. Table 250.122 lists the minimum allowable size of copper and aluminum conductors.

There are three basic components to a complete grounding system. Each is composed of one or more parts that have different, but related, functions. In order to ensure this low-impedance path back to the service, it is necessary to comply with 300.3(B) of the *NEC®*, which states, *All conductors of the same circuit and, where used, the grounded conductor and all equipment grounding conductors and bonding conductors shall be contained within the same raceway, auxiliary gutter, cable tray, cablebus assembly, trench, cable, or cord, unless otherwise permitted in accordance with 300.3(B)(1) through (B)(4).* * This becomes of extreme importance in ac applications. It is the total impedance that controls the current division among parallel paths. In 60 Hz circuits rated under 100 amperes, the circuit reactance is an insignificant part of the circuit impedance. However, reactance may be a prominent element of circuit impedance in circuits rated above 100 amperes, and reactance is the dominant element of impedance for circuits rated above 200 amperes. The reactance of an ac circuit is determined mainly by the spacing between outgoing and return conductors and is only slightly affected by conductor size. However, circuit resistance is affected by conductor size. The ratio of inductive reactance to the resistance and the relative effect of reactance on circuit impedance increases as the conductor size increases. Increased separation (spacing between grounding and phase conductors) increases not only the inductive reactance of the grounding conductor, but also the zero sequence inductive reactance of the phase conductors. Hence, on a grounded system, this conductor plus the main bonding jumper at the service and the bonding jumper between the transformer case and the transformer winding are all available as a low impedance path for fault currents. There is no dependence on the electrode earth path except for the grounded conductor if it is inadvertently interrupted. (See Figure 9–6.) If the grounded conductor, the main bonding jumper, and equipment grounding conductors are provided, properly connected, and of adequate capacity, only three conditions can prevent rapid operation of the overcurrent device: (1) a faulty or improperly rated overcurrent device; (2) a high-resistance fault, such as a fault through high-resistance contact with a ground object; or (3) a high-resistance equipment grounding conductor connecting the faulty equipment to the service equipment enclosure. The third condition can be prevented by proper selection of the grounding conductor and correct installation.

* Reprinted with permission from NFPA 70-2005.

Figure 9–6 Bonding at service disconnect enclosure as required in *NEC®* 250.92. *(Courtesy of American Iron & Steel Institute)*

Tubular metal raceway, such as rigid metal conduit, intermediate metal conduit, or electrical metallic tubing, fits the preferred conductor geometry perfectly. The phase conductors and the neutral conductors are enclosed within the metal raceway. (See Figure 9–7.) In this way, effectiveness is markedly improved. The returning ground-fault current distributes itself about the entire enclosing metal shell in such a fashion as to result in a lower round trip voltage drop. The electrical behavior during the line-to-line fault is that of a coaxial line, except for the effects of resistivity in the metal shell. All electric and magnetic fields are minimal and contained inside the shell. The external space magnetic field becomes zero.

Warning: It is extremely important that only one grounding system exist in an installation because multiple separated grounds can create unwanted ground loops that may create errors in data transmission and malfunctions. Where a ground loop exists, there are known techniques that will solve the problem. Although the single point ground is preferred, it may not always be feasible where long distances and very high frequencies are present.

Example: A dairy barn is fed underground from the farm house with a 225-ampere overcurrent device using 4/0 AWG copper phase conductors and a 1 AWG (neutral) grounded conductor. What size equipment-grounding conductor is required for this installation?

Step 1. Index, Grounding, Separate buildings, 250.32; also Grounding conductors, Sizes, Table 250.122.

Step 2. *NEC®* 250.32, Exception 2 requires an insulated or covered equipment-grounding conductor where livestock is housed; 547.9(A) contains similar requirements also.

Step 3. Table 250.122; the minimum size permitted for a 225-ampere overcurrent device.

Figure 9–7 Tubular metal raceway used as an equipment-grounding conductor. (See 250.4, 250.122(A), and Table 250.122.)

The table lists only a 200 and a 300; therefore, the 300 would apply because the 200 would not be adequate.

Answer: 4 AWG copper

Computer Model Developed

It is with a great deal of satisfaction that I can report that the excellent work that has carried this industry for the past forty years, developed and written by

R. H. "Dick" Kaufman (General Electric "GER 957A," *Some Fundamentals of Equipment Grounding Circuit Design,* IE 1058.33 November 1954, Applications and Industry, Vol. 73, Part II) and

Eustace C. Soares (Pringle Switch, *Grounding Electrical Distribution Systems for Safety,* 1966 through the 5th Edition *IAEI Soares Book on Grounding,* 1993, by J. Philip Simmons)

has been revisited.

In early 1993 at the request of J. Philip Simmons, author of the 4th and 5th editions of *IAEI Soares Book on Grounding*, the steel conduit and tubing producer members of the National Electrical Manufacturers Association (NEMA) made funding available to Georgia Institute of Technology's School of Electrical and Computer Engineering to develop a computer model. A. P. Sakis Meliopoulos, P.E., and Elias N. Glytsis, P.E., developed a model that was validated by field tests consisting of arc-voltage testing, and fault-current testing on thirteen 256-foot runs of electrical metallic tubing with both steel and die-cast set-screw and compression fittings, rigid steel conduit and intermediate steel conduit with conductors enclosed at Kearney Laboratories at McCook, Illinois. Mr. Meliopoulos stated in his final report that the techniques established by *Soares Book on Grounding* has served this industry well in providing primary design guidance. (See Figure 9–8, Figure 9–9, Figure 9–10, Figure 9–11, Figure 9–12, and Figure 9–13.)

Fortunately, computer and analytical means, which have been developed at a rapid pace, have now made it possible to develop and validate a model utilizing current data that can be used with confidence.

This new data will replace the only data available to date. Today's manufacturing processes have changed both the quality of material used to produce electrical materials and the consistency of those products. Although hand calculations were fairly accurate

to evaluate given criteria, this new model will offer a broad band of conditions and the user no longer has to extrapolate data. Mr. Meliopoulos and I presented a paper through IAS/IEEE in 1997 at the Fall Meeting, and the model is now available. The attachments to this report are excerpts from the final report. It is interesting to note that the tests and research of new materials and circuit-breaker test guidelines revealed that the criteria of 50 arc volt and 500 percent (5IP) of overcurrent device rating were overly conservative, and he recommended an arc voltage of 40 and a 4IP (4 times the rating of the overcurrent device, i.e., 30-ampere circuit-breaker, 4IP = 30 × 400% = 120) of the overcurrent device, and as stated these are still conservative numbers.

"Modeling and Testing of Steel EMT, IMC and Rigid (GRC) Conduit." Prepared by A. P. Sakis Meliopoulos and Elias N. Glytsis, School of Electrical and Computer Engineering, Georgia Institute of Technology, Atlanta, GA 30332.

For copies of this report and the free software, contact: The Steel Tube Institute of North America, http://www.steelconduit.org.

Even when the raceway does not provide the sole equipment ground but is supplemented by an equipment grounding conductor wire or a bus enclosed within, the raceway provides a parallel path and also protects the wire and bus from physical damage. Where connected as a parallel path, the raceway not only lowers the total impedance of the grounding path but carries 95 percent or more of the current. Furthermore, the raceway provides additional protection as a grounding shield; for instance, in the unlikely event that a phase conductor becomes damaged within a raceway and the conductor is exposed and comes in contact with a grounded raceway, the resulting fault current would be carried to the grounded service equipment enclosure through the main bonding jumper to the system and the overcurrent device, thus clearing the fault.

By contrast, a damaged phase conductor run as open wiring or in a nonmetallic raceway could remain exposed to accidental contact indefinitely. A fault within a nonmetallic raceway or open runs of conductors may or may not provide a low impedance return path to trip the overcurrent device protecting the circuit.

(Text continues on page 161)

Examples of Maximum Length Equipment Grounding Conductor (Steel EMT, IMC, GRC, and Copper or Aluminum Wire)
Calculated as a Safe Return Fault Path to Overcurrent Device
Based on 1997 Georgia Tech Software Version (GEMI 2.4, 1998)
With an Arc Voltage of 40 and 4 IP at 25°C Ambient
Based on a Circuit Voltage of 120 Volts to Ground
THHN/THWN Insulation

Overcurrent Device Rating Amperes (75°C)	400% (4IP) Overcurrent Device Rating Amperes	Circuit Conductor Size AWG/kcmil Copper or Aluminum	Steel EMT, IMC, GRC Trade Size	(1) Equipment Grounding Conductor Size AWG Copper or Aluminum	Length of EMT Run Calculated Maximum (In Feet)	Length of IMC Run Calculated Maximum (In Feet)	Length of GRC Run Calculated Maximum (In Feet)	Copper Grounding Conductor w/o Steel Conduit Maximum Run (In Feet)	Aluminum Grounding Conductor w/o Steel Conduit Maximum Run (In Feet)
20	80	12	1/2 (16)	—	395	398	384	—	—
20	80	12	—	12	—	—	—	300	—
20	80	10 AL	—	10 AL	—	—	—	—	293
30	120	10	1/2 (16)	—	358	383	364	—	—
30	120	10	3/4 (21)	—	404	399	386	—	—
30	120	10	—	10	—	—	—	319	—
30	120	8 AL	—	8 AL	—	—	—	—	310
40	160	8	3/4 (21)	—	407	414	395	—	—
40	160	8	1 (27)	—	447	431	418	—	—
40	160	8	—	10	—	—	—	294	—
40	160	8 AL	—	8 AL	—	—	—	—	232
60	240	6	3/4 (21)	—	350	383	363	—	—
60	240	6	1 (27)	—	404	400	382	—	—
60	240	6	—	10	—	—	—	228	—
60	240	4 AL	—	8 AL	—	—	—	—	221
100	400	3	1 1/4 (35)	—	402	397 (4)	373	—	—
100	400	3	—	8	—	—	—	229	—
100	400	1 AL	—	6 AL	—	—	—	—	222
200	800	3/0	2 (53)	—	390	389	363	—	—
200	800	3/0	—	6	—	—	—	201	—
200	800	250 AL	—	4 AL	—	—	—	—	195

(1) Per 1999 *NEC®* Table 251-18

Note: Software is not limited to above examples.

Figure 9–8 (Table derived from Software [SCA] and Testing developed at Georgia Institute of Technology and sponsored by the producers of Steel EMT, IMC, and rigid steel conduit.)

Examples of Maximum Length Equipment Grounding Conductor (Steel EMT, IMC, GRC, and Copper or Aluminum Wire) Calculated as a Safe Return Fault Path to Overcurrent Device Based on 1997 Georgia Tech Software Version (GEMI 2.4, 1998) With an Arc Voltage of 40 and 4 IP at 25°C Ambient Based on a Circuit Voltage of 277 Volts to Ground THHN/THWN Insulation

Overcurrent Device Rating Amperes (75°C)	400% (4IP) Overcurrent Device Rating Amperes	Circuit Conductor Size AWG/kcmil Copper or Aluminum	Steel EMT, IMC, GRC Trade Size	(1) Equipment Grounding Conductor Size AWG Copper or Aluminum	Length of EMT Run Calculated Maximum (In Feet)	Length of IMC Run Calculated Maximum (In Feet)	Length of GRC Run Calculated Maximum (In Feet)	Copper Grounding Conductor w/o Steel Conduit Maximum Run (In Feet)	Aluminum Grounding Conductor w/o Steel Conduit Maximum Run (In Feet)
20	80	12	1/2 (16)	—	1170	1179	1140	—	—
20	80	12	—	12	—	—	—	890	—
20	80	10 AL	—	10 AL	—	—	—	—	870
30	120	10	1/2 (16)	—	—	1135 (4)	—	—	—
30	120	10	3/4 (21)	—	1199	1182	1143	—	—
30	120	10	—	10	—	—	—	946	—
30	120	8 AL	—	8 AL	—	—	—	—	920
40	160	8	3/4 (21)	—	1208	1228	1170	—	—
40	160	8	1 (27)	—	1326	1276	1239	—	—
40	160	8	—	10	—	—	—	871	—
40	160	8 AL	—	8 AL	—	—	—	—	690
60	240	6	3/4 (21)	—	1039	1134	1075	—	—
60	240	6	1 (27)	—	1197	1186	1131	—	—
60	240	6	—	10	—	—	—	676	—
60	240	4 AL	—	8 AL	—	—	—	—	657
100	400	3	1 1/4 (35)	—	1192	1176 (4)	1107	—	—
100	400	3	—	8	—	—	—	680	—
100	400	1 AL	—	6 AL	—	—	—	—	659
200	800	3/0	2 (53)	—	1157	1155	1077	—	—
200	800	3/0	—	6	—	—	—	598	—
200	800	250 AL	—	4 AL	—	—	—	—	578

Note: Software is not limited to above examples.

(1) Per 1999 NEC® Table 251-18

Figure 9–9 (Table derived from Software [SCA] and Testing developed at Georgia Institute of Technology and sponsored by the producers of Steel EMT, IMC, and rigid steel conduit.)

Maximum Length of Electrical Metallic Tubing That May Safely Be Used as an Equipment-Grounding Circuit Conductor. Based on a Ground-Fault Current of 400% of the Overcurrent Device Rating. Circuit 120 Volts to Ground; 40 Volts Drop at the Point of Fault. Ambient Temperature 25°C

EMT Trade Size	Conductors AWG No.	Overcurrent Device Rating Amperes 75°C*	Fault Clearing Current 400% O.C. Device Rating Amperes	Maximum Length of EMT Run in Feet
1/2 (16)	3-12	20	80	395
	4-10	30	120	358
3/4 (21)	4-10	30	120	404
	4-8	50	200	332
1 (27)	4-8	50	200	370
	3-4	85	340	365
1 1/4 (35)	3-2	115	460	391
1 1/2 (41)	3-1	130	520	407
	3-2/0	175	700	364
2 (53)	3-3/0	200	800	390
	3-4/0	230	920	367
2 1/2 (63)	3-250 kcm	255	1020	406
3 (78)	3-350 kcm	310	1240	404
	3-500 kcm	380	1520	370
	3-600 kcm	420	1680	353
4 (103)	3-900 kcm	520	2080	353
	3-1000 kcm	545	2180	347

* 60°C for 20- and 30-ampere devices.

Based on 1994 Georgia Tech Model

Figure 9–10 (Table derived from Software [SCA] and Testing developed at Georgia Institute of Technology and sponsored by the producers of Steel EMT, IMC, and rigid steel conduit.)

Maximum Length of Intermediate Metal Conduit That May Safely
Be Used as an Equipment-Grounding Circuit Conductor.
Based on a Ground-Fault Current of 400% of the Overcurrent Device Rating.
Circuit 120 Volts to Ground; 40 Volts Drop at the Point of Fault.
Ambient Temperature 25°C

IMC Trade Size	Conductors AWG No.	Overcurrent Device Rating Amperes 75°C*	Fault Clearing Current 400% O.C. Device Rating Amperes	Maximum Length of IMC Run in Feet
1/2 (16)	3-12	20	80	398
	4-10	30	120	383
3/4 (21)	4-10	30	120	399
	4-8	50	200	350
1 (27)	4-8	50	200	362
	3-4	85	340	382
1 1/4 (35)	3-2	115	460	392
1 1/2 (41)	3-1	130	520	402
	3-2/0	175	700	377
2 (53)	3-3/0	200	800	389
	3-4/0	230	920	375
2 1/2 (63)	3-250 kcm	255	1020	368
3 (78)	3-350 kcm	310	1240	367
	3-500 kcm	380	1520	338
	3-600 kcm	420	1680	325
4 (103)	3-900 kcm	520	2080	320
	3-1000 kcm	545	2180	314

* 60°C for 20- and 30-ampere devices.

Based on 1994 Georgia Tech Model

Figure 9–11 (Table derived from Software [SCA] and Testing developed at Georgia Institute of Technology and sponsored by the producers of Steel EMT, IMC, and rigid steel conduit.)

Maximum Length of Galvanized Rigid Conduit That May Safely Be Used as an Equipment-Grounding Circuit Conductor.
Based on a Ground-Fault Current of 400% of the Overcurrent Device Rating.
Circuit 120 Volts to Ground; 40 Volts Drop at the Point of Fault.
Ambient Temperature 25°C

GRC Trade Size	Conductors AWG No.	Overcurrent Device Rating Amperes 75°C*	Fault Clearing Current 400% O.C. Device Rating Amperes	Maximum Length of GRC Run in Feet
1/2 (16)	3-12	20	80	384
	4-10	30	120	364
3/4 (21)	4-10	30	120	386
	4-8	50	200	334
1 (27)	4-8	50	200	350
	3-4	85	340	357
1 1/4 (35)	3-2	115	460	365
1 1/2 (41)	3-1	130	520	377
	3-2/0	175	700	348
2 (53)	3-3/0	200	800	363
	3-4/0	230	920	347
2 1/2 (63)	3-250 kcm	255	1020	356
3 (78)	3-350 kcm	310	1240	355
	3-500 kcm	380	1520	327
	3-600 kcm	420	1680	314
4 (103)	3-900 kcm	520	2080	310
	3-1000 kcm	545	2180	304

* 60°C for 20- and 30-ampere devices.

Based on 1994 Georgia Tech Model

Figure 9–12 (Table derived from Software [SCA] and Testing developed at Georgia Institute of Technology and sponsored by the producers of Steel EMT, IMC, and rigid steel conduit.)

Maximum Length of Equipment-Grounding Conductor That May Safely Be Used as an Equipment-Grounding Circuit Conductor. Based on a Ground-Fault Current of 400% of the Overcurrent Device Rating. Circuit 120 Volts to Ground; 40 Volts Drop at the Point of Fault. Ambient Temperature 25°C

Copper Equipment Grounding Conductor AWG Size***	Copper Circuit AWG Conductors	Maximum Length of Run (in feet) using Copper Equipment Ground Conductor	Aluminum Equipment Grounding Conductor AWG Size***	Aluminum Circuit AWG Conductors	Maximum Length of Run (in feet) using Aluminum Equipment Ground Conductor	Overcurrent Device Rating Amperes 75°C**	Fault Clearing Current 400% O.C. Device Rating Amperes
14	14	253	12	12	244	15	60
12	12	300	10	12	226	20	80
10	10	319	8	8	310	30	120
10	8	294	8	8	232	40	160
10	6	228	8	4	221	60	240
8	3	229	6	1	222	100	400
6	3/0	201	4	250 kcm	195	200	800
4	350 kcm	210	2	500 kcm	204	300	1200
3	600 kcm	195	1	900 kcm	192	400	1600
2	2-4/0	160	1/0	2-400 kcm	163	500	2000
1	2-300 kcm	160	2/0	2-500 kcm	161	600	2400
1/0	3-300 kcm	134	3/0	3-400 kcm	131	800	3200
2/0	4-250 kcm	114	4/0	4-400 kcm	115	1000	4000
3/0	4-300 kcm	106	250 kcm	4-500 kcm	107	1200	4800
4/0	4-600 kcm	93	350 kcm	4-900 kcm	97	1600	6400
250 kcm	5-600 kcm	78	400 kcm	5-800 kcm	79	2000	8000
350 kcm	6-600 kcm	*	600 kcm	6-900 kcm	*	2500	10,000
400 kcm	8-500 kcm	*	600 kcm	8-750 kcm	*	3000	15,000
500 kcm	8-1000 kcm	*	800 kcm	8-1500 kcm	*	4000	16,000
700 kcm	10-1000 kcm	*	1200 kcm	10-1500 kcm	*	5000	20,000
800 kcm	12-1000 kcm	*	1200 kcm	12-1500 kcm	*	6000	24,000

Based on 1994 Georgia Tech Model

* Calculations necessary
** 60°C for 20- and 30-ampere devices
*** Based on *NEC*® Chapter 9 Table 8

Figure 9–13 (Table derived from Software [SCA] and Testing developed at Georgia Institute of Technology and sponsored by the producers of Steel EMT, IMC, and rigid steel conduit.)

An equipment grounding conductor and a grounded circuit conductor are not the same electrically. It is unsafe to connect the grounded circuit conductor to the raceway or to any other noncurrent-carrying equipment at any point on the load side of the service equipment. The connections permitted to be made on the line and load sides of the service disconnecting equipment are specified in 250.142 of the *NEC®*, Parts I and II, which clarifies the preferred design method for location of these connections.

In 1998, the electronic and circuit-breaker industry introduced a new level of safety to the *NEC®*, the arc-fault circuit-breaker. Several proposals (2-128, 2-129, and 2-130) were submitted to add this new level of safety device designed to protect dwellings. The substantiation included U.S. Consumer Product Safety Commission (CPSC) fire records. They stated that there were over 40,000 fires resulting in approximately 250 deaths annually and that these figures had not decreased over the past decade. Based on a survey by a large insurance company, approximately 36 percent of these fires originated in permanent wiring, 50 percent in appliances, and 12 percent in cords. They analyzed these reports and stated that approximately 38 percent were from arcing faults and about 12 percent were from loose connections.

The solution was to add a listed arc-fault circuit interrupter (AFCI) device. The AFCI device has electronic circuitry capable of recognizing the unique current and/or voltage signatures of an arcing fault. When the arc fault is detected, it acts similarly to the GFCI device and interrupts the circuit. This technology integrates a conventional thermal circuit-breaker with an AFCI.

The *National Electrical Code®* has defined this device in 210.12. The requirement is that all new dwelling bedroom circuits be an AFCI type.

Conclusions—Why Ground Circuits and Systems?

There are many reasons to ground electrical circuits. One reason would be savings in wire cost. For instance, four wires can be used to serve the same load that a 6-wire single-phase would need. A grounded wye (3-phase) system can serve three 120-volt, single-phase loads utilizing a common neutral. Two voltages are available where a grounded circuit is employed. With a 240-volt line-to-line utilizing a grounded system, you then have 120-volt to ground. Thus, a 3-wire system can provide both 220-volt motor and appliance loads; the 120 volts can be applied to smaller appliances, convenience outlets, and lighting in all locations, including residential occupancies.

A good electrical system properly grounded and maintained limits the voltage due to lightning and line surges on the system. A good grounded system protects us from unintentional contact with high-voltage lines or the influence from higher-voltage lines run in close proximity to the lower-voltage lines. A good grounded system stabilizes the voltage-to-ground during normal operations.

Finally, and most importantly, a good grounded system provides a low-impedance path to facilitate overcurrent device operation. It makes no difference whether you use a properly sized fuse or circuit breaker. In the case of a line-to-ground fault, the overcurrent device will open. This will prevent needless loss of equipment, fires, and the loss of life and property.

An example of the horrors of ungrounded systems and what can happen was related at a recent meeting. At a manufacturing plant somewhere in the southern part of the United States operating on a 480-volt ungrounded system, production was suddenly disrupted. The first evidence came as a motor failure on the 480-volt system, then another, and in close succession, a third. The maintenance electricians made a quick inspection of the switchboard voltmeter, which measured line-to-line volts, and amperes. No unusual conditions were indicated. The system equipment continued to fail. A test voltmeter was rigged up having a full-scale calibration of 1200 volts. Upon connecting it phase-to-ground, the pointer went off the scale. A phase-to-ground potential on the 480-volt system of more than 1200 volts existed! At once the incoming service transformers were suspected of internal breakdown through high- and low-voltage windings, primary and secondary. As the last of these transformers was isolated and individually tested, it became evident they were not the problem. System equipment continued to fail, and the situation was desperate.

The frantic group went into a huddle and called in outside engineering. It was decided the only way out was to open (turn off) the main incoming service circuit-breaker, which would de-energize the entire system. At this point, one of the workmen noticed a small wisp of smoke coming from a motor starting an autotransformer; upon approaching it, he could hear a buzzing noise inside. This circuit was switched clear of the system, and the overvoltages disappeared. During the two-hour period that this arcing fault existed, between 40 and 50 motors had failed.

It was found that the autotransformer enclosing case had been bashed in and was practically in contact with the coil. The spot where arcing had taken place was evident, although not badly burned. An attempt was made to show the plant engineer the source of the trouble. A solid connection was made between the frame and the burned spot on the coil. Much to the bewilderment of the operating men, but according to the expectations of the plant engineer, no more than the 73 percent increase in voltage-to-ground on the other two phases occurred. The main ingredient of the overvoltage had been omitted (discontinuous conduction). This story is spectacular because of the magnitude of the disturbance and consequential damage. Similar occurrences of lesser extent are not uncommon; there is evidence that they occur more frequently than realized. It is characteristic of an ungrounded system to be subject to relatively severe transient overvoltages. This trouble can be avoided by properly grounding the system. Many other important benefits, as discussed in this article, are also obtained.

The term "ungrounded system" is used to identify a system where there is no intentional connection between the system conductors and ground. However, in any practical system, there always exists a capacity coupling between the system conductors and ground. Consequently, the so-called "ungrounded system" is in reality a capacitance grounded system, by virtue of distributed capacitance from the system conductor to ground. When the neutral of a system is not grounded, it is possible for destructive transient overvoltages several times normal to appear from line-to-ground during normal switching of a circuit having a line-to-ground fault. Tests have shown that overvoltages may be developed by repeated re-striking of the arc during interruption of the line-to-ground fault, particularly in low-voltage systems. Experience has proven that these overvoltages may cause failure of insulation at locations other than at the point of fault in the system; thus, line-to-ground fault on one circuit may result in damage to equipment and interruption of service on other circuits. The same condition will result from repeated strike of an arcing fault from line to ground.

In an ungrounded neutral system, the second ground fault on another phase may occur before the first fault is removed. The second fault may be on the same circuit as the original fault or on another. In any event, the resulting line-to-line fault will actuate relays or circuit breakers and trip either one or both circuits. Thus, a single ground-fault of relative unimportance

may eventually result in considerable damage because of the relatively high line-to-line fault current and interruption of one or both circuits. When the neutral of the system is not grounded, a ground-fault on one phase causes a full line-to-line voltage to appear throughout the system between the ground and the two phases, which were not-faulted phases. The voltage is 73 percent higher than normal. Usually the insulation between the line and ground is adequate to withstand line-to-line voltage. However, when this voltage is applied for long periods, it may result in failure of insulation that may have already deteriorated because of age or severe service conditions. Line-to-ground faults on ungrounded neutral systems cause a very small ground-fault current to flow through the capacitance of the cables, transformers, and other electrical equipment on the system. This current may have a magnitude of from a few amperes to as much as 25 amperes or more on larger systems. This is not, in general, enough to actuate protective devices, but it may do considerable damage if allowed to flow for a long period.

The grounded system, which has been studied, is the topic of many important meetings among the leaders of our industry over the years at IEEE meetings, National Fire Protection Association code development meetings, Canadian Electrical Code meetings, and seminars of all kinds. Proper grounding has been determined to be the safest and most economical practice to follow. The *National Electrical Code®* (*NEC®*) and *Canadian Electrical Code* (*CEC*) provide minimum standards for making the proper grounding connections on services and equipment. IEEE standards provide engineering guidelines that may be of benefit to those with unique requirements for their facility or particular installation. There are many other good documents also available.

Example: The following example illustrates how the size of the grounded (neutral) wire may be selected. A new 3-phase, 4-wire feeder is connected in an existing junction box to three single-phase plus grounded circuits, each consisting of two phase conductors and a grounded conductor. Two of these circuits are in one conduit; the third is in another. The single-phase circuit conductors are all 2 AWG THW copper. Determine the ampacity of each single-phase circuit and the required size of new service if electric discharge lighting ballasts are in the circuits, but the load is considered noncontinuous.

Step 1. Neutrals in these circuits must be counted as current-carrying conductors. (*NEC®* Article 310, Table 310.15(B)(4).) The ampacity of 2 AWG THW conductors from Table 310.16 is 115 amperes.

Step 2. Derating is not required for the circuit ampacity in the conduit with the single circuit with only three conductors. (*NEC®* Article 310.) Thus, the allowable ampacity is 115 amperes.

Step 3. In the conduit with two circuits, an 80 percent derating factor must be used, because all six wires are considered as current-carrying conductors (*NEC®* Article 310, Table 310.15(B)(2).) Each of these two circuits has a design ampacity of 80 percent × 115 = 92.

Step 4. The sum of the ampacities of all three circuits is the required ampacity of the service conductors and the service neutral (220.61).

Step 5. The value is 115 + 92 + 92 = 299. Each conductor and neutral requires not less than a 350-kcmil 75°C conductor ampacity. Reference Article 310, Table 310.16.

QUESTION REVIEW

> **Each lesson is designed purposely to require the student to apply the entire *NEC®* text, and not specific chapters or articles. It has been found that when studying for a timed, open-book examination, the student must gain proficiency in the Table of Contents, the Index, and the ability to move quickly from cover to cover to find the correct answer to each question in a timely fashion.**

1. Required grounding conductors and bonding jumpers cannot be connected solely by _____ connections.

   Answer: _____

   _____

   _____

   _____

   Reference: _____

2. The conductor permitted to bond together all isolated noncurrent-carrying metal parts of an outline lighting system is at least size ____ AWG copper.

   Answer: _____

   _____

   _____

   _____

   Reference: _____

3. A connection to a concrete-encased, driven or buried grounding electrode shall be _____.

   Answer: _____

   _____

   _____

   _____

   Reference: _____

4. The grounded conductor of an electrical branch-circuit is identified by the color _____ .

Answer: _____

_____

_____

_____

Reference: _____

Each conduit contains
three 500-kcmil conductors
500 per phase

Bonding jumper

Service equipment

Ground bus
terminal

5. *Refer to the figure above.* Each of the supply-side EMT conduits contains three 500-kcmil copper THW service conductors in parallel. As shown, only one bonding jumper is used to bond all the conduits to the grounded bus terminal. This bonding jumper must be at least size _____ AWG copper.

Answer: _____

_____

_____

_____

Reference: _____

6. A rigid metal conduit contains three circuits: two 150-ampere, 3-phase circuits, and one 300-ampere, single-phase circuit. The load side service equipment bonding jumper for this conduit must be at least size _____ AWG copper.

    Answer: _____

    _____

    _____

    _____

    Reference: _____

7. Are EMT set-screw connectors approved for grounding of service raceways or are jumpers required to bond around them?

    Answer: _____

    _____

    _____

    _____

    Reference: _____

8. Where feeder conductors are paralleled and routed in separate nonmetallic raceways, is an equipment grounding conductor required in each nonmetallic raceway or only in one nonmetallic raceway?

    Answer: _____

    _____

    _____

    _____

    Reference: _____

9. Are submersible deep-well pumps required to have an equipment grounding conductor run with the other conductors and the submersible pump motor housing grounded?

    Answer: _____

    _____

    _____

    _____

    Reference: _____

10. Is it permissible to ground the secondary of a separately derived system to the grounded terminal in the service switchboard instead of the nearest effectively grounded structural steel or nearest water pipe?

    Answer: _____

    _____

    _____

    _____

    Reference: _____

11. Where installing the grounding-electrode conductor in a raceway for physical protection, is the metal raceway required to be bonded to the grounding-electrode conductor? If the answer is yes, is it sufficient to bond it at the service panel only or must bonding be done at all terminations?

Answer: _____

_____

_____

_____

Reference: _____

12. Where installing the electrical hookups for RV sites in a recreational vehicle park, can a ground rod be driven at each RV site instead of pulling an equipment grounding conductor with a circuit conductor from the source?

Answer: _____

_____

_____

_____

Reference: _____

13. You have recently installed a separately derived system, a dry-type transformer in a commercial building. No effectively grounded structural steel was near the transformer location. The transformer secondary is grounded to a nearby metal water pipe. Is a supplemental grounding electrode required?

Answer: _____

_____

_____

_____

Reference: _____

14. When bonding the noncurrent-carrying metal parts of a recreational vehicle, what is the minimum size copper or equivalent bonding conductor?

Answer: _____

_____

_____

_____

Reference: _____

15. Does the *Code* require interior gas piping systems to be bonded to the grounding-electrode system?

Answer: _____

_____

_____

_____

Reference: _____

16. Two parallel runs of four 500-kcmil conductors each are run in rigid nonmetallic conduits to feed a 600-ampere panel. Is an equipment grounding conductor required in each conduit? What size equipment grounding conductor is required in each conduit?

    Answer: _____

    _____

    _____

    _____

    Reference: _____

17. When terminating the equipment grounding conductor in an outlet box supplied by nonmetallic sheath cables, can the equipment grounding conductors be twisted together or must they be connected with a wire connector?

    Answer: _____

    _____

    _____

    _____

    Reference: _____

18. What size aluminum equipment grounding conductor will be required to be run with the circuit conductors fed from a 60-ampere fusible switch?

    Answer: _____

    _____

    _____

    _____

    Reference: _____

19. In a run of conduit supplying a motor, it is necessary to install 4 feet (1.2 m) of flexible metal conduit where the raceway leaves the panel to get around a column. When the conduit reaches the motor, an additional 3 feet (900 mm) of flexible metal conduit is employed for convenience and flexibility. Is an equipment grounding conductor required in this metal raceway?

    Answer: _____

    _____

    _____

    _____

    Reference: _____

20. When making an installation of a high-impedance grounded neutral system, how is the neutral conductor to be sized?

Answer: _____

_____

_____

_____

Reference: _____

21. When making an installation of a hydromassage tub in a residence, how are the hydromassage bathtubs and associated electrical components to be protected?

Answer: _____

_____

_____

_____

Reference: _____

22. Underwriters Laboratory Standard UL 67 requires that backfed plug-in-type circuit breakers be secured. Does the *NEC*® require that backfed circuit breakers also be secured? What section makes this requirement?

Answer: _____

_____

_____

_____

Reference: _____

23. Where several types of grounding electrodes are available on the premises, such as an underground water line, building steel, and a concrete-encased electrode, is it permissible to jump from one electrode to another with a single properly sized grouding-electrode conductor?

Answer: _____

_____

_____

_____

Reference: _____

24. Many contractors use metal 90° elbows in nonmetallic conduit underground runs for ease of pulling the conductors and to prevent burn-through. Are these 90° bends required to be bonded? If so, how can this be accomplished from a practical standpoint when they are underground?

Answer: _____

_____

_____

_____

Reference: _____

4 750 kcmil

4 3 AWG
THWN

4 3/0 AWG
THWN

4 6 AWG
THWN

4 1/0 AWG
THWN

Switch A
100 Amp

Switch B
200 Amp

Switch C
60 Amp

Switch D
150 Amp

Underground
water pipe

Ground level

Supplementary
ground rod

25. In the figure above, what is the required size of the grounding-electrode conductor and the main grounding-electrode conductor to Switch A, Switch B, Switch C, and Switch D?

Answer: _____

_____

_____

_____

Reference: _____

Chapter Ten

WIRING METHODS

This chapter deals primarily with Chapter 3, which covers the general installation requirements and all of the general wiring methods. (See Figure 10–3.) In addition, we reference Chapter 9. Chapter 9 contains Tables 1 through 12(B). Only minor revisions have been made to Chapter 9 in the 2005 *NEC*® cycle. A new Table 2 has been added. Table 344.24, which applied to the bending radius of raceways, has been relocated as a new Chapter 9 Table 2 without modification. A reference to this table appears in each raceway article in Chapter 3. Table 4 now contains a table for each conduit and tubing type. The addition of these tables was necessary because each conduit and tubing type has a unique internal diameter (ID), with intermediate metal conduit (IMC) being the largest and rigid nonmetallic conduit Schedule 80 being the smallest standard building wiring method.

> **Conduit and tubing diameters make a difference.**
> Trade size 1/2 (16) IMC intermediate metal conduit internal diameter .0060 inch
> **Allowable 40 percent fill would permit three 8 AWG Type THWN conductors.**
> Trade size 1/2 (16) Schedule 80 rigid nonmetallic conduit internal diameter 0.526 inch
> **Allowable 40 percent fill would permit one 8 AWG Type THWN conductor.**

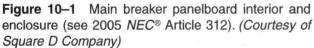

Figure 10–1 Main breaker panelboard interior and enclosure (see 2005 *NEC®* Article 312). *(Courtesy of Square D Company)*

Figure 10–2 Typical safety switch with view safety glass (see 2005 *NEC®* Article 404). *(Courtesy of Square D Company)*

Table 5 has been corrected to the true actual conductor dimensions. This table had not been corrected until now, and the insulation thickness has changed with the improvement of materials over the years.

Annex C has been added to contain the new multiple version of Tables 3 that were formerly found in Chapter 9. This new Annex C is for conduit and tubing fill tables of fixture wire and conductors of the same size.

Tables 1, 4, 5, and 8 contain conduit, tubing dimensions, (conductor) wire dimensions, and notes to determine the total cross-sectional area, including insulation to determine the required allowable trade size of conduit or tubing applicable. Tables 11(A) and 11(B) were added in the 1996 *NEC®*. They formerly appeared in Article 725. They provide power source limitations for Class 2 and 3 ac (alternating current) and dc (direct current) power sources. Tables 12(A)

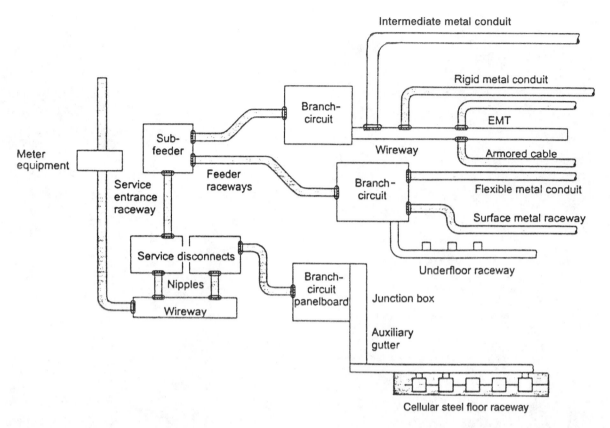

Figure 10–3 Service raceway per *NEC®* Article 230. *Note:* Location of overcurrent devices not shown. *(Courtesy of American Iron & Steel Institute)*

and 12(B) are also added to Chapter 9; they formerly appeared in Article 760. They provide power source limitations for fire alarm power limited circuits both ac (alternating current) and dc (direct current) PLFA power sources. These tables are to be used in conjunction with the appropriate articles in the *Code*.

Example: A typical 480/277-volt 400-ampere feeder is run from a distribution unit substation under a concrete slab exposed above a suspended ceiling turning down into a panelboard. Based on the length (200 feet [61 m]) of the run and the calculated load, the installer chooses to pull three 600-kcmil THWN ungrounded circuit conductors and a full-sized 600-kcmil THWN neutral (grounded conductor). If he chooses a nonmetallic raceway for a portion of this run, he recognizes that he will also have to add an equipment grounding conductor sized in accordance with Table 250.122 or a 3 AWG THWN copper. However, when he applies the tables in Chapter 9, he finds that he will need the following sizes of conduit or tubing to make this installation:

Solution:

Table 5. 600-kcmil THWN =
  0.8676 sq. in.   4 × 0.8676   = 3.47 sq. in.
Table 4.
EMT electrical metallic tubing
  40% fill as per Table 1   = 3 (78)
IMC intermediate metal conduit
  40% fill as per Table 1   = 3½ (91)
GRC galvanized rigid conduit
  40% fill as per Table 1   = 3½ (91)

Table 5. 600-kcmil THWN =
  0.8676 sq. in. 4 × 0.8676   = 3.47   sq. in.
Plus an equipment grounding
  conductor 3 AWG THWN   = 0.0973 sq. in.
                          = 3.5673 sq. in.
Table 4.
ENT electrical nonmetallic
  tubing max. size 2 (63)   = N/A
Sch. 40 rigid nonmetallic conduit
  40% fill as per Table 1   = 3½ (91)
Sch. 80 rigid nonmetallic conduit
  40% fill as per Table 1   = 4 (103)

Using these same calculations with 500-kcmil for the ungrounded and grounded conductors would require a trade size 3 (78) conduit for EMT, IMC, GRC, and PVC Schedule 40, and a trade size 3½ (91) for PVC Schedule 80. As one can see, calculations will be necessary to determine the most efficient wiring method.

The general requirements for wiring methods are found in Article 300. As you may recall in earlier chapters of this book, you have been urged several times to learn the contents of Article 300 thoroughly because it is used in some application for every installation.

The general wiring methods in Article 300 include requirements that apply to the specific wiring methods, such as nonmetallic-sheathed cable, which is covered in Article 334 (see Figure 10–4), as specific support requirements and installation criteria, such as found in 300.4(A) through (F) and the Exceptions. These requirements are mandatory and must be adhered to

Figure 10–4 Various types of hangers and supporting methods and hardware for raceways. *(Courtesy of Cooper B-Line, Inc.)*

when making an installation of any of the wiring methods and are in Chapter 3 as they apply. Chapter 3 also covers burial depths (in Table 300.5 for conductors under 600 volts) and many conditions that apply to underground installations, such as splices, taps, and the backfill material used so as not to damage the raceway or cable that you have selected as a wiring method for your installation. When a wiring method is installed in earth that is not suitable as physical protection, select fill material should be used, such as sand, or soil without rocks.

This article also covers such things as seals and protection against corrosion, indoor wet locations, raceways exposed to different temperatures, and expansion joints. We will be talking in this chapter about the expansion joint requirements for nonmetallic raceways. However, 300.7 covers all wiring methods, and where they are required to compensate for expansion or contraction (see Figure 10–5, Figure 10–6, and Figure 10-7). The electrical and mechanical continuity for metallic systems, both the raceway and the enclosures, *NEC*® 300.11, requires that these methods, the raceways, cable assemblies, boxes, cabinets, and fittings, all be securely fastened where supported on support wires above suspended ceilings (refer to Figure 10–4).

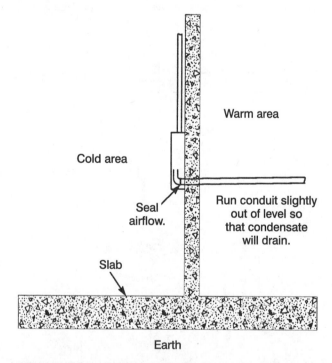

Figure 10–5 Seals are required to comply with *NEC*® 300.7 where the raceway is subjected to more than one temperature, such as shown. The system enters the building from outside.

For branch-circuit wiring, read this section very carefully. The wiring systems for feeders or services are not permitted to be secured in this manner; only certain branch-circuits are permitted. Specifically, the junction box out of the box section in Article 314 places more stringent requirements for supporting these box systems. All must be adhered to. Article 300 also tells us where a box or fitting is required and where a box only is permitted. 300.15(A) states that all fittings and connectors must be designed for the specific wiring method for which they are used and also must be listed for that purpose. You are urged to look up the definition for a branch-circuit in Article 100 at this point. A very important requirement for all wiring methods is found in 300.21. The requirement aims to limit the spread of fire or products of combustion and requires that the wiring system be installed so as to minimize the spread of fire and products of combustion and that all penetrations through floors, walls, and ceilings be sealed in an approved manner to accomplish this purpose. 300.22, which contains the provisions for the installation and uses of electrical wiring and equipment in ducts, plenums, and other air-handling spaces, deals with safety concerns in determining the wiring method to be used in electrical installations and should be studied carefully before making the final selections. Part II of Article 300 explains the requirements for installations over 600 volts nominal. 300.50 covers the underground installations and references, which are the general wiring requirements for installations over 600 volts in the table and section in 300.50. Other general articles that relate to wiring methods are found in Chapter 3. However, temporary wiring is covered in Chapter 5. Article 590 covers the temporary wiring requirements needed to provide power for the workers making the installation and constructing the facility during the construction phases of a job. Article 590 also covers other types of temporary wiring, such as temporary wiring concerned with holiday tree lighting, trade fairs, emergencies and tests, and experimental and developmental work. Be aware of Article 590. Article 310 names the conductors to be used for general wiring and all of the requirements related to these conductors. The requirements do not apply to conductors that form an integral part of equipment, such as motors, motor controllers and similar equipment, or conductors specifically provided for elsewhere in the *Code*. Examples would be Article 400 and Article 402 for flexible cords and fixture wires. Other sections dealing with conductors would be Chapter 7 and

Table 352.44(A) Expansion Characteristics of PVC Rigid Nonmetallic Conduit Coefficient of Thermal Expansion = 6.084 × 10⁻⁵ mm/mm/°C (3.38 × 10⁻⁵ in./in./°F)

Temperature Change (°C)	Length Change of PVC Conduit (mm/m)	Temperature Change (°F)	Length Change of PVC Conduit (in./100 ft)	Temperature Change (°F)	Length Change of PVC Conduit (in./100 ft)
5	0.30	5	0.20	105	4.26
10	0.61	10	0.41	110	4.46
15	0.91	15	0.61	115	4.66
20	1.22	20	0.81	120	4.87
25	1.52	25	1.01	125	5.07
30	1.83	30	1.22	130	5.27
35	2.13	35	1.42	135	5.48
40	2.43	40	1.62	140	5.68
45	2.74	45	1.83	145	5.88
50	3.04	50	2.03	150	6.08
55	3.35	55	2.23	155	6.29
60	3.65	60	2.43	160	6.49
65	3.95	65	2.64	165	6.69
70	4.26	70	2.84	170	6.90
75	4.56	75	3.04	175	7.10
80	4.87	80	3.24	180	7.30
85	5.17	85	3.45	185	7.50
90	5.48	90	3.65	190	7.71
95	5.78	95	3.85	195	7.91
100	6.08	100	4.06	200	8.11

Table 352.44(B) Expansion Characteristics of Reinforced Thermosetting Resin Conduit (RTRC) Coefficient of Thermal Expansion = 2.7 × 10⁻⁵ mm/mm/°C (1.5 × 10⁻⁵ in./in./°F)

Temperature Change (°C)	Length Change of RTRC Conduit (mm/m)	Temperature Change (°F)	Length Change of RTRC Conduit (in./100 ft)	Temperature Change (°F)	Length Change of RTRC Conduit (in./100 ft)
5	0.14	5	0.09	105	1.89
10	0.27	10	0.18	110	1.98
15	0.41	15	0.27	115	2.07
20	0.54	20	0.36	120	2.16
25	0.68	25	0.45	125	2.25
30	0.81	30	0.54	130	2.34
35	0.95	35	0.63	135	2.43
40	1.08	40	0.72	140	2.52
45	1.22	45	0.81	145	2.61
50	1.35	50	0.90	150	2.70
55	1.49	55	0.99	155	2.79
60	1.62	60	1.08	160	2.88
65	1.76	65	1.17	165	2.97
70	1.89	70	1.26	170	3.06
75	2.03	75	1.35	175	3.15
80	2.16	80	1.44	180	3.24
85	2.30	85	1.53	185	3.33
90	2.43	90	1.62	190	3.42
95	2.57	95	1.71	195	3.51
100	2.70	100	1.80	200	3.60

Figure 10–6 (Reprinted with permission from NFPA 70-2005, the *National Electrical Code*®, Copyright © 2004–, National Fire Protection Association, Quincy, MA 02269. This reprinted material is not the complete and official position of the National Fire Protection Association on the referenced subject, which is represented only by the standard in its entirety.)

Example

380 ft. of conduit is to be installed on the outside of a building exposed to the sun in a single straight run. It is expected that the conduit will vary in temperature from 0°F in the winter to 140°F in the summer (this includes the 30°F for radiant heating from the sun). The installation is to be made at a conduit temperature of 90°F. From the table, a 140°F temperature change will cause a 5.7 in. length change in 100 ft. of conduit. The total change for this example is 5.7" × 3.8 = 21.67" which should be rounded to 22". The number of expansion couplings will be 22 ÷ coupling range (6" for E945, 2" for E955). If the E945 coupling is used, the number will be 22 ÷ 6 = 3.67 which should be rounded to 4. The coupling should be placed at 95 ft. intervals (380 ÷ 4). The proper piston setting at the time of installation is calculated as explained above.

$$0 = \left[\frac{140 - 90}{140} \right] 6.0 = 2.1 \text{ in.}$$

Insert the piston into the barrel to the maximum depth. Place a mark on the piston at the end of the barrel. To properly set the piston, pull the piston out of the barrel to correspond to the 2.1 in. calculated above.

See the drawing below.

Figure 10–7 Rigid nonmetallic conduit. For aluminum raceway, 21.7 × 0.1 = 2.17 inches an expansion fitting may be required. *(Courtesy of Carlon, a Lamson & Sessions Company)*

Chapter 8 of the *NEC®*, which deal with limited voltage wire, such as in Article 725, 760, and so on. Article 310 must be used in every installation made. The requirements in 310.1 through 310.15 are general requirements for the use and installation of conductors.

> *Note:* When using the applicable tables, read each heading and each column carefully. Read the ambient temperature correction factors below the column, and remember that the notes (footnotes) under each table are a part of those tables and are mandatory requirements.

This is important when studying for a test, because many test questions will require you to apply these additional parts of the table to get the correct answer. For an example, the obelisk note under each table applies to general wiring methods. However, notice that it says unless specifically otherwise provided for or permitted in this *Code*. Motors, for instance, generally are not required to comply with this note, because Article 430 has specific requirements for branch-circuit conductors for motors. Study these tables carefully.

The next important part of Article 310 is 310.15 (B)(6). Table 310.15(B)(6) has less restrictive ampacity requirements for dwelling services and feeders.

> *Note:* These only apply to single-phase, 120/240 volt, 3-wire systems and do not apply to any other type systems.

Example: NEC® 392.9(A)(1)

Width selection for cable tray containing 600-volt multiconductor cables, sizes 4/0 AWG and larger only. Cable installation is limited to a single layer. The sum of the cable diameters (Sd) must be equal to or less than the usable cable tray width.

Cable tray width is obtained as follows:

A—Width required for 4/0 AWG and larger multiconductor cables [*see diagram top of page 177*]:

Item No.	List Cable Sizes	(D) List Cable Outside Diameter	(N) List Number of Cables	Multiply (D) × (N) Subtotal of Sum of Cable Diameters (Sd)
1	3/C–500 kcmil	2.25 inches	4	9.00 inches
2	3/C–250 kcmil	1.76 inches	3	5.78 inches
3	3/C–4/0 kcmil	1.55 inches	10	15.50 inches

The sum of the diameters (Sd) of all cables (add total Sd for items 1, 2, and 3):

9.00 inches + 5.78 inches + 15.50 inches = 30.28 inches (Sd)

A cable tray with a usable width of 30 inches is required. For approximately 15% percent more, a 36-inch wide cable tray could be purchased that would provide for some future cable additions.

Notes:
1. Cable sizes used in this example are a random selection.
2. Cables—copper conductor with cross-linked polyethylene insulation and a PVC jacket. (These cables could be ordered with or without an equipment grounding conductor.)

NEC® 392.9(A)(1)

Cross Section of the Cables and the Cable Tray

(Courtesy of Cooper B-Line, Inc.)

3. Total cable weight per foot for this installation:
     61.4 lb/ft (without equipment grounding
          conductors)
     69.9 lb/ft (with equipment grounding con-
          ductors)

   This load can be supported by a load symbol
"B" cable tray–75 lb/ft.

Example: NEC® 392.9(A)(2)
   Width selection for cable tray containing 600-
volt multiconductor cables, sizes 3/0 AWG and
smaller. Cable tray allowance fill areas are listed
in Column 1 of Table 392.9.

Method 1:
The sum of the total areas for items 1, 2, 3, and 4:

   3.34 sq. in. + 3.04 sq. in. + 6.02 sq. in. + 16.00
   sq. in. = 28.40 sq. inches

   From Table 392.9, Column 1, a 30-inch (750
mm) wide tray with an allowable fill area of 35
square inches must be used. The 30-inch (750
mm) cable tray has the capacity for additional
future cables (6.60 square inches additional allowable fill area can be used).

Method 2:
The sum of the total areas for items 1, 2, 3, and 4
multiplied by (6 sq. in./7 sq. in.) = cable tray width
required:

   3.34 sq. in. + 3.04 sq. in. + 6.02 sq. in. + 16.00
   sq. in. = 28.40 sq. inches

$$\frac{28.40 \text{ sq. inches} \times 6 \text{ sq. inches}}{7 \text{ sq. inches}} = 24.34 \text{ inch}$$
                                        cable tray width
                                        required

Use a 30-inch (750 mm) cable tray.

NEC® 392.9(A)(2)
30" Usable cable tray width

Cross Section of the Cables and the Cable Tray

(Courtesy of Cooper B-Line, Inc.)

Notes:

1. The cable sizes used in this example are a random selection.
2. Cables—copper conductors with cross-linked polyethylene insulation and a PVC jacket. These cables could be ordered with or without an equipment grounding conductor.
3. Total cable weight per foot for this installation: 31.9 lb/ft. (Cables in this example do not contain equipment grounding conductors.) This load can be supported by a load symbol "A" cable tray—50 lb/ft.

Example: *NEC® 392.9(A)(3)*

Width selection for cable tray containing 600-volt multiconductor cables, sizes 4/0 AWG and larger (single layer required) and 3/0 AWG and smaller. These two groups of cables must have dedicated areas in the cable tray.

Cable tray width is obtained as follows:

A—Width required for 4/0 AWG and larger multiconductor cables:

Item No.	List Cable Sizes	(D) List Cable Outside Diameter	(N) List Number of Cables	Multiply (D) × (N) Subtotal of Sum of Cable Diameters (Sd)
1	3/C–500 kcmil	2.26 inches	3	6.78 inches
2	3/C–4/0 kcmil	1.55 inches	4	6.20 inches

Total cable tray width required for items 1 and 2:

6.78 inches + 6.20 inches = 12.98 inches

B—Width required for 3/0 AWG and smaller multiconductor cables:

Total cable tray width required for items 3, 4, and 5:

(3.20 sq. in. + 4.00 sq. in. + 3.20 sq. in.) (6 sq. in. / 7 sq. in.)1= (10.4 sq. in.) (6 sq. in. / 7 sq. in.)1 = 8.92 inches

Actual cable tray width is A "Width" (12.98 inches) + B "Width" (8.92 inches) = 21.90 inches

A 24-inch (600 mm) wide cable tray is required. The 24-inch (600 mm) cable tray has the capacity for additional future cables (3.1 inches or 3.6 square inches allowable fill can be used).

Notes:

1. This ratio is the inside width of the cable tray in inches divided by its maximum fill area in square inches from Table 392.9, Column 1.
2. The cable sizes used in this example are a random selection.
3. Cables—copper conductor with cross-linked polyethylene insulation and a PVC jacket.
4. Total cable weight per foot for this installation: 40.2 lb/ft (Cables in this example do not contain equipment grounding conductors.) This load can be supported by a load symbol "A" cable tray—50 lb/ft.

NEC® 392.9(A)(3)

Cross Section of the Cables and the Cable Tray

(Courtesy of Cooper B-Line, Inc.)

Example: NEC® Section 392.9(B)

50 percent of the cable tray usable cross-sectional area can contain type PLTC cables.

4 inches × 6 inches × .500 = 12 square inches
allowable fill area

2/C 16 AWG 300-volt shielded instrumentation cable O.D. = .224 inches

Cross-Sectional Area = 0.04 square inch

$$\left(\frac{12 \text{ sq. in.}}{0.04 \text{ sq. in. cable}}\right) = 300 \text{ cables can be installed in this cable tray}$$

$$\left(\frac{300 \text{ cables}}{26 \text{ cables/row}}\right) = 11.54 \text{ rows can be installed in this cable tray}$$

Notes:
1. The cable sizes used in this example are a random selection.
2. Cable—copper conductors with PVC insulation, aluminum/mylar shielding, and PVC jacket.

Following these general application Articles 300 and 310, the remaining articles in Chapter 3 are specific articles. Article 392 covers cable trays (again see Figure 10–4). Cable trays are a support system and not a wiring method.

Note: As you begin to study this part of Chapter 3, refer to the definitions in Article 100 for raceway. A fine print note will alert you to the types of raceways. There is a distinct difference between raceways and cables and other wiring methods. Study that definition or refer to it as needed. A cable tray is neither a raceway nor a cable. It is merely a support system for other wiring methods.

392.3 Uses Permitted. *Cable tray shall be permitted to be used as a support system for service conductors, feeders, branch circuits, communications circuits, control circuits, and signaling circuits. Cable tray installations shall not be limited to industrial establishments. Where exposed to direct rays of the sun, insulated conductors and jacketed cables shall be identified as being sunlight resistant. Cable trays and their associated fittings shall be identified for the intended use.*

(A) Wiring Methods. *The wiring methods in Table 392.3(A) shall be permitted to be installed in cable tray systems under the conditions described in their respective articles and sections.*

(B) In Industrial Establishments. *The wiring methods in Table 392.3(A) shall be permitted to be used in any industrial establishment under the conditions*

NEC® 392.9(B)

Cross Section of the Cables and the Cable Tray

(Courtesy of Cooper B-Line, Inc.)

described in their respective articles. In industrial establishments only, where conditions of maintenance and supervision ensure that only qualified persons service the installed cable tray system, any of the cables in 392.3(B)(1) and (B)(2) shall be permitted to be installed in ladder, ventilated trough, solid bottom, or ventilated channel cable trays.

(C) Equipment Grounding Conductors. Metallic cable trays shall be permitted to be used as equipment grounding conductors where continuous maintenance and supervision ensure that qualified persons service the installed cable tray system and the cable tray complies with provisions of 392.7.

(D) Hazardous (Classified) Locations. Cable trays in hazardous (classified) locations shall contain only the cable types permitted in 501.10, 502.10, 503.10, 504.20, and 505.15.

(E) Nonmetallic Cable Tray. In addition to the uses permitted elsewhere in 392.3, nonmetallic cable tray shall be permitted in corrosive areas and in areas requiring voltage isolation.

392.4 Uses Not Permitted. Cable tray systems shall not be used in hoistways or where subject to severe physical damage. Cable tray systems shall not be used in ducts, plenums, and other air-handling spaces, except as permitted in 300.22, to support wiring methods recognized for use in such spaces.

* Reprinted with permission from NFPA 70-2005.

Example: Application of 392.13. Ampacity of Type MV and Type MC Cables (2001 Volts or Over) in Cable Trays—(b) Single Conductor cables. These single conductor cables can be in-

stalled in a cable tray cabled together (triplexed, quadruplexed, and so on) if desired. Where the cables are installed according to the requirements of 392.12, the ampacity requirements are as shown in the accompanying chart.

Refer to the articles in Chapter 3 for the specific types of raceways and cables as needed. In addition to the references for raceways in Chapter 3, you will find references related to Chapter 9 tables and examples, such as the reference found in 342.22, the numbers of conductors in conduit. When sizing the number of conductors in a raceway, conduit, tubing, and so on, it is necessary to refer to the permitted percent fills specified in the tables in Chapter 9, Table 1 for percentage fill and Table 4 for the conduit dimensions. Similar references are found in 344.22 for rigid metal conduit, 352.22 for rigid nonmetallic conduit, 358.22 for electrical metallic conduit, and 348.22 for flexible metallic conduit. Other similar references are found for other raceway types. The number of conductors allowed in a given raceway must be calculated in square inches, based on the insulation and wire size listed in Chapter 9, Table 5 and Table 5-A for compact strand aluminum. For bare conductors, Chapter 8 dimensions are used. Where conductors of all the same size are installed in a single raceway, Tables C1 through C12A in Annex 3 may be permitted.

The following example will help you learn to apply Chapter 9.

Example: What size PVC rigid nonmetallic conduit is required for a multiple branch-circuit

Sec. No.	Cable Sizes	Solid Unventilated Cable Tray Cover	Applicable Ampacity Tables *	Amp. Table Values By	Special Conditions
(1)	1/0 AWG and Larger	No Cover Allowed **	310.69 and 310.70	0.75	—
(1)	1/0 AWG and Larger	Yes	310.69 and 310.70	0.70	—
(2)	1/0 AWG and Larger in Single Layer	No Cover Allowed **	310.69 and 310.70	1.00	Maintained Spacing of 1 Cable Diameter
(2)	Single Conductor in Triangle Config. 1/0 AWG and Larger	No Cover Allowed **	310.71 and 310.72	1.00	Maintained Spacing of 2.15 × 1 Conductor O.D.

* The ambient ampacity correction factors must be used.
** At a specific position where it is determined that the tray cables require mechanical protection, a single cable tray cover of six feet (1.8 m) or less in length can be installed.

supplying a cooling tower located adjacent to a hospital containing three 6 AWG THWN stranded copper conductors, three 4 AWG THWN copper conductors, four 1/0 AWG THW copper conductors, and one 2 AWG bare copper conductor?

Step 1. Rigid nonmetallic conduit, Article 352. Number of conductors, 352.22; Ref. Table 1, Chapter 9.

Step 2. Chapter 9, Notes 1, 2, Table 1 and Notes. Read the notes to the table carefully because they are a part of the table and, where applicable, they are mandatory.

Step 3. Chapter 9 Table 1. Over two conductors permit a 40 percent fill of the raceway.

Step 4. Table 4. Dimensions and percent are of conduit and tubing in square inches. Therefore, select square-inch dimensions from Tables 5 and 8 for the correct calculations.

Step 5.
Table 5.
3 6 AWG THWN = .0507 × 3 = .1521 sq. inch
3 4 AWG THWN = .0824 × 3 = .2472 sq. inch
4 6 AWG THWN = .2223 × 4 = .8892 sq. inch

Table 8.
1 2 AWG BARE = .067 × 1 = .067 sq. inch
                        Total = 1.3555 sq. inches

Table 4.
Not lead covered over two conductors in a raceway 40 percent. Trade size 2 (53) Rigid PVC Schedule 40 raceway is 1.316 square inches. Trade size 2½ (63) raceway is 1.878. Therefore, 2 (53) is not large enough, and it is necessary to install 2½ (63) rigid nonmetallic conduit for this installation.

Note: Trade size 2 (53) (IMC) intermediate metal conduit (Article 342) would permit a 40 percent fill of 1.452 square inches. Therefore, trade size 2 (53) IMC could be used.

Article 314 covers outlet boxes, pull boxes, junction boxes, conduit bodies, and fittings. A conduit body is what is normally referred to in the trade as a Tee, LB, or as a condulet. Use Article 314 carefully, and as you determine the number of conductors for a given box, the sizing requirements will be found in 314.16.

Note: A deduction must be added for most boxes with the integral-type clamps, such as nonmetallic outlet boxes. If uncertain, check with the manufacturer. In a test, normally the test material will tell you whether or not there

NEC® 314.16(A)(2)

14-2 AWG Type NM cable
12-2 AWG Type NM cable

are clamps to be calculated. Study 314.16(A) and (B) and the tables carefully.

Example: What size outlet box with integral clamps is required to accommodate a 15-ampere snap switch to control lighting and a 20-ampere duplex receptacle? The switch has a 2-wire w/GRD 14 AWG nonmetallic-sheathed cables feeding it with another 14 AWG NM 2-wire w/GRD feeding the luminaire (light fixture). The receptacle is fed with a 2-wire w/GRD 12 AWG nonmetallic-sheathed cable and a 2-wire w/GRD 12 AWG nonmetallic sheath cable goes on to feed additional duplex receptacles in the same area.

Solution: 314.16(A)(1) and (2): The minimum size of box in cubic inches is based on the following calculations:

The required deduction for each type fitting (i.e., clamps, hickeys, and the like) and all equipment grounding conductors require one deduction for each based on the largest conductor entering the box; 314.16(A) (1), (2), and Table 314.16(B).

In this example, a 12 AWG is the largest, therefore, a 2.25 cubic inch deduction is required for each.

Equipment grounding conductors = 2.25
Cable clamps = 12 AWG 2.25 × 1 = 2.25

Each device requires two deductions each, based on the size of conductor connected to the device.

Receptacle = 12 AWG 2 × 2.25 = 4.50
Switch = 14 AWG 2 × 2.00 = 4.00

Pull and junction boxes. Minimum size as required in *NEC®* 314.28.

Conductors entering the box, one deduction each.

12 AWG $2.25 \times 4 = 9.00$
14 AWG $2.00 \times 4 = 8.00$

Answer: The minimum size box permitted would be a 30 cubic inch box.

In addition, sizing of pull boxes and junction boxes will be included in most examinations. These requirements are found in 314.28.

Example: Four trade size 3 (78) rigid steel conduits or EMT, straight pull. Three trade size 2 (53) rigid steel conduits or EMT, angle pull.

Recommended minimum spacing for 3 (78) conduits, 4¾ inches; for 2 (53) conduits, 3⅜ inches; and between a 3 (78) and a 2 (53), conduit, 4 inches.

For depth of box (A):

Three trade size 2 (53) conduits, angle pull:

$A = (6 \times 2) + 2 + 2$ = 16 inches

Spacing for two rows of conduit:

Edge of box to center of trade size 3 (78) conduit	= 2⅜ inches
Center to center, trade size 3 (78) to trade size 2 (53) conduit	= 4
Minimum distance, D	= 8½
Total	= 14⅞ inches

Required depth: Use 16 inches.

For length of box (B):

Four trade size 3 (78) conduits, straight pull:
$B = 8 \times 3$ = 24 inches
Trade size 2 (53) conduit, angle pull:
$B = A$ = 16 inches

Required length: Use 24 inches.

For width of box (C):

Recommended spacing for trade size 3 (78) conduits	= 4¾ inches
$C = 4 \times 4¾$ inches	= 19 inches

Required width: Use 19 inches.

Angle pull distance (D):

$D = 6 \times 2$ inches   = 12 inches

Required distance: Use 12 inches.

This section should also be studied, and you must know where to find it quickly so that you can determine the size, outlet, or junction box as required (see Figure 10–8).

Article 312 covers cabinet, cutout boxes, and meter socket enclosures. Most panelboards, switches, disconnecting switches, conductors, and so on, are enclosed in a cabinet or cutout box (see Figure 10–9). When determining the requirements for cabinets and cutout boxes, refer to Article 312.

Figure 10–8 A typical pull box (junction manufactured and sized in accordance with Article 314). *(Courtesy of Allied Tube and Conduit)*

Figure 10–9 Typical cabinet manufactured in accordance with Article 312. *(Courtesy of Unity Manufacturing)*

Article 366 covers auxiliary gutters. Auxiliary gutters are not raceways or cables. An auxiliary gutter is an extension used to supplement wiring spaces in meter enclosures, motor control centers, distribution centers, switchboards, and similar wiring systems, and may enclose conductors or bus bars, but is not permitted for switches, overcurrent devices, appliances, or other similar equipment. An auxiliary gutter, although very similar in appearance to a wireway, has a different application and use. Be aware of Article 366 for auxiliary gutters (see Figure 10–10).

Article 404 covers switches. In this article you may find requirements for all types of switches and some installation requirements, such as accessibility and grouping. Article 408 covers switchboards and panelboards. This article will undoubtedly be included as a part of most examinations and includes many of the requirements for panelboards and switches. Of particular importance is 110.26(F), which applies to

Figure 10–10 This figure was given to a panel of *NEC®* experts and the question was asked, "Which of these sketches are auxiliary gutters?" The consensus of that group was that only the center sketch is an auxiliary gutter (Article 366), the other two are wireways (Article 376).

equipment covered in Articles 408 and 404 for the dedicating working space above and below equipment. This section will also undoubtedly be referred to in most examinations. You need not try to memorize this, because it contains specific requirements for this equipment, and your proficiency in using this as a ref-

erence document will adequately guide you through these wiring method articles. Be familiar with the article names as listed in the Table of Contents, and it will greatly assist you in using this book as a reference or as a tool in the field, and will greatly increase your proficiency in preparing for an examination.

QUESTION REVIEW

> Each lesson is designed purposely to require the student to apply the entire *NEC* text, and not specific chapters or articles. It has been found that when studying for a timed, open-book examination, the student must gain proficiency in the Table of Contents, the Index, and the ability to move quickly from cover to cover to find the correct answer to each question in a timely fashion.

1. When installing Type SER service-entrance cable in a building from the service panel to a dryer outlet, what are the support requirements for this cable?

   Answer: _____

   _____

   _____

   _____

   Reference: _____

2. When installing conduit bodies that are legibly marked with their cubic inch capacity, how can one determine the number of conductors and splices and taps permitted?

   Answer: _____

   _____

   _____

   _____

   Reference: _____

3. Is trade size ⅜ (12) liquidtight flexible metal conduit permitted for enclosing motor leads in accordance with 430.245(B)?

   Answer: _____

   _____

   _____

   _____

   Reference: _____

4. Structural bar joists are often spaced at up to 5-foot (1.5 m) intervals. Support is difficult to achieve when installing metal raceways on these bar joists. Is it permitted to mount an outlet box on a bar joist and extend intermediate metal conduit (IMC) 5 feet (1.5 m) to the first support?

Answer: _____

_____

_____

_____

Reference: _____

5. When installing electrical nonmetallic tubing (ENT), is it permissible to install this material through metal studs without securing the ENT every 3 feet (900 mm) as required by the *Code*?

Answer: _____

_____

_____

_____

Reference: _____

6. What is the support requirements for Type MC cable?

Answer: _____

_____

_____

_____

Reference: _____

7. What are the support requirements and methods of securing Type NM cable?

Answer: _____

_____

_____

_____

Reference: _____

8. Can nonmetallic wireways be installed in hazardous Class 1 or Class 2 locations?

Answer: _____

_____

_____

_____

Reference: _____

9. What wiring methods are permitted for installing a branch-circuit routed through the bar joists of a metal building for a 4160-volt, 3-phase motor?

Answer: _____

_____

_____

_____

Reference: _____

10. A contract to install a branch-circuit under a bank president's desk to serve a calculator is awarded to a contractor. The bank has asked that there be no drilled holes in the walls. Can the conductors be laid on the floor under the carpet? If so, what type of wiring method should be used?

Answer: _____

_____

_____

_____

Reference: _____

11. When installing wireway metallic or nonmetallic, either type, how do you derate the conductors when exceeding the number permitted?

Answer: _____

_____

_____

_____

Reference: _____

12. When installing Type AC cable on wood studs, routing them parallel to the framing members, what distance clearance must be maintained where the Type AC cable is likely to be penetrated by screws or nails?

Answer: _____

_____

_____

_____

Reference: _____

13. Is it necessary to support electrical metallic tubing (EMT) within 3 feet (900 mm) of each coupling or fitting?

Answer: _____

_____

_____

_____

Reference: _____

14. How do the installation requirements for intermediate metal conduit (IMC) and that of rigid metal conduit (RMC) differ? What installation requirement is permitted by one wiring method that is not permitted by the other?

Answer: _____

_____

_____

_____

Reference: _____

15. What is the minimum size conductor that can be run in parallel, generally? (Do not consider exceptions.)

Answer: _____

_____

_____

_____

Reference: _____

16. Where installing electrical nonmetallic tubing for the branch-circuit conductors in a five-story building, does the material from which the walls, floors, and ceilings are constructed need to be considered?

Answer: _____

_____

_____

_____

Reference: _____

17. Christmas tree lights have been installed on the local courthouse. The mayor suggested that we leave them up throughout the year. Is this permitted? How long are they permitted to remain installed?

Answer: _____

_____

_____

_____

Reference: _____

18. When installing a run of conduit, it is necessary to install a piece of liquidtight flexible metal conduit 6 feet (1.8 m) long through a plenum. Is this permitted by the *Code*? How many feet of liquidtight flexible metal conduit can you install in a plenum?

Answer: _____

_____

_____

_____

Reference: _____

19. Where making an installation of rigid nonmetallic conduit, Schedule 80, along the outside of a building to feed outdoor floodlights and studying the *Code*, you find that in 300.7(B) expansion joints must be provided where necessary. How much expansion or contraction must rigid nonmetallic conduit have before an expansion fitting is required?

Answer: _____

_____

_____

_____

Reference: _____

20. A column panel is a narrow panel that is installed inside or flush with an I-beam. It is primarily used in warehouses and areas where the panel is subjected to moving traffic and warehousing storage, such as forklifts and palletizing. When installing a column-type panel, an auxiliary gutter is installed above the panel, and the neutral connections are made up at the top of the auxiliary gutter and not brought down into the panelboard. Is this an acceptable wiring method?

Answer: _____

_____

_____

_____

Reference: _____

21. When wiring a single-family dwelling to save space, a trade size 2 (53) rigid conduit is installed from the top of the panelboard into the attic space and nonmetallic-sheathed cables are pulled within that trade size 2 (53) conduit. Is this a permitted wiring method?

Answer: _____

_____

_____

_____

Reference: _____

22. A small lake cabin is to be wired. This cabin will contain wood heating, no air conditioning, and only six single-pole branch-circuits for lighting and receptacle outlets. Using only six switches in this panel, are you therefore exempt from using a main circuit breaker or a single disconnecting switch as required in Chapter 2 for services?

Answer: _____

_____

_____

_____

Reference: _____

23. A location in a small commercial office building appears to be dangerous. The light switches located on the wall are installed in a standard three-gang switch box. Each light switch controls a row of fixtures, and each is fed from a different phase in the panelboard. The light switch on the right is fed from B-phase of a 480-volt, 277-volt panelboard. The center switch is fed from B-phase, and the switch on the right is fed from C-phase, making the potential between each switch at 480 volts. This appears to be dangerous. Which section in the *Code* prohibits this installation?

Answer: _____

_____

_____

_____

Reference: _____

24. Two new circuits are to be added in an existing old house with knob-and-tube wiring. The owner has asked that this extension be made with the same wiring method that is now in the house, concealed knob-and-tube. The attic space has had 6 inches (150 mm) of blown insulation installed; the wires are covered by the insulation. The owner has stated that you should pull the insulation back, make the extension, and then relocate the insulation over the wiring. Is this wiring method still acceptable in the *Code*?

Answer: _____

_____

_____

_____

Reference: _____

25. What is the minimum size single conductor cable permitted within cable trays to be used as ungrounded circuit conductors?

Answer: _____

_____

_____

_____

Reference: _____

Chapter Eleven

OBJECTIVES

Studying this chapter along with the 2005 *NEC*® gives the reader a basic introduction to the branch-circuit and feeder requirements in the *Code* that apply to all installations for above and below 600 volts.

After studying this chapter, you should know:

- The definitions and the application of terms that are applicable and that are unique to the 2005 *NEC*®

- The new *Code* articles that have been relocated to Chapter 4 and are addressed as they apply to the *NEC*®

- The utilization equipment requirements found in this chapter

- The flexible cords and fixture wire requirements and applications

- The switch types, requirements, and applications

- The new Article 406 covering all requirements for receptacles, cord connectors, and attachment plugs (caps)

- The switchboard and panelboard requirements and applications

- The basic requirements for luminaires (lighting fixtures), motors, transformers, and other general-use equipment commonly found in all installations

GENERAL-USE UTILIZATION EQUIPMENT

The equipment covered in Chapter 4 is equipment found generally in all occupancies—residential, commercial, and industrial. These are things that utilize electricity to operate, and include items such as luminaires (lighting fixtures), motors, and appliances. They are not considered as special equipment, which is dealt with in Chapter 6. Chapter 4 starts out with two articles, Articles 400 and 402, dealing with wire. One might think that these articles are misplaced and should be found in Chapter 3. However, flexible cords, cables, and fixture wires are most generally associated with general-purpose equipment, such as flexible cords for appliances and motors and fixture wiring for luminaires (lighting fixtures). The application for these articles is not the same as general wiring. *NEC*® 400.7 and 400.8 cover the uses permitted and not permitted. Cords are not generally used as a fixed-wiring method; there are exceptions for limited length and use applications. It would not be acceptable to connect many types of equipment with a flexible cord—for example, water heaters, gas furnaces, luminaires (lighting fixtures), except as permitted in 410.30(A). Many appliances, however, are listed with a flexible cord attached or are designed to have a flexible cord installed—for example, dishwashers, disposals, washing machines, dryers, and ranges. Article 402 is used even less in normal applications than is Article 400. Apply the rules carefully and in the proper application.

Article 410 covers luminaires (lighting fixtures) of all types. Article 406—Receptacles, Cord Connectors, and Attachment Plugs (Caps) was added to the 2002 *NEC*®. This article is a compilation of all the requirements that were formerly found in Article 210 and Part L of Article 410. This was done in an attempt to make the *Code* more user friendly. Many had believed that receptacles and plugs should haven been a part of Article 404 (switches). Although they are not utilization equipment, they have always been located in Article 410, and those just beginning to study the *Code* should mark this article or make notes to remind themselves where this information is located. Notes can be beneficial to quickly find items in the *Code*. However, do not mark up your *Code* book so that there are so many notes that they are difficult to distinguish from each other. Some testing agencies do not permit the book used for the examination to be marked up. (See Figure 11–1, Figure 11–2, and Figure 11–3.)

As you study Article 410, it will quickly become apparent that this article is divided into sections that cover only specific types of luminaires (fixtures)—that is, incandescent, high discharge including fluorescent below 1000 volts and less (refer to Figure 6–4), high

Figure 11–1 Compliance with 406.8. *(Courtesy of TayMac Corporation)*

discharge including fluorescent greater than 1000 volts, and recessed fixtures and lampholders. 410.4 covers the requirements for installing fixtures in specific locations. This section contains the only requirements found in the *NEC®* for locating fixtures in bathrooms and clothes closets. Many believe that the *Code* contains requirements for protecting luminaires (lighting fixtures) in bathrooms with ground-fault protection. The *NEC®* contains no such requirements.

1. The maximum wattage permitted for a medium base incandescent lamp (standard light bulb) is _____ watts.
2. The maximum wattage permitted for a mogul-base incandescent lamp is _____ watts.
3. A special base or other means must be used for incandescent lamps of _____ watts and larger.

Answers: 1. 300; 2. 1500; 3. 1501.
Reference: Section 410.53

Article 411—Lighting Systems Operating at 30 Volts or Less: This is a very short article with only seven sections, and the requirements are very easy to understand. However, one must be aware that it exists; otherwise misapplication of Article 410 may result.

Figure 11–2 Examples of common receptacles, cord connectors, and attachment plugs (caps) covered in Article 406.

LANGUAGE *NEC®* 406.8 Receptacles in Damp or Wet Locations*

406.8 Receptacles in Damp or Wet Locations.

(A) Damp Locations. A receptacle installed outdoors in a location protected from the weather or in other damp locations shall have an enclosure for the receptacle that is weatherproof when the receptacle is covered (attachment plug cap not inserted and receptacle covers closed).

An installation suitable for wet locations shall also be considered suitable for damp locations.

A receptacle shall be considered to be in a location protected from the weather where located under roofed open porches, canopies, marquees, and the like, and will not be subjected to a beating rain or water runoff.

(B) Wet Locations.

(1) 15- and 20-Ampere Outdoor Receptacles in a Wet Location. 15- and 20-ampere, 125- and 250-volt receptacles installed in a wet location shall have an enclosure that is weatherproof whether or not the attachment plug cap is inserted.

(2) Other Receptacles. All other receptacles intsalled in a wet location shall comply with (B)(2)(a) or (B)(2)(b).

(a) A receptacle installed in a wet location, where the product intended to be plugged into it is not attended while in use, shall have an enclosure that is weatherproof with the attachment plug cap inserted or removed.

(b) A receptacle installed in a wet location, where the product intended to be plugged into it will be attended while in use (e.g., portable tools) shall have an enclosure that is weatherproof when the attachment plug is removed.

(C) Bathtub and Shower Space. A receptacle shall not be installed within or directly over a bathtub or shower stall.

(D) Protection for Floor Receptacles. Standpipes of floor receptacles shall allow floor-cleaning equipment to be operated without damage to receptacles.

(E) Flush Mounting with Faceplate. The enclosure for a receptacle installed in an outlet box flush-mounted in a finished surface shall be made weatherproof by means of a weatherproof faceplate assembly that provides a watertight connection between the plate and the finished surface.

PURPOSE To ensure receptacles remain weatherproof when equipment is used in damp or wet locations.

WET LOCATIONS

"Std Wp cover"

"Raintite" while in use

Figure 11–3 Weatherproof cover on the left meets 406.8. The weatherproof cover on the right meets the general rule 406.8. *(Courtesy of TayMac Corporation)* (*Reprinted with permission from NFPA 70-2005, the *National Electrical Code®* Copyright © 2004–, National Fire Protection Association, Quincy, MA 02269. This reprinted material is not the complete and official position of the National Fire Protection Association on the referenced subject, which is represented only by the standard in its entirety.)

Article 422 covers appliances. "**Appliance**" is defined in Article 100 as *Utilization equipment, generally other than industrial, that is normally built in standardized sizes or types and is installed or connected as a unit to perform one or more functions such as clothes washing, air conditioning, food mixing, deep frying, and so forth.* This article gives some special requirements for calculating the branch-circuits supplying appliances, but Articles 210 and 220 govern generally. 422.11 should be studied carefully because specific overcurrent protection requirements are contained in this section for appliances.

NEC® 422.12 states that central heating equipment must be supplied by an "**individual branch circuit**." What is an individual branch circuit?

Answer: An individual branch circuit is defined in Article 100 as a branch circuit that supplies only one utilization equipment.

Would it be permissible to connect a waste (garbage) disposal and a built-in dishwasher with flexible cords? Both cords would be a minimum of 42 inches (1.1 m).

Answer: NEC® 422.16(B)(2) would permit the built-in dishwasher to be connected with a cord 3 feet (900 mm) to 4 feet (1.2 m) long. 422.16(B)(1), however, would prohibit the 42-inch (1.1 m) cord on the waste disposal. This section limits the cord for disposals to 18 to 36 inches (450 to 900 mm).

Note: Portable dishwashers are supplied by the manufacturer with a cord and plug attached. This cord is evaluated as a part of the listing, and length is not limited by 422.16.

Fixed electric space heating equipment is covered in Article 424. It is important to understand that this article covers electrical resistance furnaces, baseboard heaters, heating cable, duct heaters, and boilers, resistance or electrode types. It does not cover the typical gas-fired furnace common in many residential occupancies throughout the United States, which are appliances and covered in Article 422. It also does not cover heat pumps; they are covered in Article 440.

Article 430 covers motors, motor circuits, and controllers (see Figure 11–4). This article technically amends several articles in Chapter 1 through Chapter 4—210, 220, and 310. Although Article 90 states that

Chapter 4 is a part of the general requirements, each of these articles contains a special allowance to which Article 430 applies.

Example: Motor branch-circuit conductors ampacities are as per Article 310. However, 240.4(D) limiting 14 AWG to a 15-ampere overcurrent device, 12 AWG to a 20-ampere device, and 10 AWG to a 30-ampere device does *not* apply. 430.6 only references the table and not 240.4(D).

Read *NEC®* 430.6 carefully.

The motor nameplate current is *not* to be used; the values in Tables 247 through 250 are to be used instead.

Motor-circuit conductors are determined in Article 430 Part II, branch-circuit and overcurrent protection in Part III. Article 430 also contains special tap rules for motors in 430.28. Where the values for branch-circuit, short-circuit, and ground-fault protective devices do not correspond to standard sizes or ratings, the next higher standard size, rating, or possible setting is permitted.

Careful study is required for Part IX of Article 430, which covers the disconnecting means, the location, type, size, and so on.

Article 440 covers air-conditioning and refrigeration equipment (see Figure 11–5). The rules for hermetically sealed units are covered in this article; belt-driven units are referenced to Article 430 for the rules.

Example: In a steel mill, a number of small 3-horsepower, 3-phase, 480-volt motors are supplied by a 400-ampere feeder. What is the minimum size 10-foot (3.0 m) tap conductor permitted to be used to feed each motor?

Step 1. 240.4(E), which permits tap conductors to be protected against overcurrent in accordance with 210.19(A)(3), 240.5(B)(2), 240.21, 368.17(B), 368.17(C), and 430.53(D).

Step 2. 240.21(F) references 430.28 and 430.53(D) for motor taps.

Step 3. *NEC®* 430.28, 400 amperes ÷ 1000% = 40 amperes

Step 4. Table 310.16

Answer: 8 AWG THWN

Figure 11–4 Cutaway of a standard totally enclosed fan-cooled squirrel-cage motor.

Figure 11–5 Typical air handler (fan coil unit).

Article 409—Industrial Control Panels, new for the 2005 NEC®

These control panels are for general use; they operate at 600 volts or less. They are general manufactured or shop fabricated to UL 508A, the safety standard for industrial control panels. They are defined as an assembly of a systematic and standard arrangement of two or more components such as motor controllers, overload relays, fused disconnect switches, and circuit breakers and related control devices such as pushbutton stations, selector switches, timers, switches, control relays, and the like with associated wiring, terminal blocks, pilot lights, and similar components. The industrial control panel does not include the equipment that it is designed to operate.

Note: Some would permit a 10 AWG conductor to be used in this calculation as taken from the 90° column because a 3-horsepower motor (Table 430.250) would only draw 4.8 amperes, and the temperature would not be exceeded.

Article 450 covers transformers and transformer vaults (see Figure 11–6). The secondary conductors are always tap conductors, and the rules for protecting tap conductors are found in 240.21. Autotransformers (often referred to as buck or boost transformers) are covered in this article; however, branch-circuit and feeder rules are found in 210.9.

This chapter also contains several articles for special use types of general equipment, such as phase converters (Article 455), which are used often in rural farm-type installations; capacitors (Article 460) now generally found in many types of equipment and used to improve the efficiency of many services as a power factor correction; and generators (Article 445) often used to supplement or provide emergency or standby power; and others.

As your studies take you through Chapter 4 of the NEC®, be aware that the requirements in Article 410 cover luminaires (lighting fixtures), lampholders, and lamps. However, Article 410 does not apply to signs. Article 422, which covers the appliances, generally has specific requirements for branch-circuits, installation of appliances, and the control and protection of appliances. Space heating, motors, and air-conditioning and refrigeration equipment articles also have general provisions and installation provisions within those articles. Study them carefully for correct application. Many examination questions are derived from Chapter 4.

Figure 11–6 Typical dry-type transformer.

QUESTION REVIEW

Each lesson is designed purposely to require the student to apply the entire *NEC®* text, and not specific chapters or articles. It has been found that when studying for a timed, open-book examination, the student must gain proficiency in the Table of Contents, the Index, and the ability to move quickly from cover to cover to find the correct answer to each question in a timely fashion.

THREE-PHASE MOTOR AS FOLLOWS:

Type:	Squirrel cage, high reactance	Duty:	Continuous
Voltage:	460 volts, 3-phase	Start:	High reactance
HP:	30	Code letter:	None
Current:	Nameplate is the same as *NEC®* current Tables 430.247–250	Service factor:	None

Refer to the description of the 3-phase motor above for Questions 1–3.

1. What *NEC®* section requires that the Tables 430.247–250 be used instead of actual current rating marked on the motor nameplate?

   Answer: _____

   _____

   _____

   _____

   Reference: _____

2. What size inverse-time circuit breaker is needed to provide the maximum branch-circuit, short-circuit, ground-fault protection?

   Answer: _____

   _____

   _____

   _____

   Reference: _____

3. A dual element (time delay) fuse can be sized a maximum of _____ percent when used as branch-circuit protection.

   Answer: _____

   _____

   _____

   _____

   Reference: _____

4. A 120-volt single-phase store sign panel supplies only the sign as follows: Hours-on 24 hours a day, three ¼ horsepower motors, 24-0.80 ampere ballasts. Each ungrounded conductor in the subfeeder to the sign panel must be at least _____ AWG *aluminum* THW.

Answer: _____

_____

_____

_____

Reference: _____

5. Where small conductors are tapped to larger conductors as permitted in the *NEC*®, the tap cannot be larger than the _____ size of the conductor being tapped.

Answer: _____

_____

_____

_____

Reference: _____

In Questions 6 through 9, calculate the ampere rating of the fuse size needed for protection of the following copper conductors where no motors are connected:

6. 12 AWG THHN _____ ampere rated fuse

Answer: _____

_____

_____

_____

Reference: _____

7. 8 AWG THW _____ ampere rated fuse

Answer: _____

_____

_____

_____

Reference: _____

8. 3 AWG THW _____ ampere rated fuse

Answer: _____

_____

_____

_____

Reference: _____

9. 3/0 AWG THWN _____ ampere rated fuse

   Answer: _____

   _____

   _____

   _____

   Reference: _____

10. For a transformer 600 volts or less that has no secondary overcurrent protection, the primary overcurrent protection shall not exceed _____ percent of the primary current.

    Answer: _____

    _____

    _____

    _____

    Reference: _____

11. The maximum branch-circuit and ground-fault protection using time-delay fuses for typical motors are sized at _____ percent of the motor's full-load current.

    Answer: _____

    _____

    _____

    _____

    Reference: _____

12. A motor pulls 22 amperes at 230 volts, and the feeder circuit is 150 feet (45.7 m) in length. If a 10 AWG copper wire is desired, what would be the voltage drop?

    Answer: _____

    _____

    _____

    _____

    Reference: _____

13. What is the percentage of voltage drop on the circuit described in Question 12?

    Answer: _____

    _____

    _____

    _____

    Reference: _____

14. What is the correct size conductor to maintain 3 percent or less voltage drop in Question 12?

    Answer: _____

    _____

    _____

    _____

    Reference: _____

15. Overload protection (heaters, relays, time delay fuses, thermal overloads) for typical motors having a service factor of 1.15 is generally sized at _____ percent of the motor's full-load current.

    Answer: _____

    _____

    _____

    _____

    Reference: _____

16. A motor controller that is installed with the expectation of its being submerged in water occasionally for short periods shall be installed in a rated enclosure type _____ AWG.

    Answer: _____

    _____

    _____

    _____

    Reference: _____

17. Where recessed high-intensity discharge fixtures are installed _____ and operated by remote ballasts, both the fixture and the ballast require thermal protection.

    Answer: _____

    _____

    _____

    _____

    Reference: _____

18. Any pipe or duct system foreign to the electrical installation must not enter a transformer vault. The _____ piping is not considered foreign to the vault.

    Answer: _____

    _____

    _____

    _____

    Reference: _____

19. A 10 kVA dry-type transformer rated at 480 volts can be installed on a building column and shall not be required to be _____.

    Answer: _____

    _____

    _____

    _____

    Reference: _____

20. A capacitor is located indoors. It must be enclosed in a vault if it contains more than a minimum of _____ gallon(s) of flammable liquid.

    Answer: _____

    _____

    _____

    _____

    Reference: _____

21. A single-phase hermetic refrigerant motor-compressor has a rated load current of 24 amperes and a branch-circuit selection current of 30 amperes. The branch-circuit conductors are copper, Type TW. They operate at 80°F, and they are the only conductors in the conduit to this compressor. The smallest possible branch-circuit conductors must be at least size ____ AWG.

    Answer: _____

    _____

    _____

    _____

    Reference: _____

22. A 50-horsepower, 3-phase, squirrel-cage motor with full-voltage reactor starting has no code letter. The calculated size of the fuse to protect the branch-circuit of the motor would be sufficient for the starting current of the motor. The nontime delay fuse to protect the circuit of this motor shall be sized at _____ percent of full-load current.

    Answer: _____

    _____

    _____

    _____

    Reference: _____

23. A dry-type transformer is to be installed indoors. If rated more than ____ kilovolt amperes, the transformer must be installed in a fire-resistant transformer room.

    Answer: _____

    _____

    _____

    _____

    Reference: _____

24. An electric resistance heater is rated for 2400 watts at 240 volts. What power is consumed when the heater is operated at 120 volts?

    Answer: _____

    _____

    _____

    _____

    Reference: _____

25. A 230-volt, single-phase circuit has 10 kilowatts of load and 50 amperes of current. The power factor is _____ percent.

    Answer: _____

    _____

    _____

    _____

    Reference: _____

Chapter Twelve

OBJECTIVES

Studying this chapter along with the 2005 *NEC®* gives the reader a thorough review of the 2005 *NEC®*.

After studying this chapter, you should know:

- The definitions and the application of terms that are applicable and that are unique to the 2005 *NEC®*

- The *Code* requirements in which special equipment and occupancies are addressed as they apply to the *NEC®*

- The differences between Class I, Class II, and Class III systems as they relate to hazardous occupancies

- The differences between U.S. Division classification and the International Zone classification methods

- The wiring methods allowed in special occupancies

- Healthcare facilities and the requirements for those occupancies

- The definitions and the application of terms that are applicable and that are unique to Article 517 of the 2005 *NEC®*

SPECIAL EQUIPMENT AND OCCUPANCIES

Special occupancies are covered in Chapter 5, Articles 501 through 555. (See Figure 12–1.) It is very important to be aware that the *NEC®* covers installation guidelines for wiring and equipment. The classification of areas and operational rules are covered in other standards.

Articles 500 through 517 contain hazardous (classified) locations within those occupancies. When an electrical system, or any portion of it, is to be installed in an area containing a hazardous atmosphere, protection against possible explosions becomes a major design factor. All the basic design considerations still apply to determination of conductor sizes and general layout of the overall system. In addition, the portion of the system in or passing through an area where flammable gases, combustible dusts, or ignitable fibers may exist must have equipment enclosures specially designed to prevent electrical sparks and arcs from igniting the surrounding atmospheres and causing serious fires and/or explosions. For this portion of the design, the electrical engineers must be doubly careful to select equipment specifically approved for the conditions encountered.

The *NEC®* describes three classifications of hazardous locations where an unconfined spark could result in explosion or fire. Each of these classifications identifies conditions that may exist either under normal operations; under occasional, but frequent conditions resulting from maintenance operations; or under conditions of accidental breakdown that may cause the release of flammable or readily combustible materials.

Warning: *NEC®* 90.3 Chapter 5 only modifies and augments Chapters 1 through 4. Corrosion, capacity, suitability, and service are not considered. Therefore, it is the responsibility of the designer to take these factors into account along with the hazardous location requirements that are covered in Chapter 5.

NEC® Articles 500 through 506, 510, 511, 513, 514, 515, 516, and 517 contain rules covering the type of wiring and apparatus to be installed in hazardous locations. Because the actual degree of hazard is frequently evaluated by local inspection and fire department authorities, they should be consulted during the design stage of the wiring system that may be in the hazardous area.

Class I

Those areas in which flammable gases or vapors are or may be present in the air in quantities sufficient to produce explosive or ignitable mixtures are Class I areas.

Figure 12–1 Special occupancy hazardous (classified) fuel storage facility. *(Courtesy Crouse Hinds, Division of Cooper Industries)*

Class II

Those areas that are hazardous because of the presence of combustible dust are Class II areas.

Class III

Those areas that are hazardous because of the presence of easily ignitable fibers or flyings, but in which such fibers or flyings are not likely to be in suspension in the air in quantities sufficient to produce ignitable mixtures, are Class III areas.

Division I in the Normal Situation

This hazard would be expected to be present in every-day production operations or during frequent repair and maintenance activity (see Figure 12–2, Figure 12–3, and Figure 12–4).

Division II in the Abnormal Situation

The material is expected to be confined within closed containers or closed systems and will be present only through accidental rupture, breakage, or unusual faulty operation (see Figure 12–2, Figure 12–3, and Figure 12–4).

Groups

The gases and vapors of Class I locations are broken into four groups by the codes: A, B, C, and D. These materials are grouped according to their explosion pressures and other flammable characteristics. Class II dust locations are groups E, F, and G. These groups are classified according to the ignition temperature and the conductivity of the hazardous substance (see Figure 12–2, Figure 12–3, and Figure 12–4).

Hazardous (Classified) Locations in Accordance with Article 500, *National Electrical Code*® — 2005

Class I Flammable Gases or Vapors

Division 1
- Exists under normal conditions
- May exist because of:
  - repair operations
  - maintenance operations
  - leakage
- Released concentration because of:
  - breakdown of equipment
  - breakdown of process
  - faulty operation of equipment
  - faulty operation of process that causes simultaneous failure of electrical equipment

Division 2
- Liquids and gases are in closed containers or the systems are:
  - handled
  - processed
  - used
- Concentrations are normally prevented by positive mechanical ventilation.
- Adjacent to a Class I, Division 1 location

Group A: Atmospheres containing acetylene.

Group B: Atmospheres such as butadiene, ethylene oxide, propylene oxide, acrolein, or hydrogen (or gases or vapors equivalent in hazard to hydrogen, such as manufactured gas).

Group C: Atmospheres such as cyclopropane, ethyl ether, ethylene, or gases or vapors equivalent in hazard.

Group D: Atmospheres such as acetone, alcohol, ammonia, benzine, benzol, butane, gasoline, hexane, laquer solvent vapors, naptha, natural gas, propane, or gases or vapors equivalent in hazard.

Figure 12–2 Class I flammable gases or vapors

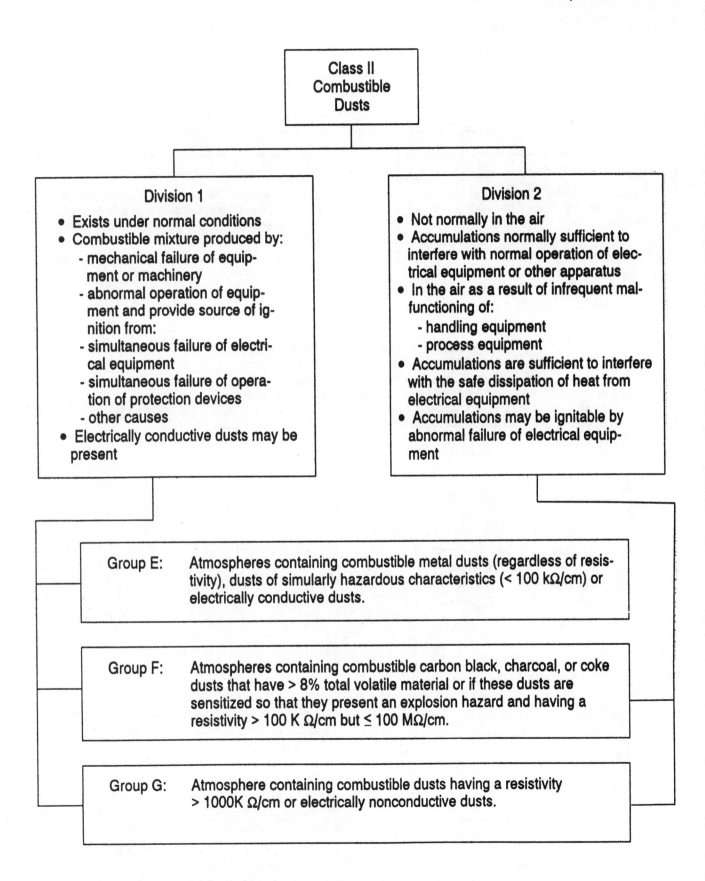

Figure 12–3 Class II combustible dusts

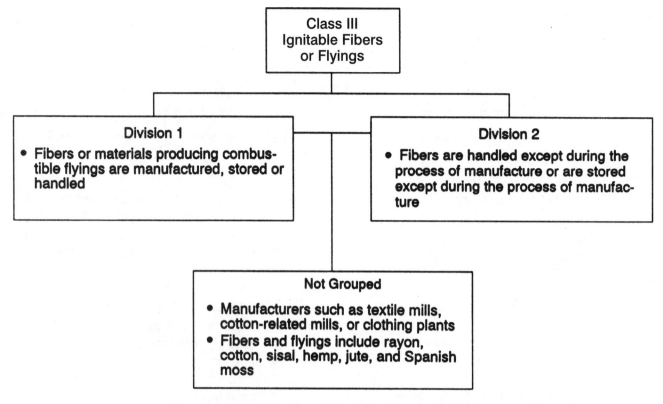

Figure 12–4 Class III ignitable fibers or flyings

Seals

Special fittings are required in Class I locations to minimize the passage of gases and vapors or, in the case of an explosion, prevent the passage of flames. They are also required in Class II locations to prevent the passage of combustible dusts (see Figure 12–5).

Article 500

Articles 500 through 503

These articles explain in detail the requirements for the installation of wiring or electrical equipment in hazardous locations. These articles, along with other applicable regulations, local governing inspection authorities, insurance representatives, and qualified engineering/technical assistance, should be your guides to the installation of wiring or electrical equipment in any hazardous or potentially hazardous location.

Class I Wiring Methods

The minimum wiring methods are specified in *NEC®* 501.10(A) for Class I Division 1 locations. *NEC®* 501.10(B) covers Class I Division 2 locations.

Class I, Division 1 Wiring Methods

NEC® 501.10(A) requires wiring methods of *NEC®* Article 344—Threaded Rigid Metal Conduit (RMC); Article 342—Threaded Steel Intermediate Metal Conduit (IMC); or Article 332—Type MI cable (see Figure 12–6) with termination fittings approved for the location. All boxes, fittings, and joints shall be threaded for a connection to conduit or cable terminations and shall be explosionproof. Threaded joints must be made up with at least five threads fully engaged.

> **MI cable must be installed and supported in a manner to avoid tensile stress at the terminations.**

When it is necessary to use flexible connections as provided in *NEC®* 501.10, flexible fittings approved for Class I locations must be used. *NEC®* 501.140, however, permits flexible cord or connection between portable lighting equipment or other portable utilization equipment and a fixed portion of its supply circuit with rigid requirements: Extra-hard usage cord must be employed and must, in addition to the conductors in the circuit, employ a grounding conductor complying

EYD
Drain seal
Trade size 1/2–1
(16)–(27)

EYS—Vertical seal

CHICO "A"
compound

CHICO "X"
fiber dam

CHICO A-P
INTRAPAK®
premeasured
sealing compound
and water in
plastic mixing
pouch

CHICO A
sealing compound

Figure 12–5 Typical seal fitting, cutaway of a seal fitting that has been poured, and examples of approved pouring compound. *(Courtesy of Crouse Hinds, Division of Cooper Industries)*

Figure 12–6 MI cable installed as per *NEC*® Article 332. *(Courtesy of Pyrotenax USA Inc., photos by Robert Stewart Associates)*

with *NEC*® 400.23. It must be connected to terminals or supply conductors in an approved manner and be supported by clamps or other suitable means in such manner that there will be no tension on the terminal connections, and must be provided with seals where the cord enters boxes, fittings, or enclosures of explosion-proof type.

Class I, Division 2 Wiring Methods

NEC® 501.10(B) requires somewhat less stringent wiring requirements. However, it is equally as important to wire areas classified as Class I, Division 2 correctly. The Class I, Division 2 minimum wiring methods are *NEC*® Article 342—Threaded Steel Intermediate Conduit; *NEC*® Article 344—Threaded Rigid Metal Conduit; *NEC*® Article 368—Enclosed Gasketed Busways; *NEC*® Article 376—Enclosed Gasketed Wireways; PLTC cable in accordance with the provisions of *NEC*® Article 725 and Types MI, MC, MV, PC, or SNM cable with approved fitting and terminations.

> **Types PLTC, MI, MC, MV, TC, and SNM are permitted to be installed in cable trays and shall be installed in a manner to avoid tensile stress at the termination fittings.**

Boxes and joints are not required to be explosion-proof in these areas, except as required in *NEC*® 501.105(B)(1), 501.115(B)(1), and 501.150(B)(1). When provisions must be made for limited flexibility, such as at motor terminals, flexible metal conduit with approved fittings, *NEC*® Article 320—Type AC Armored Cable with approved fittings, *NEC*® Article 350—Liquidtight Flexible Metal Conduit with approved fittings, and Article 356—Liquidtight Flexible Nonmetallic Conduit with approved fittings, or flexible cord approved for hard uses provided with approved bushed fittings may be used. The additional conductor for grounding must be included in the flexible cord unless other accessible means for grounding are provided.

Wiring that cannot release sufficient energy to ignite a specific ignitable atmospheric mixture by opening, shorting, or grounding may be permitted in this area using any of the methods suitable for general wiring applications.

Sealing and Drainage Requirements

NEC® 501.15 covers the minimum requirements for sealing and drainage in Class I, Divisions 1 and 2. Fittings approved for the location and purpose must be used with approved sealing compounds. Sealing compounds approved for Type MI cable must also exclude moisture and other fluids from the cable insulation. In addition, see the requirements of *NEC*® 501.15(D).

> Seals are provided in conduit and cable systems to minimize the passage of gases and vapors and prevent the passage of flames through the wiring method from one electrical system or enclosure to another. Type MI cable provides this inherently and prevents the passage of gases and vapors through the insulation.

The specific requirements for installing the seals are found in *NEC*® 501.15(A) through (E) and the manufacturer's instructions. Installation of these seals is a critical part of this installation and improperly made will not provide the protection needed to isolate accidental explosions. Raceways in Class I hazardous locations are required to be provided with sealing fittings to prevent the passage of gases, vapors, or flames through the raceway from one portion of an electrical installation to another, and to prevent pressure piling or cascading. Pressure piling may occur in these locations when an explosion in some part of the system, such as a switch enclosure, is not effectively blocked or sealed off. The first explosion may be made harmless if properly confined by a seal to the switch enclosure. If allowed to enter the raceway, it builds up great pressure and another more powerful explosion may follow. The second explosion builds up additional pressures in the gas-filled conduit, and a greater explosion results. This may continue in very long raceway runs not provided with intermittent seals until some part of the system ruptures. The hot or burning gases could then escape and ignite flammable gases in the surrounding atmosphere, causing a general explosion.

Proper sealing is one of the most important requirements for safety in a Class I hazardous location. Explosive vapors are prevented from passing to a non-hazardous area where they could be ignited and carry the flame back to the explosive vapor location. In the manner of a pilot light on a gas range, arcs, sparks, and hot particles from an arcing device are also prevented from entering the raceway by the seal. Sealing fittings are designed to permit the pouring of an approved water-base sealing compound into the fitting around and between the conductors (refer to Figure 12–5).

> Conduit seals are limited to a 25 percent conductor fill unless specifically identified for a higher percentage fill. Where raceways are filled to 40 percent, check with the manufacturer for conductor fill limits.

The raceway should first be blocked at the bottom of the sealing fitting with a packed material to prevent the sealing mixture from draining out of the fitting. The liquid mixture solidifies after a few hours and becomes very hard. Proper mixing of the compound and liquid is essential; otherwise, the seal would not prevent the pressure piling. Even though it may prevent the passage of vapors, arcs, and hot particles, all voids under, between, and over conductors must be filled. If the sealing mixture is to perform its function of preventing the passage of gases, vapors, or flames through the raceway, complete elimination of the pressure piling can be achieved by installing effective seals in every raceway regardless of size where it enters or leaves any junction box at regular intervals, in all raceway runs, and at points at which they presently are required by the *NEC®*. In the absence of tests for various gases, determine the proper spacing of seals. In long conduit runs, it is not unreasonable to specify that they be installed at 50-foot (15.25 m) intervals in all conduit runs within a Class I Division 1 location. If installed in accordance with the *NEC®* requirements, a rigid steel conduit system with suitable enclosure and seals will effectively contain an internal explosion and prevent the escape of burning gases until they are cooled sufficiently to prevent the ignition of flammable vapors in the surrounding atmosphere. The gases are cooled by expansion as they escape through the machine-grounded joints of predetermined width and through threaded joints. Ground joints, unless properly maintained, are not as reliable as threaded joints. A single loose screw in a cover of a ground joint, or scratches in ground services, will render an otherwise explosion-proof enclosure ineffective. Threaded joints, on the other hand, are reliable because the gases escape only by following the threads. The *NEC®* requires that at least five full threads be engaged at each coupling. Effective seals must be placed in a raceway as required by the *NEC®* within 18 inches (450 mm) of an arcing device or at or near the point where the raceway leaves

or enters a hazardous area. In the case of a raceway 2 inches (50 mm) in diameter or larger, seals are required within 18 inches (450 mm) of any junction box containing splices, taps, or terminal connections. No coupling or fitting is permitted between the seal and the boundary between the hazardous and nonhazardous areas. In many cases, special panelboards and controls are designed with factory seals that are an integral part of the equipment designed for Class I locations. Because such seals effectively isolate the arc-producing segments of the assembly, additional seals are not generally required in the conduit leaving these panelboards. Incandescent and fluorescent fixtures in Class I locations are also available with integral seals.

> Seals are **not** intended to prevent the passage of gases or vapors at a continuous pressure. Even at differences in pressure across the seal, there may be a slow passage of gases or vapors.

> Leakage and the passage or propagation of flames through the interstices of stranded conductors can be reduced through the use of compact stranding or sealing of the individual stranding.

Although wiring methods are less restrictive in Class I, Division 2 locations, sealing is still required as specified in *NEC®* 501.15(B). In each conduit run passing from a Class I, Division 2 location into an unclassified area, a sealing fitting is permitted to be on either side of the boundary of such location, but so designed and installed as to minimize the amount of gas or vapor that may enter the conduit system within the Class II location from being communicated to the conduit beyond the seal. Rigid metal conduit and threaded steel intermediate conduit shall be used between the sealing fitting and the point at which the conduit leaves the Division 2 location, and a threaded connection shall be used at the sealing fitting. No fitting or coupling is permitted between the sealing fitting and the point in which the conduit leaves the Division 2 location. Conduit systems terminating to an open raceway in an outdoor unclassified area are not required to be sealed between the point at which the conduit leaves the classified location and enters the open raceway. The minimum requirements for the location of sealing fittings are found in *NEC®* 501.15(A), (B), (C), (D), (E), and (F).

Class II, Division 1 Wiring Methods

Class II, Division 1 502.10(A)(1) and (2) permit threaded rigid metal conduit, threaded steel intermediate metal conduit, or Type MI cable with termination fittings approved for the location and installed to avoid tensile stress at the fittings. The fittings and boxes with threaded bosses and close-fitting covers shall have no openings through which dust might enter or sparks or burning material might escape. In locations where combustible dusts of an electrically conductive nature are likely, the fittings and boxes that are designed to contain taps, joints, or terminal connections must be approved for Class II locations. Flexible connections shall be made with dusttight flexible connectors; liquidtight flexible metal conduit with approved fittings; liquidtight flexible nonmetallic conduit with approved fittings; or flexible cord approved for extra-hard usage and provided with bushed fittings. Flexible cords must comply with 502.10(A)(2)(4) and 502.140, where flexible connections subject to oil or other corrosive conditions shall be of a type approved for the condition or shall be protected by a suitable sheath.

> FPN: See 502.30(B) for grounding requirements where flexible conduit is used. [502.10(A)(2), reprinted with permission from NFPA 70-2005.]

Class II, Division 2 Wiring Methods

Class II, Division 2 502.10(B)(1) and (2) permit the use of rigid metal conduit, intermediate metal conduit, electrical metallic tubing, dusttight wireways, or Types MI, MC, or SNN cable with approved termination fittings, ITC, PLTC, or TC cable installed in ventilated channel-type cable trays in a single layer with a space of not less than the larger cable diameter between the two adjacent cables is also permitted. Wiring in nonincendive circuits is permitted using any of the methods suitable for wiring in ordinary locations.

Sealing Requirements for Class II, Divisions 1 and 2

Combustible and explosive dusts constitute a hazard in Class II locations. Such dusts do not travel for great distance within a conduit. Therefore, seals are not required if the conduit running between a dust-ignition-proof enclosure and an enclosure that is not dust-ignitionproof consists of either a horizontal section not less than 10 feet (3.0 m) in length or a vertical section not less than 5 feet (1.5 m) in length extending downward from the dust-ignitionproof enclosure through the raceway. Where such conduit runs are not provided, seals must be used to prevent the entrance of dust into ignitionproof enclosures through the raceway. Sealing fittings and poured sealing compound may be used in both Division 1 and 2. They are not required to meet the same stringent requirements of those in 501.5. The term "dust-ignitionproof" is defined in *NEC®* 500.2.

> **Class III Divisions 1 and 2**
> *Live uninsulated parts shall not be exposed.* [503.25]

Wiring Methods for Class III, Division 1 Locations

The permitted wiring methods are rigid metal conduit, intermediate metal conduit (IMC), electrical metallic tubing (EMT), rigid nonmetallic conduit, Type MC, MI, or SNM cable with approved termination fittings or dusttight wireways. Flexible connections must be made with dusttight flexible fittings and installed in accordance with 503.10(A)(2) using liquidtight flexible metal conduit with approved fittings, liquidtight flexible nonmetallic conduit with approved fittings, or flexible cord in conformance with 503.140.

Boxes and fittings must be dusttight.

Class III, Division 2 Locations

Wiring methods approved for Class III Division 1 locations are acceptable and, in addition, in open wiring on insulators may be used in accordance with all of the provisions of Article 398, not subject to physical damage in compartments or locations solely for storage, containing no machinery.

Grounding and Bonding Class I, II, and III, Divisions 1 and 2

> **In accordance with Article 250**

Bonding jumpers with proper fittings or other approved means of bonding shall be used and apply to all intervening raceways, fittings, boxes, enclosures, and so on, between classified locations and the point of grounding for service equipment or point of grounding of a separately derived system, except as otherwise permitted.

> **The locknut-bushing and double-locknut types shall not be depended upon for bonding purposes.**

Equipment grounding conductors shall be installed in accordance with 250.120. Where flexible conduit is used, it must be installed with internal or external bonding jumpers in parallel with each conduit and complying with 250.102, with exceptions in Chapter 5.

> Seals are not required in Class III locations.

Article 504 covers "intrinsic safe systems." Intrinsic safety is an explosion-prevention design technique applied to electrical equipment and the wiring of hazardous locations where flammable or combustible material is present. The technique is based on limiting electrical and thermal energy levels below that which is required to ignite the specific hazardous atmospheric mixture present. Intrinsically safe wiring shall not be capable of releasing sufficient electrical or thermal energy under normal or abnormal conditions to cause ignition to a specific flammable or combustible atmospheric mixture in its most easily ignitable concentration.

The entire intrinsically safe system must be installed in accordance with control drawings, as required in 504.10. The control drawing identification must be marked on all equipment and apparatus. Generally, intrinsically safe systems may be permitted using any wiring method suitable for unclassified locations. However, sealing shall be provided in accordance with 504.70, and separation of intrinsically safe systems shall be provided in accordance with 504.30.

General wiring methods are permitted, in other words, any of the wiring methods in Chapter 3.

> **Warning:** Corrosion and physical protection have not been considered, only circuit safety. It is the designer's responsibility to consider all facets of the installation for performance and safety.

It should be noted that intrinsically safe systems always fail in the safe position. Should a wire or cable be crushed or cut accidentally, the operation of the equipment must fail safe in every instance. No arcing or sparking will occur due to the design and nature of intrinsically safe systems. Designers generally require physical protection, such as metal raceway, to protect intrinsically safe conductors and to provide assurance from costly accidental shutdowns and operational stoppages. *NEC®* 504.70 requires that conduits and cables be sealed in accordance with the hazardous location in which they are installed, such as Article 501 for Class I hazardous flammable vapors and gases. Sealing of intrinsically safe systems must be provided as specified in 501.25. For flammable or combustible dusts, sealing is required as specified in 502.25. Seals are not required for enclosures that contain only intrinsically safe apparatus.

Article 505 introduces the European zone system as a new alternate method as an alternative for engineers. It provides a new level of the classification of Class I hazardous (classified) locations. This alternate method uses the zone concept to meet the requirements for electrical equipment and wiring for all voltages in locations where fire or explosive hazards may exist due to flammable gases, vapors, and flammable liquids.

Article 505 modifies the general rules of the *Code* as it applies to the electric wiring and equipment in locations classified as Class I, Zone 0, Zone 1, and Zone 2.

The 2005 *NEC®* introduces the second half of the European zone system for classifying hazardous locations. Article 505, which was introduced into the *NEC®* a few years earlier, was the European alternative for Class I locations and first introduction to the Zone classification system to the United States. This new Article 506—Zone 20, 21, and 22 provides the alternative classification scheme for Class II Locations, where flammable dusts, fibers, and flyings are likely to be present. The new article covers the requirements for the Zone classification system as an alternative to the division classification system covered in Article 500, Article 502, and Article 503 for electrical and electronic equipment and wiring for all voltages. Article 506 divides hazardous (classified) locations where fire and explosion hazards may exist due to combustible dusts, or ignitable fibers, or flyings into Zone 20, Zone 21, and Zone 22. Zone 20, Zone 21, and Zone 22 area classifications are based on a modified IEC area classification system.

Combustible metallic dusts are not covered by the requirements of Article 506.

The varying rules for hazardous classified locations as covered in Articles 500 through 504 are applied to the specific locations as covered in Articles 510 through 517 with amendments to those rules. For example, Article 511 covers commercial garages. The standard for classifying these areas is NFPA 88. The hazardous concerns are the presence of Class I classified liquids, gases, and vapors in dangerous concentrations.

The difficulty in interpreting the rules is that Chapters 1 through 4 apply generally. Article 511 amends those general rules, but references Articles 500 and 501.

Article 517 covers the installation requirements for health care facilities. This article contains many extracts from NFPA 99, which covers other than installation requirements for these locations.

Chapter 5 also covers many other articles related to special occupancies from places of assembly (Article 518) to mobile homes (Article 550).

Chapter 6 covers special equipment. This is equipment that is not found in all occupancies, such as Article 600—Electric Signs and Outline Lighting. They are not required, nor are they found on all occupancies. However, when installed they do introduce a hazard if not safely installed; therefore, the *NEC®* contains specific requirements for sign installation.

Examination Study

Questions related to Chapter 6, Special Equipment, can generally be answered quickly because these articles are usually short and cover very specific requirements. The most important study advice is to learn the names of these articles for quick reference.

- Article 600—Signs and Outline Lighting
- Article 605—Office Furnishings
- Article 610—Cranes and Hoists
- Article 620—Elevators, Dumbwaiters, Escalators, Moving Walks, Wheelchair Lifts, and Stairway Chairlifts
- Article 625—Electric Vehicle Charging System Equipment. This article places installation requirements for electrically powered automobiles in the *Code*. This article does not cover other charging systems, such as for golf carts, forklifts, etc. It does cover individual residential-type chargers and the proposed commercial quick-charging stations that may some day appear at rest stops along the interstate highway system. It is said that someday soon you will be able to stop for coffee and get your electric car charged in a few minutes.
- Article 630—Electric Welders
- Article 640—Sound Recording and Similar Equipment
- Article 645—Information Technology Equipment (formerly Computer/Data Processing Equipment)
- Article 647—Sensitive Electronic Equipment
- Article 650—Pipe Organs
- Article 660—X-ray Equipment
- Article 665—Induction and Dielectric Heating Equipment
- Article 668—Electrolytic Cells
- Article 669—Electroplating
- Article 670—Industrial Machinery
- Article 675—Electrical-Driven or Controlled Irrigation Machinery
- Article 680—Swimming Pools, Fountains, and Similar Equipment
- Article 685—Integrated Electrical Systems
- Article 690—Solar Photovoltaic Systems
- Article 692—Fuel Cell Systems
- Article 695—Fire Pumps. This article has added installation requirements into the *Code* that were formerly found in NFPA 20.

QUESTION REVIEW

> **Each lesson is designed purposely to require the student to apply the entire *NEC*® text, and not specific chapters or articles. It has been found that when studying for a timed, open-book examination, the student must gain proficiency in the Table of Contents, the Index, and the ability to move quickly from cover to cover to find the correct answer to each question in a timely fashion.**

1. Name three types of optical fiber cable.

   Answer: _____

   _____

   _____

   _____

   Reference: _____

2. Fixed wiring over a Class I location in a commercial garage cannot be enclosed in nonmetallic-sheathed cables. True or False?

   Answer: _____

   _____

   _____

   _____

   Reference: _____

3. Are seals required for an outdoor propane-dispensing unit located 50 feet (15.24 m) from the office where the branch-circuit supplying the unit originates? Where are the seals required to be located?

   Answer: _____

   _____

   _____

   _____

   Reference: _____

4. Emergency electrical systems are those systems legally required and classed as emergency by Article 700 *NEC*®. True or False?

   Answer: _____

   _____

   _____

   _____

   Reference: _____

5. In a commercial garage, Class _____, Division _____ wiring methods must be met for an electrical outlet installed 12 inches (300 mm) above the floor if there is no mechanical ventilation.

Answer: _____

_____

_____

_____

Reference: _____

6. According to the *NEC*®, transfer equipment for emergency systems shall be designed and installed to prevent _____ of normal and emergency sources of power.

Answer: _____

_____

_____

_____

Reference: _____

7. The sealing compound in a completed conduit seal for a Class I location must be at least _____ inch(es) (_____ mm) thick.

Answer: _____

_____

_____

_____

Reference: _____

8. Bare conductors are field-connected to fixed terminals of different phases on an outdoor, 13.8-kV circuit. According to the *National Electrical Code*®, the air separation between these conductors shall be a minimum of _____ inch(es) (_____ mm).

Answer: _____

_____

_____

_____

Reference: _____

9. In a hospital's general care areas, the number of receptacles required to be in a patient-bed location is a minimum of _____.

Answer: _____

_____

_____

_____

Reference: _____

10. A 120-volt, single-phase store sign panel supplies only the sign as follows: Hours-on 24 hours a day, three ⅓ horsepower motors, 36-0.80 ampere ballasts. Each ungrounded conductor in the subfeeder to the sign panel must be at least _____ AWG copper THWN.

    Answer: _____

    _____

    _____

    _____

    Reference: _____

11. Portable gas tube signs for interior use can be connected with a supply cord having a maximum length of _____ feet (____ m).

    Answer: _____

    _____

    _____

    _____

    Reference: _____

12. Electric discharge (neon) tubing shall be a _____ and _____ so as **not** to cause steady overvoltage on the transformer.

    Answer: _____

    _____

    _____

    _____

    Reference: _____

Answer Questions 13 through 20 about Type S fuses True or False.

13. The S indicates "size rejection."

    Answer: _____

    _____

    _____

    _____

    Reference: _____

14. A 15-ampere Type S fuse will fit into a 20-ampere Type S adapter.

    Answer: _____

    _____

    _____

    _____

    Reference: _____

15. A 20-ampere Type S fuse will fit into a 20-ampere Type S adapter.

    Answer: _____

    _____

    _____

    _____

    Reference: _____

16. A 20-ampere Type S fuse will fit into a 15-ampere Type S adapter.

    Answer: _____

    _____

    _____

    _____

    Reference: _____:

17. A 30-ampere Type S fuse will fit into a 20-ampere Type S adapter.

    Answer: _____

    _____

    _____

    _____

    Reference: _____:

18. A 25-ampere Type S fuse will fit into a 30-ampere Type S adapter.

    Answer: _____

    _____

    _____

    _____

    Reference: _____

19. A 6½-ampere Type S adapter will accept a 6¼-ampere Type S fuse.

    Answer: _____

    _____

    _____

    _____

    Reference: _____

20. A 6¼-ampere Type S adapter will accept a 3-ampere, 4-, 4½-, 5-, 5⁶⁄₁₀-, and 6¼-ampere Type S fuse.

    Answer: _____

    _____

    _____

    _____

    Reference: _____

21. The *NEC*® requires that the available fault current be marked when series rated systems are installed. (a) Who is required to install that marking? (b) What must the marking state? (c) Who is responsible for providing this information?

    Answer: _____

    _____

    _____

    _____

    Reference: _____

22. A 1½-horsepower, single-phase motor that has an efficiency of 80 percent, operates at 230 volts and has an input current of _____ amperes. (1 hp = 746 watts)

    Answer: _____

    _____

    _____

    _____

    Reference: _____

23. A 30-horsepower, wound-rotor induction motor with no code letter is to be installed with 460-volt, 3-phase, alternating current. Disregarding all the exceptions, the nontime delay fuse for short-circuit protection of the motor branch-circuit must be rated at a maximum _____ amperes.

    Answer: _____

    _____

    _____

    _____

    Reference: _____

24. The following 480-volt, 3-phase, 3-wire, intermittent-use equipment is in a commercial kitchen: two 5000-watt water heaters, four 3000-watt fryers, and two 6000-watt ovens. Each ungrounded conductor in the feeder circuit for this kitchen equipment must be sized to carry a minimum calculated load of _____ amperes.

Answer: _____

_____

_____

_____

Reference: _____

25. A 240/480-volt, 3-phase power panelboard supplies only one 15,000 VA, 480-volt, 3-phase balanced resistive load. Each ungrounded conductor in the subfeeder to this power panel has a total net calculated load of _____ amperes.

Answer: _____

_____

_____

_____

Reference: _____

Chapter Thirteen

OBJECTIVES

Studying this chapter along with the 2005 *NEC®* gives the reader a basic introduction to the grounding requirements in the *Code* that apply to special systems, all limited voltage, and communications installations.

After studying this chapter, you should know:

Special Systems (Conditions)

- The definitions and the application of terms that are applicable and that are unique to the 2005 *NEC®*
- The *Code* requirements in which special systems or conditions are addressed as they apply to the *NEC®*
- The differences between emergency systems and legally required or optional systems
- The differences between power-limited and non–power-limited circuits and equipment
- The basics of optical fiber cable and raceways
- The interconnected electrical power production sources (commonly called "co-generation")

Communications Systems

- The definitions and the application of terms that are applicable and that are unique to Chapter 8 of the 2005 *NEC®*
- The *Code* requirements in which special systems or conditions are addressed as they apply to Chapter 8 systems
- The network-powered broadband communications systems

SPECIAL CONDITIONS AND COMMUNICATION SYSTEMS

Special Conditions

The first three articles in Chapter 7 cover Emergency Systems in Article 700, Legally Required Standby Systems in Article 701, and Optional Standby Systems

Class 1 Circuit. The portion of the wiring system between the load side of the overcurrent device or power-limited supply and the connected equipment. The voltage and power limitations of the source are in accordance with 725.21.

Class 2 Circuit. The portion of the wiring system between the load side of a Class 2 power source and the connected equipment. Due to its power limitations, a Class 2 circuit considers safety from a fire initiation standpoint and provides acceptable protection from electric shock.

Class 3 Circuit. The portion of the wiring system between the load side of a Class 3 power source and the connected equipment. Due to its power limitations, a Class 3 circuit considers safety from a fire initiation standpoint. Since higher levels of voltage and current than for Class 2 are permitted, additional safeguards are specified to provide protection from an electric shock hazard that could be encountered.

> 725.41(A) *FPN No. 2: Table 11(A) and Table 11(B) in Chapter 9 provide the requirements for listed Class 2 and Class 3 power sources.*

III. Class 2 and Class 3 Circuits

725.41 Power Sources for Class 2 and Class 3 Circuits.

(A) Power Source. The power source for a Class 2 or a Class 3 circuit shall be as specified in 725.41(A)(1), (A)(2), (A)(3), (A)(4), or (A)(5)

Figure 725.41 Class 2 and Class 3 Circuits.

Reprinted with permission from NFPA 70-2005.   *(continued)*

The power source types for a Class 2 or a Class 3 circuit shall be as specified below:

725.41(A)
(1) A listed Class 2 or Class 3 transformer
(2) A listed Class 2 or Class 3 power supply
(3) Other listed equipment marked to identify the Class 2 or Class 3 power source

Exception No. 1: Thermocouples shall not require listing as a Class 2 power source.

Exception No. 2: Limited power circuits of listed equipment where these circuits have energy levels rated at or below the limits established in Chapter 9, Table 11(A) and Table 11(B).

> *FPN: Examples of other listed equipment are as follows:*
> *(1) A circuit card listed for use as a Class 2 or Class 3 power source where used as part of a listed assembly*
> *(2) A current-limiting impedance, listed for the purpose, or part of a listed product, used in conjunction with a non–power-limited transformer or a stored energy source, for example, storage battery, to limit the output current*
> *(3) A thermocouple*
> *(4) Limited voltage/current or limited impedance secondary communications circuits of listed industrial control equipment*

(4) Listed information technology (computer) equipment limited power circuits.

> *FPN: One way to determine applicable requirements for listing of information technology (computer) equipment is to refer to UL 1950–1995, Standard for Safety of Information Technology Equipment, Including Electrical Business Equipment. Typically such circuits are used to interconnect information technology equipment for the purpose of exchanging information (data).*

(5) A dry cell battery shall be considered an inherently limited Class 2 power source, provided the voltage is 30 volts or less and the capacity is equal to or less than that available from series connected No. 6 carbon zinc cells.

725.41*(B) Interconnection of Power Sources.* *Class 2 or Class 3 power sources shall not have the output connections paralleled or otherwise interconnected unless listed for such interconnection.*

Reprinted with permission from NFPA 70-2005.

building official, or the like. The *NEC*® does not make these requirements; it does not require exit lights in buildings; it does not require smoke detectors. However, when these items are required to be installed, the *NEC*® governs and the installation requirements must conform to these articles as appropriate.

Requirements of Systems above 600 Volts

- Article 490—Equipment, over 600 Volts Nominal applies to the general requirements for equipment over 600 volts.

- The wiring method requirements are located in Article 300 Part II, 300.31 through 300.50.

- Overcurrent requirements are covered in Article 240 Part IX 240.100 and 240.101.

- Workspace and accessibility requirements are located in Article 110 Part III.

- Tunnel installations over 600 volts nominal are located in Part IV of Article 110.

- The grounding requirements are covered in Article 250 as appropriate.

Article 720 covers the *few* circuits and equipment operating at less than 50 volts that are not already covered in the *Code* by Articles 411, 551, 650, 669, 690, 725, and 760.

Article 725 covers power-limited circuits of Class 1, 2, and 3. It is very important that one understand the parameters of each class of wiring system. The power is limited by the power source, generally a transformer.

NEC® 725.8 covers the mechanical execution of work as follows:*

725.8 Mechanical Execution of Work. Class 1, Class 2, and Class 3 circuits shall be installed in a neat and workmanlike manner. Cables and conductors installed exposed on the outer surface of ceiling and sidewalls shall be supported by the building structure in such a manner that the cable will not be damaged by normal building use. Such cables shall be supported by straps, staples, hangers, or similar fittings designed and installed so as not to damage the cable. The installation shall also conform with 300.4(D).

and Other Special Systems in Article 702. The determination for which type system is installed in a building or facility is made by others—fire marshal,

* Reprinted with permission from NFPA 70-2005.

These circuits can often be run as open wiring without physical protection with specific cable types for the location. They can also be run with standard types of building wire where placed in the proper wiring method.

> For example: Class II conductors run through a plenum must be run as Type CL2P, CMP, or CL3P, or with appropriate building wire run in RMC, IMC, EMT Type MI cable or specific types of solid sheathed MC cable. (See Section 300.22.)

Article 760 covers fire alarm systems, both the power-limited and the non–power-limited. This article is very complete in its requirements for these systems. The NPLFA (Nonpower-Limited Fire Alarm) wiring methods are in accordance with Chapters 1 through 4 and 300.15(A), with the exceptions provided in 760.26 through 760.30. PLFA (Power-Limited Fire Alarm) wiring methods are covered in 760.51 and are similar to those in Article 725 as shown previously.

> For example: PLFA conductors run through a plenum must be run as Type FPLP, MPP, or CMP, or with appropriate building wire run in RMC, IMC, EMT Type MI cable or specific types of solid sheathed MC cable. (See Section 300.22.)

NEC® 760.8 covers the mechanical execution of work for fire alarm circuits, and equipment shall be installed in a neat and professional manner. Cables shall be supported by the building structure in such a manner that the cable will not be damaged by normal building use.

Article 770 covers the installation requirements for optical fiber cables and raceways. The question always comes up: Why are installation requirements in the *NEC®* for this system, which is not electrical? There is concern for static and resistance to the spread of fire. There are different types of fiber-optic cables. They are grouped into three types, as defined in 770.5. The composite types are hybrid and can carry both current and noncurrent. All types are capable of controlling current flow.*

NEC® 770.8 covers mechanical execution of work for optical fiber cables and raceways, and equipment shall be installed in a neat and professional manner. Cables shall be supported by the building structure in such a manner that the cable will not be damaged by normal building use.

Article 780—Closed-Loop and Programmed Power Distribution (Smart House) applies to premise power distribution systems jointly controlled by a signaling between the energy-controlling equipment and utilization equipment. All equipment, including the conductors, is required to be listed and identified. It will apply only for signaling circuits not exceeding 24 volts; the current required shall not exceed 1 ampere where protected by an overcurrent device or an inherently limited power source. The article also contains specific safety requirements for this new system.

Listed hybrid cable is permitted that may contain power, communications, and signaling conductors under a common jacket. The jacket must be applied so as to separate the power conductors from the communications and signaling conductors. The individual conductors of a hybrid cable shall conform to the *Code* provisions applicable to their current, voltage, and insulation rating. The signaling conductors shall not be smaller than 24 AWG copper. The power, communications, and signaling conductors of listed hybrid cable are permitted to occupy the same cabinet, panel, or outlet box (or similar enclosure housing the electrical terminations of electric light or power circuits) only if connectors specifically listed for hybrid cable are employed.

Chapter 8 is a stand-alone chapter in that 90.3 states:

> *Chapter 8 covers communications systems not subject to the requirements of Chapter 1 through 7 except where the requirements are specifically referenced in Chapter 8.*

This chapter is divided into four articles.

> ***800.1 Scope.*** *This article covers telephone, telegraph (except radio), outside wiring for fire alarm and burglar alarm, and similar central station systems; and telephone systems not connected to a central station system but using similar types of equipment, methods of installation, and maintenance.*
>
> *FPN No. 1: For further information for fire alarm, sprinkler waterflow, and sprinkler supervisory systems, see Article 760.*
>
> *FPN No. 2: For installation requirements of optical fiber cables, see Article 770.*
>
> *FPN No. 3: For installation requirements for network-powered broadband communications circuits, see Article 830.*
>
> *(continued)*

> **810.1. Scope.** *This article covers antenna systems for radio and television receiving equipment, amateur radio transmitting and receiving equipment, and certain features of transmitter safety. This article covers antennas such as multi-element, vertical rod, and dish, and also covers the wiring and cabling that connects them to equipment. This article does not cover equipment and antennas used for coupling carrier current to power line conductors.*
>
> **820.1. Scope.** *This article covers coaxial cable distribution of radio frequency signals typically employed in community antenna television (CATV) systems.*
>
> **830.1. Scope.** *This article covers network-powered broadband communications systems that provide any combination of voice, audio, video, data, and interactive services through a network interface unit.*
>
> *FPN No. 1: A typical basic system configuration includes a cable supplying power and broadband signal to a network interface unit that converts the broadband signal to the component signals. Typical cables are coaxial cable with both broadband signal and power on the center conductor, composite metallic cable with a coaxial member for the broadband signal and a twisted pair for power, and composite optical fiber cable with a pair of conductors for power. Larger systems may also include network components such as amplifiers that require network power.*
>
> *FPN No. 2: See 90.2(B)(4) for installations of broadband communications systems that are not covered.*
>
> Reprinted with permission from NFPA 70-2005.

Articles 800, 810, and 820 do not reference Table 300.5 (wiring method burial requirements). Therefore, the table cannot be applied. There are no burial depth requirements for these communication cables. The reason is that the *NEC®* is primarily a safety document. Communications generally pose no safety hazards by interruption of service. Therefore, the installers often bury these cables for service and economic reasons. However, they pose no safety hazards, so the *NEC®* does not set any requirements. Local requirements or amendments may apply.

NEC® 800.8, 820.8, and 830.8 cover the mechanical execution of work for communications circuits, and equipment shall be installed in a neat and professional manner. Cables shall be supported by the building structure in such a manner that the cable will not be damaged by normal building use.

Article 830 covers requirements for "Network-Powered Broadband Communications Systems." This is a new generation communication system and provides multimedia services in any combination of audio, voice, video, data, and interactive services. This new system may operate at higher levels of power and recognizes two levels of power (see *NEC®* Table 830.4).

Network Power Source	Low	Medium
Maximum power rating (volt-amperes)	100	100
Maximum voltage rating (volts)	100	150

It is because of these higher operating voltages that Article 830 contains burial depth requirements. *NEC®* 830.12(A) through (D) and Table 830.12 contain requirements similar to those in Table 300.5 and generally require cables to be buried at 18 inches (450 mm) below grade. When the cables are enclosed in a nonmetallic raceway 12 inches (300 mm) and in RMC or IMC, the burial depth is 6 inches (150 mm). The notes at the bottom of the table provide additional information that can be applied if necessary.

Chapter 8 is unique. Study it carefully. Study the scope, as shown throughout this chapter, of each article carefully to be sure that you are answering a question from the right article and section of this chapter.

QUESTION REVIEW

> Each lesson is designed purposely to require the student to apply the entire *NEC®* text, and not specific chapters or articles. It has been found that when studying for a timed, open-book examination, the student must gain proficiency in the Table of Contents, the Index, and the ability to move quickly from cover to cover to find the correct answer to each question in a timely fashion.

1. Does the *Code* address grounding CATV service to a dwelling unit?

   Answer: _____

   _____

   _____

   _____

   Reference: _____

2. After you install a satellite dish at a local residence, the inspector informs you that you must bury the cable from the satellite dish to the house in accordance with Table 300.5 of the *NEC®*. Was that inspector correct?

   Answer: _____

   _____

   _____

   _____

   Reference: _____

3. Does the *Code* require grounding the metal sheath of CATV cable entering a dwelling?

   Answer: _____

   _____

   _____

   _____

   Reference: _____

4. According to the *NEC®*, open conductors for communication equipment on a building shall be separated at least _____ feet (_____ mm) from *lightning* conductors, where practicable.

   Answer: _____

   _____

   _____

   _____

   Reference: _____

5. In a highly hazardous location where propane is used, the customer states that the wiring to the control circuit is of sufficient low voltage that a spark cannot be generated from this circuit. The devices and wiring have a control drawing number on them. What type wiring is most likely employed in this installation where there is not enough voltage to generate a spark sufficient to cause an explosion?

Answer: _____

_____

_____

_____

Reference: _____

6. A motor control center is to be installed in a local factory. Does the *NEC®* cover this type of installation? If so, which article is applicable?

Answer: _____

_____

_____

_____

Reference: _____

7. You have been asked to install a small transformer that will transform 480 volts to 240/120 volts. The inspector states that this must be wired as a separately derived system. What is a separately derived system? How do you wire this transformer?

Answer: _____

_____

_____

_____

Reference: _____

8. You recently encountered a receptacle with an orange triangle on the face of it. You were told that this was an isolated ground receptacle. Where can this be found in the *NEC®*?

Answer: _____

_____

_____

_____

Reference: _____

9. You recently encountered a rental storage facility that is an area in a building to store old movie films. Where would the wiring requirements for this area be found in the *NEC®*?

Answer: _____

_____

_____

_____

Reference: _____

10. The inspector informed you that a bathroom area was defined in the *NEC*®. Where is this definition located?

    Answer: _____

    _____

    _____

    _____

    Reference: _____

11. In wiring a large church, an installation of a pipe organ is required. Are pipe organs covered in the *NEC*®?

    Answer: _____

    _____

    _____

    _____

    Reference: _____

12. In making an installation, it was noted that the clearance between two large switchboards was rather tight, only about 3 feet (900 mm). With the doors open, there was not room to pass. Are there any requirements in the *NEC*® to cover the requirements for working clearances?

    Answer: _____

    _____

    _____

    _____

    Reference: _____

13. After you completed wiring a new home, the inspector came and told you to cover the panelboard, that if the sheetrockers got their spackling material or paint on the panel, it was a code violation. Is this correct?

    Answer: _____

    _____

    _____

    _____

    Reference: _____

14. You recently mounted a new service on a new mobile home on a privately owned lot. The inspector turned it down and said this was not permitted. Which section of the *NEC*® did you violate?

    Answer: _____

    _____

    _____

    _____

    Reference: _____

15. What is an optional standby system?

    Answer: _____

    _____

    _____

    _____

    Reference: _____

16. Is color-coding required in the *NEC*®? You know that the neutral conductor is required to be white and the grounding conductor is required to be green, but what about the ungrounded conductors?

    Answer: _____

    _____

    _____

    _____

    Reference: _____

17. You recently installed 14 AWG THHN conductors to supply a general purpose branch-circuit supplying duplex receptacles in an office area. You connected the 14 AWG THHN to a 20-ampere breaker. Table 310.16 allows 25 amperes on a 14 AWG conductor. Was this in accordance with the *NEC*®?

    Answer: _____

    _____

    _____

    _____

    Reference: _____

18. In making an installation to an outdoor ballfield, you proposed a common neutral to be run with multiple branch-circuit conductors. How many branch-circuit conductors can be supplied with a common neutral?

    Answer: _____

    _____

    _____

    _____

    Reference: _____

19. You have been asked to wire a dairy barn. Are there any specific requirements in the *NEC*® for this type of installation?

    Answer: _____

    _____

    _____

    _____

    Reference: _____

20. You recently observed a local electrician installing wiring under a carpet in a bank. Is this wiring method permitted? Which article covers wiring under carpets?

Answer: _____

_____

_____

_____

Reference: _____

21. More buildings are increasingly requiring lightning arrestors to be installed on the service. Which section of the *NEC*® covers this installation?

Answer: _____

_____

_____

_____

Reference: _____

22. Are temporary wiring methods covered in the *National Electrical Code*®? You regularly install a temporary meter loop to supply power to the construction crews until the permanent power system has been installed. Where is this covered in the *NEC*®?

Answer: _____

_____

_____

_____

Reference: _____

23. A 480-volt, 3-phase circuit with a rated load of 150 kW draws 210 amperes. The power factor is _____ percent.

Answer: _____

_____

_____

_____

Reference: _____

24. A _____-horsepower, 208-volt, 3-phase motor operating at 91 percent efficiency has an input current of 141 amperes. (1 hp = 746 watts)

Answer: _____

_____

_____

_____

Reference: _____

25. A feeder supplying a second building requires a maximum conductor resistance of 0.02 ohm. The minimum size conductor must have a resistance of no more than _____ ohm(s) per 1000 feet (305 m). The total length of the circuit conductor is 190 feet (58 m).

Answer: _____

_____

_____

_____

Reference: _____

Chapter Fourteen

ELECTRICAL AND *NEC®* QUESTION REVIEW

Now you are about to begin the final chapter of this book. You should have a good understanding of the *Code* arrangement and its content. You should now understand that the chapters are arranged in the order in which you use them in most common installations, from Article 90—Introduction; Chapter 1—general requirements; Chapter 2—wiring and protection; Chapter 3—wiring methods and materials; Chapter 4—general use equipment; Chapters 5, 6, and 7—special occupancies, uses, and equipment; and Chapter 8—communication systems. The final chapter in the *NEC®* is Chapter 9, which has the tables. Furthermore, you should now understand this arrangement and know the articles contained in each chapter. For instance, you should understand that the article covering rigid metal conduit (RMC) would automatically be found in Chapter 3, because rigid metal conduit is a wiring method. You should understand that the conditions for installation of a luminaire (light fixture) would be found in Chapter 4, because luminaires (light fixtures) are general-use equipment. Or you should understand that hospital installation requirements would be found in Chapter 5, because a hospital is a special occupancy and those are found in Chapter 5, such as mobile homes, hazardous materials and spray booths, commercial garages, and so on. You should automatically remember that a sign would be found in Chapter 6, special equipment, because although signs are common, they are not general equipment found in all types of installations. Signs are special equipment and are found in Chapter 6. These things must be understood before you take the examination. It is necessary so that you might use the *National Electrical Code®* efficiently as a reference document. Memorization is not recommended. However, with a good, clear understanding, one can refer to the Table of Contents in the front of the book and quickly find the article number if you have an understanding of where that article should be found in the *Code*. You should then be able to go to the Index and find the proper reference so that you might quickly look up the question and prove the answer. Once this has been done, you should have no problem passing any competency exam given throughout the country that is based on your knowledge of the *National Electrical Code®*.

With that competency achieved and your understanding of the *National Electrical Code®* arrangement ensured, the next important task you must improve is the understanding of what each question is asking. You must read the question through, then determine exactly what the question is asking. You will find key words, and with those key words you can simplify the question generally so it can be quickly researched using the Table of Contents, the Index, and your knowledge of the *Code* arrangement.

Example: As an example, in the following question, select the key words and simplify the question:

"May Type NM cable be used for a 120/140-volt branch-circuit as temporary wiring in a building under construction where the cable is supported on insulators at intervals of not more than 10 feet (3 m)?"

Now, you may rephrase the question slightly different in your mind, or you may look for the key words. NM cable is a key word, temporary wiring is a key, and supported on insulators at intervals of not more than 10 feet (3 m) is a key. So,

"Is NM cable permitted as temporary wiring where supported on insulators?"

Now, with the three key words in mind, we must remember that they are not asking where this wiring method is acceptable, they are telling you that the installation would be temporary wiring. If we go to the Table of Contents in the front of the *NEC*®, we find that temporary wiring systems is found in Article 590. As we turn to Article 590 and we look in the scope under 590.1, we find out that, yes, lesser methods are acceptable. As we research that article quickly, we find in 590.4(C) that this open wiring cable is an acceptable method, provided that it is supported on insulators at intervals of not more than 10 feet (3 m), so the answer would be yes. References would be *NEC*® 590.1 and 590.4(C). As you see in this case when selecting the key words or phrases, using the Table of Contents was sufficient and it was not necessary to go to the Index.

Example: Let us look at another example. The question is:

"Is liquidtight flexible nonmetallic conduit permitted to be used in circuits in excess of 600 volts?"

As we look at that question, we see that liquidtight flexible nonmetallic conduit is the key phrase. Then we look at permitted use as a key phrase and 600 volts as the final key phrase.

Again, we can go to the Table of Contents and find that liquidtight flexible nonmetallic conduit is found in Article 356. As we turn to Article 356, we see uses permitted in 356.10, uses not permitted in 356.12. Immediately as we scan down the uses permitted and the uses not permitted; we find in Section 356.12(4) that liquidtight flexible nonmetallic conduit is not permitted for circuits in

excess of 600 volts. However, you will note that there is an exception for electrical signs, so the answer is: "Yes, for electrical signs by exception. No, generally."

If you will take each question and break it down so that you are looking at the key phrase or key word, you will find it much easier to reference the correct *Code* article and section, and you will not waste the valuable time needed to complete your examination. All questions will not be as easy as these two examples to identify the key phrases, but in all questions there are key phrases or key words, and with the practice you will receive during the following exercises in Chapter 14, you should greatly improve your competency and speed when preparing for an examination, which will allow you ample time to get through the examination and still have some time to do additional research on the tough questions. What you will find by following these examples is that there will be fewer tough questions in the exam. Good luck!

2005 Highlighted Code Changes

There were several thousand proposals submitted to change the 2002 *NEC*® and over 2,000 public comments. The *NEC*® Technical Correlating Committee continued with appointed task groups to make the *Code* more friendly and easier to interpret. Several new articles were added.

Articles New for the 2005 *NEC*®

Article 80 has been relocated as a new addendum G. There is a new **Article 353** for high density polyethylene (HDPE) conduit. **Article 366**—Auxiliary Gutters and **Article 368**—Busways have been rewritten and renumbered to conform to the numbering style in Chapter 3. A new **Article 409**—Industrial Control Panels has been added. A new **Article 506** is added to cover the Zone alternative system for hazardous dusts. Temporary Installations have been relocated in the 2005 *NEC*® and is renumbered as **Article 590**. A new **Article 682** has been added to cover bodies of water not covered by Article 680. This will include fish farms and similar bodies of water.

The following tests include questions that highlight a few of the most important *NEC*® changes. See a good edition analysis of the 2005 changes.

PRACTICE EXAMINATION

The following test was developed and administered by a nationally known seminar presenter and author at a meeting of *Code* authorities. Although this test is not representative of national electricians' tests, it does provide a good exercise in researching the *NEC*®. You will find the questions to be somewhat tricky in some cases. However, if you can pass this examination within one hour, you should be able to pass most national examinations. It will exercise your knowledge of the *NEC*®, your ability to use the document as a reference, and your ability to research variable difficult questions.

1. Equipment enclosed in a case or cabinet that is provided with a means of sealing or locking so that live parts cannot be made accessible without opening the enclosure is said to be
   A. guarded.
   B. protected.
   C. sealable equipment.
   D. lockable equipment.

   Answer: _____   Reference: _____

2. The letter(s) _____ indicate(s) two insulated conductors laid parallel within an outer nonmetallic covering.
   A. D
   B. M
   C. T
   D. II

   Answer: _____   Reference: _____

3. Class 1 circuit conductors shall be protected against overcurrent
   A. in accordance with the values specified in Table 310.16 through Table 310.31 for 14 AWG and larger.
   B. shall not exceed 7 amperes for 18 AWG.
   C. shall not exceed 10 amperes for 16 AWG.
   D. and derating factors do not apply.
   E. all of the above

   Answer: _____   Reference: _____

4. Enclosures for switches or circuit breakers shall not be used as
   A. junction boxes.
   B. auxiliary gutters.
   C. raceways.
   D. all of the above

   Answer: _____   Reference: _____

5. A plug fuse of the Edison base has a maximum rating of _____ amperes.
   A. 20
   B. 30
   C. 40
   D. 50
   E. 60

   Answer: _____   Reference: _____

6. Non–power-limited fire-protective signaling circuit conductors shall be
   A. solid copper.
   B. bunch-tinned stranded copper.
   C. bonded stranded copper.
   D. any of the above

   Answer: _____   Reference: _____

7. When conduit or tubing nipples having a maximum length not to exceed _____ inches (_____ mm) are installed between boxes and similar enclosures, the fill shall be permitted to 60 percent.
   A. 6 (150)
   B. 12 (300)
   C. 24 (600)
   D. 30 (750)

   Answer: _____   Reference: _____

8. Circuit breakers shall be
   A. capable of being opened by manual operation.
   B. capable of being closed by manual operation.
   C. trip free.
   D. all of the above
   E. A and B only

   Answer: _____   Reference: _____

9. For a circuit operating at less than 50 volts, standard lampholders having a rating not less than _____ watts shall be used.
   A. 300
   B. 550
   C. 660
   D. 770

   Answer: _____   Reference: _____

10. 18 AWG TFF wire is rated _____ amperes.
    A. 14
    B. 10
    C. 8
    D. 6

    Answer: _____   Reference: _____

11. A 10 AWG solid copper wire has a cross-section area of _____ square inches (_____ mm²).
    A. .008 (5.26)
    B. 1.21 (3.984)
    C. .006 (4.25)
    D. .109 (70.41)

    Answer: _____   Reference: _____

12. Conductors of light or power can occupy the same enclosure or raceway with conductors of power-limited fire-protective signaling circuits.
    A. True
    B. False

    Answer: _____ Reference: _____

13. Explanatory material in the *National Electrical Code®* is in the form of
    A. footnotes.
    B. fine print notes.
    C. obelisks and asterisk.
    D. red print.

    Answer: _____ Reference: _____

14. Straight runs of trade size 1¼ (35) rigid metal conduit using threaded couplings can be secured at not more than _____ foot (_____ m) intervals.
    A. 5 (1.5)
    B. 10 (3)
    C. 12 (3.7)
    D. 14 (4.3)

    Answer: _____ Reference: _____

15. An isolating switch is one that
    A. is not readily accessible to persons unless special means for access are used.
    B. is capable for interrupting the maximum operating overload current of a motor.
    C. is intended for use in general distribution and branch-circuits.
    D. is intended for isolating an electrical circuit from the source of power.

    Answer: _____ Reference: _____

16. Where damage to remote-control circuits of safety-control equipment would produce a hazard, all conductors of this Class 1 circuit shall be installed in _____ or otherwise suitably protected from physical damage.
    I.   mineral-insulated or metal clad cable
    II. rigid metallic conduit
    III. rigid nonmetallic conduit
    A. I only
    B. II only
    C. III only
    D. I, II, or III

    Answer: _____ Reference: _____

17. _____ on equipment to be grounded shall be removed from contact surfaces to assure good electrical continuity.
    A. Conductive coatings
    B. Nonconductive coatings
    C. Manufacturers instructions
    D. all of the above

    Answer: _____ Reference: _____

18. A _____ conductor is one having one or more layers of nonconducting materials that are not recognized by this *Code*.
    A. noninsulating
    B. bare
    C. covered
    D. none of these

    Answer: _____ Reference: _____

19. To guard live parts, _____ accessible to qualified persons only.
    A. isolate in a room
    B. locate on a balcony
    C. enclose in a cabinet
    D. any of these

    Answer: _____ Reference: _____

20. Conductors larger than 4/0 AWG are measured in
    A. inches.
    B. circular mils.
    C. square inches.
    D. none of the above

    Answer: _____ Reference: _____

21. Measurement of preference: The metric system used in the 2005 *NEC*® is known as the
    A. English metric system.
    B. International system of units.
    C. SOPT metric conversion.
    D. hard metric conversion.

    Answer: _____ Reference: _____

22. The voltage of a circuit is defined by the *Code* as the _____ root-mean-square (effective) difference of potential between any two conductors in the circuit.
    A. lowest
    B. greatest
    C. average
    D. nominal

    Answer: _____ Reference: _____

23. Doorbell wiring, rated as Class 2, in a residence _____ run in the same raceway with light and power conductors.
    A. is permitted with 600-volt insulation
    B. shall not be
    C. shall be
    D. is permitted if insulation is equal to highest installed

    Answer: _____ Reference: _____

24. Equipment or materials to which has been attached a symbol or other identifying mark acceptable to the authority having jurisdiction is known as
    A. listed.
    B. labeled.
    C. approved.
    C. rated.

    Answer: _____ Reference: _____

25. The grounded conductor of a branch-circuit shall be identified by a _____ color.
    I.   gray
    II. continuous green
    III. continuous white
    IV. green with yellow stripe
    A. I and III
    B. II and IV
    C. III only
    D. II only

    Answer: _____ Reference: _____

26. Circuits for lighting and power shall not be connected to any system containing
    A. hazardous material.
    B. trolley wires with ground returns.
    C. poor wiring methods.
    D. dangerous chemicals or gases.

    Answer: _____ Reference: _____

27. The minimum clearance between an electric space heating cable and an outlet box shall not be less than _____ inches (_____ mm).
    A. 8 (200)
    B. 12 (300)
    C. 18 (450)
    D. 6 (150)

    Answer: _____ Reference: _____

28. The ampacity requirements for a disconnecting means of X-ray equipment shall be based on _____ percent of the input required for the momentary rating of the equipment if greater than the long-time rating.
    A. 125
    B. 100
    C. 50
    D. none of the above

    Answer: _____ Reference: _____

29. Open conductors installed outside shall be separated from open conductors of other circuits by not less than _____ inches (_____ mm).
    A. 4 (100)
    B. 6 (150)
    C. 8 (200)
    D. 10 (250)
    E. 12 (300)

    Answer: _____   Reference: _____

30. The minimum headroom of working spaces about motor control centers shall be
    A. 3½ feet (1 m).
    B. 5 feet (1.5 m).
    C. 6 feet, 6 inches (2 m).
    D. 6 feet, 3 inches (1.8 m).

    Answer: _____   Reference: _____

31. Where fixed multioutlet assemblies are employed in locations where a number of appliances are likely to be used simultaneously, each foot or fraction thereof shall be considered as an outlet of not less than _____ VA.
    A. 180
    B. 200
    C. 60
    D. 35

    Answer: _____   Reference: _____

32. Soft-drawn or medium-drawn copper lead in conductors for television equipment antenna systems shall be permitted where the maximum span between points of support is less than _____ feet (_____ m).
    A. 35 (11)
    B. 30 (9.1)
    C. 20 (6)
    D. 10 (3)

    Answer: _____   Reference: _____

33. For a one-family dwelling, the service disconnect shall not be less than _____.
    A. 60
    B. 100
    C. 150
    D. 200

    Answer: _____   Reference: _____

34. Which of the following machines shall be provided with speed-limiting devices or other speed-limiting means?
    I.   series motors
    II. induction
    III. self-excited DC motors
    A. I only
    B. II only
    C. III only
    D. I and III

    Answer: _____   Reference: _____

35. Where within _____ feet (_____ m) of any building or other structure, open wiring on insulators shall be insulated or covered.
    A. 4 (1.2)
    B. 3 (900 mm)
    C. 6 (1.8)
    D. 10 (3)

    Answer: _____   Reference: _____

36. High-voltage conductors in tunnels shall be installed in
    A. rigid metal conduit.
    B. MC cable.
    C. other metal raceways.
    D. any of the above

    Answer: _____   Reference: _____

37. For garages and outbuildings on residential property, a _____ suitable for use on branch-circuits shall be permitted as the disconnecting means.
    A. snap switch
    B. set of three-way or four-way snap switches
    C. set of two-way or four-way snap switches
    D. A or B
    E. none of the above

    Answer: _____   Reference: _____

38. Utilization equipment fastened in place connected to a branch-circuit with other loads shall not exceed _____ percent of the branch-circuit rating.
    A. 50
    B. 60
    C. 80
    D. 100

    Answer: _____   Reference: _____

39. The maximum amperage rating of a 4 inch × ½ inch busbar is _____ amperes.
    A. 500
    B. 1000
    C. 700
    D. 2000

    Answer: _____   Reference: _____

40. For temporary wiring over 600 volts, nominal, _____ shall be provided to prevent access of other than authorized and qualified personnel.
    I. fencing
    II. barriers
    III. signs
    A. I only
    B. II only
    C. III only
    D. I or II

    Answer: _____   Reference: _____

41. Each continuous-duty motor _____ horsepower or less, not permanently installed, is nonautomatically started, and is within sight of the controller shall be permitted to be protected against overload by the branch-circuit protective device.
    A. ⅛
    B. ½
    C. ¾
    D. 1

    Answer: _____   Reference: _____

42. The neutral (grounded) conductor in a mobile home shall be insulated from the equipment grounding system to which of the following location(s)?
    A. range and oven
    B. distribution panel
    C. clothes dryer
    D. all of the above

    Answer: _____   Reference: _____

43. The bottom of sign and outline lighting enclosures shall not be less than _____ feet (_____ m) above areas accessible to vehicles.
    A. 12 (3.7)
    B. 14 (4.25)
    C. 16 (4.9)
    D. 18 (5.5)

    Answer: _____   Reference: _____

44. The conductor used to ground the outer cover of a coaxial cable shall be
    I.   insulated.
    II. 14 AWG minimum.
    III. guarded from physical damage when necessary.
    A. I only
    B. II only
    C. III only
    D. I, II, III

    Answer: _____   Reference: _____

45. Where energized live parts are exposed, the minimum clear workspace shall not be less than _____ feet (_____ m) high for over 600 volts.
    A. 3 (900 mm)
    B. 5 (1.8)
    C. 3½ (1)
    D. 6¼ (1.9)
    E. 6½ (2)

    Answer: _____   Reference: _____

46. Conductors _____ AWG or larger supported on solid knobs shall be securely tied thereto by tie wires having an insulation equivalent to that of the open wire.
    A. 14
    B. 12
    C. 10
    D. 8
    E. 6

    Answer: _____   Reference: _____

47. A branch-circuit that supplies a number of outlets for lighting and appliances is known as a
    A. general-purpose branch-circuit.
    B. a multipurpose branch-circuit.
    C. a utility branch-circuit.
    D. none of the above

    Answer: _____   Reference: _____

48. The minimum size conductor that can be used for an overhead feeder from a residence to a remote garage is
    A. 10 AWG cu.
    B. 12 AWG cu.
    C. 6 AWG al.
    D. 10 AWG al.

    Answer: _____   Reference: _____

49. Two or more _____, 120-volt small-appliance branch-circuits 15- or 20-ampere receptacle outlets in dwelling unit, kitchen, dining room, breakfast room, pantry, or similar dining areas are required in all dwelling units.
    A. 15-ampere
    B. 20-ampere
    C. 30-ampere
    D. 15- or 20-ampere

    Answer: _____   Reference: _____

50. Each kitchen countertop surface in a dwelling unit shall be supplied by not fewer than _____ small-appliance branch-circuit(s).
    A. one
    B. two
    C. three
    D. no minimum

    Answer: _____   Reference: _____

2005 *CODE* CHANGES QUIZ 1

> **Each lesson is designed purposely to require the student to apply the entire *NEC®* text, and not specific chapters or articles. It has been found that when studying for a timed, open-book examination, the student must gain proficiency in the Table of Contents, the Index, and the ability to move quickly from cover to cover to find the correct answer to each question in a timely fashion.**

1. Which article in the *NEC®* is applicable when correctly installing the electrical systems for commercial fish ponds and sewer ponds?

   Answer: _____   Reference: _____

2. High-density polyethylene conduit as an assembly with conductors enclosed. Where is high-density polyethylene conduit covered in the *Code*?

   Answer: _____   Reference: _____

3. Where are the requirements for installing temporary wiring on a construction job for the workers only during construction?

   Answer: _____   Reference: _____

4. An area including a basin with one or more of the following—a toilet, a tub, or a shower—is defined in the *NEC®* as a _____.

   Answer: _____   Reference: _____

5. Hazardous locations determined to be neither Class I, Division 1; Class I, Division 2; Class I, Zone 0; Class I, Zone 1; Class I, Zone 2; Class II, Division 1; Class II, Division 2; Class III, Division 1; Class III, Division 2; nor any combination thereof are _____.

   Answer: _____   Reference: _____

6. The 2002 *NEC®* Article 80 contained administration and enforcement rules. Was this material deleted from the 2005 *NEC®*?

   Answer: _____   Reference: _____

7. The main bonding jumper is used to bond the grounded conductor to the grounding electrode and equipment grounding. What is the system bonding jumper used for?

   Answer: _____   Reference: _____

8. The conductor used to connect the grounding electrode(s) to the equipment grounding conductor, to the grounded conductor, or to both, at the service, at each building or structure where supplied by a feeder(s) or branch-circuit(s) from a common service, or at the source of a separately derived system is called the _____.

   Answer: _____   Reference: _____

9. An enclosure identified for use in underground systems, provided with an open or closed bottom, and sized to allow personnel to reach into, but not enter, for the purpose of installing, operating, or maintaining equipment or wiring or both is defined as a _____.

   Answer: _____ Reference: _____

10. A kitchen not located within a dwelling is defined as an area with a sink and permanent facilities for food preparation, and cooking. What section of the *Code* places special requirements for these areas?

    Answer: _____ Reference: _____

11. All 125-volt, 15- or 20-ampere receptacles located in other than dwelling units outdoors in public places must have _____ for personnel.

    Answer: _____ Reference: _____

12. Where is the allowable minimum bending radius for all conduit and tubing found?

    Answer: _____ Reference: _____

13. Article 830 covers network-powered broadband communications systems that are covered by the *NEC*®. Those not covered are found in Section _____ .

    Answer: _____ Reference: _____

14. Article 820 covers CATV systems and radio distribution systems. How is the maximum number of coaxial cables permitted in conduit calculated?

    Answer: _____ Reference: _____

15. GFCI protection is required for 15A and 20A, 125V, receptacles in aircraft hangers. True or False?

    Answer: _____ Reference: _____

2005 *CODE* CHANGES QUIZ 2

1. Abandoned power cables, communication cables, and connecting and interconnection cables are required to be removed under raised floors in information technology (IT) rooms unless contained in _____.

   Answer: _____ Reference: _____

2. Contact conductors and exposed short lengths of conductors at resistors, collectors, and other equipment for cranes or hoists are not required to be _____.

   Answer: _____ Reference: _____

3. Outdoor or under-chassis line-voltage (120 volts, nominal, or higher) wiring exposed to moisture or physical damage shall be protected by _____ or _____ unless they are routed closely against the frame and equipment enclosure of mobile or manufactured homes.

   Answer: _____ Reference: _____

4. A service for a carnival would have to comply with all the requirements of Article _____ and Sections _____.

   Answer: _____   Reference: _____

5. The wiring methods for Zone 20 are found in Section _____.

   Answer: _____   Reference: _____

6. Resistors commonly found in electrical control cabinets and other electrical apparatus are covered in _____.

   Answer: _____   Reference: _____

7. The disconnecting means for motor circuits rated 600 volts, nominal, or less shall have an ampere rating not less than _____ percent of the full-load current rating of the motor, except for specific listed nonfused motor-circuit switches.

   Answer: _____   Reference: _____

8. Receptacles shall not be installed within or directly over _____ or _____.

   Answer: _____   Reference: _____

9. Conductors in vertical poles used as raceways shall be supported per Section _____.

   Answer: _____   Reference: _____

10. Flexible cords and cables must conform to the description provided in _____.

    Answer: _____   Reference: _____

11. Open wiring on insulators shall not be permitted in other than industrial or agricultural installations and where concealed by the building structure or where the voltage exceeds _____.

    Answer: _____   Reference: _____

12. Do the ampacity adjustment factors apply to conductors in cellular metal raceways?

    Answer: _____   Reference: _____

13. Nonmetallic cable tray shall be permitted only in corrosive areas and in areas requiring voltage isolation. True or False?

    Answer: _____   Reference: _____

14. Busways may not be installed where subject to _____ or severe _____.

    Answer: _____   Reference: _____

15. If I ask for 10 lengths of 27 IMC, what should I receive from the supplier?

Answer: _____   Reference: _____

2005 *CODE* CHANGES QUIZ 3

1. Integrated gas spacer cable is not permitted to be used as interior wiring, exposed or in contact with buildings or _____.

    Answer: _____   Reference: _____

2. _____ receptacles or covers are required in psychiatric and pediatric general-care areas where accessible to patients.

    Answer: _____   Reference: _____

3. The _____ cannot be used as the sole equipment grounding conductor or as an effective ground-fault current path.

    Answer: _____   Reference: _____

4. The processes of purging or pressurizing supplying enclosures in Class 1 hazardous locations to an acceptable level are acceptable protection techniques. What NFPA standard covers these two protection techniques?

    Answer: _____   Reference: _____

5. Single conductors specified in Table 310.13 must only be installed in a recognized wiring method of Chapter 3 or be part of a Chapter 3 wiring method except in accordance with _____.

    Answer: _____   Reference: _____

6. Where grounded conductors of different systems are installed in the same raceway, cable, box, auxiliary gutter, or other type of enclosure, each grounded conductor shall be identified by system identification that distinguishes each system. This means of identification shall be permanently posted at each _____.

    Answer: _____   Reference: _____

7. An attachment plug and multioutlet receptacle assembly where used as permitted assembly shall consist of an attachment plug, a flexible cord, and a multioutlet receptacle strip. Each assembly must be energized from a receptacle outlet. The assembly must be listed, and the total length of the assembly shall not exceed _____ feet?

    Answer: _____   Reference: _____

8. In mobile or manufactured homes, receptacle outlets shall not be permitted to be installed within a _____ or _____ or _____ position in any countertop or _____ electric baseboard heaters, unless provided for in the listing or manufacturer's instructions.

    Answer: _____   Reference: _____

9. The conduit or tubing not smaller than _____ trade size for neon secondary circuit conductors, over 1000 volts, nominal, shall contain no more than _____ conductor.

   Answer: _____   Reference: _____

10. Which article covers the requirements for electric cutting process equipment?

    Answer: _____   Reference: _____

11. Is a disconnect switch required to be inside of the motor? Or can the disconnect be located elsewhere if it can be locked in the off position?

    Answer: _____   Reference: _____

12. How is the size of an equipment bonding jumper determined?

    Answer: _____   Reference: _____

13. Do raceways to be installed by directional boring equipment have to be listed for the purpose?

    Answer: _____   Reference: _____

14. What cables do not have to be marked as "sunlight resistant" when they are used in installations exposed to direct sunlight?

    Answer: _____   Reference: _____

15. Is it permissible to have 220-volt lighting circuits in a dwelling?

    Answer: _____   Reference: _____

16. Would a 120-volt receptacle installed outdoors at a dwelling for a sewer lift pump require GFCI protection?

    Answer: _____   Reference: _____

17. What is the trade size of a 103 intermediate conduit?

    Answer: _____   Reference: _____

18. Is a green or bare equipment grounding conductor required to be installed with the circuit conductors installed in electrical metallic tubing?

    Answer: _____   Reference: _____

19. Where are the installation requirements for industrial control cabinets found?

    Answer: _____   Reference: _____

20. The TVSS shall be connected to each _____ conductor.

    Answer: _____   Reference: _____

2005 *CODE* CHANGES QUIZ 4

1. Is it permissible for a switch in a patient care area, located outside the patient vicinity, to be supplied by a switch loop consisting of Type MC cable?

   Answer: _____

   _____

   _____

   _____

   Reference: _____

2. What is the difference between a lighting fixture and a luminaire?

   Answer: _____

   _____

   _____

   _____

   Reference: _____

3. Does the 2005 *NEC*® contain adoptive ordinance language similar to other model codes?

   Answer: _____

   _____

   _____

   _____

   Reference: _____

4. Has the 2005 *Code* been changed or modified in any way to make it a more suitable document for international use?

   Answer: _____

   _____

   _____

   _____

   Reference: _____

5. Does 90.4 address signaling and communications systems as part of the term "electrical installations"?

Answer: _____

_____

_____

_____

Reference: _____

6. Are informative product standards referenced in the *NEC®*? If so, where?

Answer: _____

_____

_____

_____

Reference: _____

7. Are metric measurements the measurement system of preference in the 2005 *Code*?

Answer: _____

_____

_____

_____

Reference: _____

8. Where are the requirements for receptacles, cord connectors, and attachment plugs found in the 2005 *NEC®*?

Answer: _____

_____

_____

_____

Reference: _____

9. Would the definition of a garage in Article 100 include a building where electric vehicles are stored?

Answer: _____

_____

_____

_____

Reference: _____

10. Would 110.22 literally not require identification of a disconnecting means that was not required by the *Code*?

    Answer: _____

    _____

    _____

    _____

    Reference: _____

11. Is panic hardware on electrical room doors required?

    Answer: _____

    _____

    _____

    _____

    Reference: _____

12. Where can the requirements for the specific wiring methods for electric discharge luminaires (lighting fixtures) supported independent of the outlet box be found?

    Answer: _____

    _____

    _____

    _____

    Reference: _____

13. What does the new term "luminaire" used throughout the *Code* refer to?

    Answer: _____

    _____

    _____

    _____

    Reference: _____

14. Does the *Code* prohibit receptacles from being installed in a face-up position in a dwelling-unit laundry room countertop?

    Answer: _____

    _____

    _____

    _____

    Reference: _____

15. Are dwelling unit bedroom lighting circuits required to be AFCI protected?

    Answer: _____

    _____

    _____

    _____

    Reference: _____

16. Receptacles are usually installed in appliance garages on dwelling-unit kitchen countertops. Can this receptacle be considered as one of the required receptacles for the countertop?

    Answer: _____

    _____

    _____

    _____

    Reference: _____

17. Is GFCI protection required for electrically heated floors in bathrooms, hydromassage bathtub, spa, and hot tub locations?

    Answer: _____

    _____

    _____

    _____

    Reference: _____

18. Is it permissible to use a gray-colored conductor as a grounded (neutral) conductor?

    Answer: _____

    _____

    _____

    _____

    Reference: _____

19. Are branch-circuits over 600 volts permitted in residential and commercial occupancies where there are no maintenance or supervision provisions?

    Answer: _____

    _____

    _____

    _____

    Reference: _____

20. Is a service conductor for an overhead service drop permitted to be bare?

Answer: _____

_____

_____

_____

Reference: _____

2005 CODE CHANGES QUIZ 5

1. *NEC*® 230.46 allows service-entrance conductors, including directly buried conductors, to be spliced. Are service-lateral conductors permitted to be to be spliced?

Answer: _____

_____

_____

_____

Reference: _____

2. Is cable tray a wiring method for all service conductors?

Answer: _____

_____

_____

_____

Reference: _____

3. *NEC*® 230.46 permits service-entrance conductors to be spliced using clamped or bolted connections. Could a crimp connection or exothermic welding be used for this purpose?

Answer: _____

_____

_____

_____

Reference: _____

4. Would 230.70 allow the use of a shunt trip switch at the location specified in 230.70(A)(1) as the service disconnecting means if the circuit breaker were located inside the building?

Answer: _____

_____

_____

_____

Reference: _____

5. A mobile home is equipped with a chassis bonding connection as required in 550.16(C)(1). The metal gas piping is bonded with a factory-installed bonding jumper connected to the mobile home frame at a point remote from the chassis bonding connection. Does this installation meet all the requirements?

   Answer: _____

   _____

   _____

   _____

   Reference: _____

6. Does the *Code* specifically allow a molded case switch to be used as a motor controller?

   Answer: _____

   _____

   _____

   _____

   Reference: _____

7. The service disconnecting means is located outside a building. Can it be located 50 feet (15 m) away or within sight of the building?

   Answer: _____

   _____

   _____

   _____

   Reference: _____

8. Where can requirements for separately derived systems that operate at 120 volts line-to-line and 60 volts line-to-ground be found?

   Answer: _____

   _____

   _____

   _____

   Reference: _____

9. What section of the *Code* permits solar photovoltaic systems, fuel cell systems, or interconnected electric power production sources to be connected to the supply-side of the service disconnect?

   Answer: _____

   _____

   _____

   _____

   Reference: _____

10. There is presently no maximum rating or setting for overcurrent devices protecting fire pumps and related equipment. How would overcurrent devices for fire pumps be sized?

Answer: _____

_____

_____

_____

Reference: _____

11. Is ground-fault circuit-interrupter protection required for all single-phase, 15- and 20-ampere receptacles in commercial kitchens?

Answer: _____

_____

_____

_____

Reference: _____

12. Where in the *Code* are the branch-circuit overcurrent devices supplying a motel room required to be readily accessible to the motel room occupant?

Answer: _____

_____

_____

_____

Reference: _____

13. Are circuit breakers permitted to be used as a switch? Are they permitted for switching HID luminaires (lighting fixtures) if they are only marked SWD?

Answer: _____

_____

_____

_____

Reference: _____

14. Are there presently any *Code* requirements for the removal or termination of abandoned low-voltage cables in buildings or structures?

Answer: _____

_____

_____

_____

Reference: _____

15. Where in the *Code* does it require that temporary electrical power and lighting installations for holiday decorative lighting be limited to 90 days?

    Answer: _____

    _____

    _____

    _____

    Reference: _____

16. Are underground water pipe supplementary grounding electrodes required to meet the requirements of 250.56 if they are rod pipe or plate types of electrodes?

    Answer: _____

    _____

    _____

    _____

    Reference: _____

17. Is it permissible to connect a grounding electrode to equipment such as parking light poles?

    Answer: _____

    _____

    _____

    _____

    Reference: _____

18. What is the difference in performance between a surge arrester and a surge suppressor?

    Answer: _____

    _____

    _____

    _____

    Reference: _____

19. In walls and ceilings constructed of wood or other combustible surface material, boxes shall be flush with the finished surface or project therefrom. Is gypsum board (sheetrock) considered a combustible surface material?

    Answer: _____

    _____

    _____

    Reference: _____

20. A luminaire (lighting fixture) that weighs more than _____ shall be supported independently of the outlet box unless the outlet box is listed for the weight to be supported.

Answer: _____

_____

_____

_____

Reference: _____

2005 *CODE* CHANGES QUIZ 6

1. Type AC cable shall be permitted to be unsupported where the cable is not more than _____ from the last point of support for connections within an accessible ceiling to luminaire(s) (lighting fixture[s]) or equipment.

Answer: _____

_____

_____

_____

Reference: _____

2. Does the *Code* require a raceway to be sealed where it passes from the interior to the exterior of a building where condensation is known to be a problem?

Answer: _____

_____

_____

_____

Reference: _____

3. Does the *Code* allow cables to be used as a means of support for other cables?

Answer: _____

_____

_____

_____

Reference: _____

4. Is liquidtight flexible metal conduit permitted to be used in ducts, plenums, and other air-handling spaces as long as no one run exceeds 6 feet (1.8 m) in length?

Answer: _____

_____

_____

_____

Reference: _____

5. Which article of the *Code* covers temporary installations?

    Answer: _____

    _____

    _____

    _____

    Reference: _____

6. What is an arc-fault circuit interrupter as per *Code*?

    Answer: _____

    _____

    _____

    _____

    Reference: _____

7. Do the ampacity adjustment factors for cables apply to all Type MC or AC cable?

    Answer: _____

    _____

    _____

    _____

    Reference: _____

8. Is the messenger-supported wiring method permitted where subject to physical damage?

    Answer: _____

    _____

    _____

    _____

    Reference: _____

9. Does the *Code* allow Type AC cable to be run unsupported in lengths not exceeding 6 feet (1.8 m) between luminaires (light fixtures) in accessible ceilings?

    Answer: _____

    _____

    _____

    _____

    Reference: _____

10. New hybrid-type vehicles incorporate gasoline engines and electric motors. Would Article 625 or Article 511 govern repair facilities for these vehicles?

Answer: _____

_____

_____

_____

Reference: _____

11. Could intermediate metal conduit (IMC) be used as an exposed vertical riser to support a lighted display case in a department store?

Answer: _____

_____

_____

_____

Reference: _____

12. Does the *Code* permit threadless connectors to be used on the threaded end of rigid metal conduit (RMC)?

Answer: _____

_____

_____

_____

Reference: _____

13. Are flat cable assemblies Type FC cable permitted to be installed where subject to physical damage?

Answer: _____

_____

_____

_____

Reference: _____

14. Does the minimum wire bending space at terminals in Table 312.6(B) apply to aluminum conductors?

Answer: _____

_____

_____

_____

Reference: _____

15. Would the metal faceplate (cover) grounding requirements apply to fan speed controls or spring-wound timers?

    Answer: _____

    _____

    _____

    _____

    Reference: _____

16. Are flash protection warning markings required for switchboards, panelboards, and motor control centers that are likely to require servicing while energized?

    Answer: _____

    _____

    _____

    _____

    Reference: _____

17. Is nonmetallic-sheathed cable Type NM permitted as open runs in dropped or suspended ceilings in other than one- and two-family and multifamily dwellings?

    Answer: _____

    _____

    _____

    _____

    Reference: _____

18. A service drop point of attachment is anchored to a trade size 2 (53 designator) rigid metal conduit riser 3 feet (900 m) above the roofline. There is a threaded coupling on the riser between the highest strap and the weather-head. Is this a *Code* violation?

    Answer: _____

    _____

    _____

    _____

    Reference: _____

19. Could the required emergency disconnect for gasoline dispensers serve as the maintenance disconnect required?

    Answer: _____

    _____

    _____

    _____

    Reference: _____

20. Do the dedicated space requirements for switchboards and panelboards now apply to a fusible switch as well?

Answer: _____

_____

_____

_____

Reference: _____

2005 *CODE* CHANGE QUIZ 7

1. Does the AFCI branch-circuit protection for bedroom circuits require the use of AFCI circuit breakers?

Answer: _____

_____

_____

_____

Reference: _____

2. Could an individual refrigerator branch-circuit be considered in the small appliance branch-circuit calculation? Or would this individual circuit be included as part of the general lighting load?

Answer: _____

_____

_____

_____

Reference: _____

3. Why is Type AC cable not permitted to be used as a service-entrance wiring method?

Answer: _____

_____

_____

_____

Reference: _____

4. Would the bonding jumper connection to metal gas piping be required to be accessible after installation?

Answer: _____

_____

_____

_____

Reference: _____

5. Could Type NM cable be run unsupported in lengths not exceeding 4½ feet (1.4 m) between luminaires (lighting fixtures) in an accessible ceiling?

   Answer: _____

   _____

   _____

   _____

   Reference: _____

6. Is nonferrous intermediate metal conduit (IMC) manufactured?

   Answer: _____

   _____

   _____

   _____

   Reference: _____

7. Is electrical nonmetallic tubing (ENT) permitted outside exposed to the direct rays of the sun?

   Answer: _____

   _____

   _____

   _____

   Reference: _____

8. Does the *Code* allow nonmetallic-sheathed cable (NM) to be installed in a complete raceway system?

   Answer: _____

   _____

   _____

   _____

   Reference: _____

9. Is there a maximum number of bends allowed between pull points—for example, conduit bodies and boxes for liquidtight flexible metal conduit?

   Answer: _____

   _____

   _____

   _____

   Reference: _____

10. Would a raceway, used as physical protection for direct burial conductors of over 600 volts emerging from the ground, be required to be listed for the purpose?

Answer: _____

_____

_____

_____

Reference: _____

11. Luminaires (lighting fixtures) shall not be used as a raceway for circuit conductors unless _____ for use as a raceway.

Answer: _____

_____

_____

_____

Reference: _____

12. What is Type NUCC? Is it required to be listed? Are the conductors inside NUCC required to be listed? Is the raceway required to be listed?

Answer: _____

_____

_____

_____

Reference: _____

13. Is Type ENT permitted where subject to physical damage?

Answer: _____

_____

_____

_____

Reference: _____

14. Cable trays shall be supported at intervals _____.

Answer: _____

_____

_____

_____

Reference: _____

15. What is the maximum water level for a swimming pool?

   Answer: _____

   _____

   _____

   _____

   Reference: _____

16. Is electrical metallic tubing (EMT) an equipment grounding conductor in accordance with the *Code* or is a separate equipment grounding conductor required?

   Answer: _____

   _____

   _____

   _____

   Reference: _____

17. Do Tables 352.44(A) and (B) take into account heating of PVC conduit by direct sunlight, or are the tables only based on ambient temperature?

   Answer: _____

   _____

   _____

   _____

   Reference: _____

18. Are nonmetallic wireways required to be listed?

   Answer: _____

   _____

   _____

   _____

   Reference: _____

19. A service riser contains a 6 AWG bare neutral conductor. Is this permitted by the *Code*?

   Answer: _____

   _____

   _____

   _____

   Reference: _____

20. Where in the *Code* does it require that voltage drop for branch-circuits not exceed 1.5 percent?

Answer: _____

_____

_____

_____

Reference: _____

2005 *CODE* CHANGE QUIZ 8

1. Is liquidtight flexible nonmetallic conduit permitted to enclose conductors above 600 volts?

Answer: _____

_____

_____

_____

Reference: _____

2. The equipment bonding jumper for liquidtight flexible nonmetallic conduit must be installed in accordance with _____.

Answer: _____

_____

_____

_____

Reference: _____

3. Are metal wireways required to be listed?

Answer: _____

_____

_____

_____

Reference: _____

4. The ampacity of the conductors from the generator terminals to the first distribution device(s) containing over-current protection shall not be less than _____ of the nameplate current rating of the generator.

Answer: _____

_____

_____

Reference: _____

5. Single-phase cord-and-plug connected room air conditioners shall be provided with factory-installed _____ protection. The protection shall be an integral part of the attachment plug or be located in the power supply cord within 12 inches (300 mm) of the attachment plug.

Answer: _____

_____

_____

Reference: _____

6. Is each motor circuit required to include coordinated protection to automatically interrupt overload and fault currents in the motor, the motor-circuit conductors, and the motor control apparatus?

Answer: _____

_____

_____

Reference: _____

7. Unless protected by ground-fault circuit-interrupter protection for personnel, the secondary winding of the isolation transformer connected to the impedance heating elements shall not have an output voltage greater than _____ volts ac.

Answer: _____

_____

_____

Reference: _____

8. Are snap switches and control switches required to be grounded?

Answer: _____

_____

_____

Reference: _____

9. If a motor were accessible from a raised platform, would the *Code* allow the disconnect to be mounted higher than 6 feet, 7 inches (2 m) above the platform?

Answer: _____

_____

_____

Reference: _____

10. Is it acceptable to install more than one neutral under a terminal in a panelboard?

    Answer: _____

    _____

    _____

    _____

    Reference: _____

11. Exposed vertical risers from industrial machinery or fixed equipment shall be permitted to be supported at intervals not exceeding 20 feet (6 m). If the conduit is made up with threaded couplings, the conduit is firmly supported at the top and bottom of the riser, and no other means of intermediate support is readily available. What types of conduit are permitted to be used?

    Answer: _____

    _____

    _____

    _____

    Reference: _____

12. May a 25-ampere branch-circuit supply two or more outlets for fixed electric space-heating equipment?

    Answer: _____

    _____

    _____

    _____

    Reference: _____

13. The essential electrical distribution system for other than health care facilities shall be a _____ system.

    Answer: _____

    _____

    _____

    _____

    Reference: _____

14. Where Class II, Group E dusts are present in hazardous quantities, there are only _____ locations.

    Answer: _____

    _____

    _____

    _____

    Reference: _____

15. Overcurrent protection shall not be required for conductors from a battery rated less than _____ if the battery provides power for starting, ignition, or control of prime movers.

    Answer: _____

    _____

    _____

    _____

    Reference: _____

16. All circuits and circuit modifications shall be legibly identified as to _____ or _____ on a circuit directory located on the face or inside of the panel door in the case of a panelboard, and at each switch on a switchboard.

    Answer: _____

    _____

    _____

    _____

    Reference: _____

17. As presently worded, does 430.245(B) require a motor lead to be a solid conductor if larger than a 10 AWG ?

    Answer: _____

    _____

    _____

    _____

    Reference: _____

18. A ballast in a fluorescent luminaire (lighting fixture) that is used for egress lighting and is energized only during an emergency shall not have _____.

    Answer: _____

    _____

    _____

    _____

    Reference: _____

19. Would the *Code* allow the specified disconnecting means to be mounted on a removable cover of air-conditioning or refrigeration equipment if the disconnect were supplied from a flexible wiring method?

    Answer: _____

    _____

    _____

    _____

    Reference: _____

20. A separable connector or a plug-and-receptacle combination in the supply line to an oven or cooking unit shall be approved for the _____ of the space in which it is located.

Answer: _____

_____

_____

_____

Reference: _____

2005 *CODE* CHANGES QUIZ 9

1. Can the electric "service" supplying a building or structure be supplied from a private generator?

Answer: _____

_____

_____

_____

Reference: _____

2. Where are the requirements for tunnel installations over 600 volts covered?

Answer: _____

_____

_____

_____

Reference: _____

3. A dedicated space equal to the depth and width of the equipment from the floor to a height of _____ or to the structural ceiling whichever is _____.

Answer: _____

_____

_____

_____

Reference: _____

4. Grounding and bonding requirements for tunnel installations over 600 volts are covered in *NEC*® _____.

Answer: _____

_____

_____

_____

Reference: _____

5. Generally, grounded conductors 6 AWG or smaller shall be identified by a continuous white or gray finish or by _____ on other than green insulation along its entire length.

Answer: _____

_____

_____

_____

Reference: _____

6. Is a receptacle installed in an accessory building such as a storage building at grade level associated with a dwelling required to have a ground-fault circuit-interrupter protection for personnel?

Answer: _____

_____

_____

_____

Reference: _____

7. What is an arc-fault circuit interrupter?

Answer: _____

_____

_____

_____

Reference: _____

8. May the service conductors originate at a generator where there is no utility serving the premises?

Answer: _____

_____

_____

_____

Reference: _____

9. Other electrical equipment associated with the electrical installation located above or below shall not be allowed to extend beyond the front of electrical equipment. True or False?

Answer: _____

_____

_____

_____

Reference: _____

10. The requirements for dedicated electrical space are located in Article _____.

    Answer: _____

    _____

    _____

    _____

    Reference: _____

11. Equipment requirements for over 600 volts nominal are located in Article _____.

    Answer: _____

    _____

    _____

    _____

    Reference: _____

12. An insulated grounded conductor of 6 AWG or smaller shall be identified by a continuous white or gray or _____.

    Answer: _____

    _____

    _____

    _____

    Reference: _____

13. Receptacles located in garages or accessory buildings having a floor located at or below grade and are limited to storage areas, work areas, or areas of similar use are required to be of the _____ type.

    Answer: _____

    _____

    _____

    _____

    Reference: _____

14. All branch-circuits supplying 125-volt, single-phase, 15- and 20-ampere branch-circuits installed in bedrooms shall be protected by _____.

    Answer: _____

    _____

    _____

    _____

    Reference: _____

15. Are the required two small-appliance circuits permitted to serve more than one kitchen?

    Answer: _____

    _____

    _____

    _____

    Reference: _____

16. In dwelling units, at least one receptacle shall be located not more than _____ feet (_____ mm) of the outside edge of each basin in bathrooms.

    Answer: _____

    _____

    _____

    _____

    Reference: _____

17. In guest rooms of hotels, motels, and similar occupancies, at least _____ of the receptacles required by 210.52(A) shall be _____.

    Answer: _____

    _____

    _____

    _____

    Reference: _____

18. An additional load of _____ shall be included for each 2 feet (600 mm) of track lighting or fraction thereof in all occupancies except dwelling units or guest rooms of hotels and motels.

    Answer: _____

    _____

    _____

    _____

    Reference: _____

19. Outside feeders and branch-circuits not exceeding 300 volts-to-ground over residential property and driveways, and commercial areas not subject to truck traffic, shall have a minimum clearance from ground of _____ feet (_____ m).

    Answer: _____

    _____

    _____

    _____

    Reference: _____

20. For a one-family dwelling, the feeder disconnecting means shall have a rating of not less than _____, 3-wire.

Answer: _____

_____

_____

_____

Reference: _____

2005 *CODE* CHANGES QUIZ 10

1. Which section of the *Code* permits one additional set of service-entrance conductors to supply a separate meter for public or common area loads not permitted to be supplied from an individual dwelling unit?

Answer: _____

_____

_____

_____

Reference: _____

2. For a one-family dwelling, the service disconnecting means shall have a rating of not less than _____, 3-wire.

Answer: _____

_____

_____

_____

Reference: _____

3. Do the overcurrent protection requirements in 240.4(D) for small conductors apply to motor or motor-operated appliances?

Answer: _____

_____

_____

_____

Reference: _____

4. Tap conductor is defined in *NEC*® _____.

Answer: _____

_____

_____

_____

Reference: _____

5. Where a tap is located outdoors except for the point of termination, the conductors are protected from physical damage and terminate in a single overcurrent device, the disconnecting means is readily accessible either outside or inside the building nearest the point of entrance, and the overcurrent device is integral or immediately adjacent to the disconnect, the length shall not exceed _____ feet (_____ m).

Answer: _____

_____

_____

_____

Reference: _____

6. Where in the *Code* is a "Supervised Industrial Installation" defined?

Answer: _____

_____

_____

_____

Reference: _____

7. Why is GFCI protection for personnel using 120-volt, 15- and 20-ampere rated receptacles, especially outdoors, not required for carnival and exhibition shows?

Answer: _____

_____

_____

_____

Reference: _____

8. I have been told that a tap conductor could not be tapped from a conductor that had been tapped from another conductor. Where is this stated in the *NEC*®?

Answer: _____

_____

_____

_____

Reference: _____

9. An individual 15-ampere branch-circuit is installed for refrigeration equipment in the kitchen of a dwelling unit per 210.52(B)(1), Exception No. 2. Does *NEC*® 220.52(A) require additional 1500 volt-amperes for this 15-ampere circuit?

Answer: _____

_____

_____

_____

Reference: _____

10. Do all the receptacle outlets in a hotel or motel room located conveniently for permanent furniture layout have to be accessible?

Answer: _____

_____

_____

_____

Reference: _____

11. If outlets end up behind nonstationary furniture such as a recliner chair, bed table, or desk such that they would have to be moved in order to plug into a receptacle outlet, does this mean they are not considered accessible?

Answer: _____

_____

_____

_____

Reference: _____

12. When should a transfer-switch switch the grounded conductor as well as the ungrounded conductors between a utility service and a generator service?

Answer: _____

_____

_____

_____

Reference: _____

13. Can a receptacle be located below the rim of a pedestal sink in a bathroom, and if so, how far below before it would not be considered as the required receptacle outlet for the bathroom basin?

Answer: _____

_____

_____

_____

Reference: _____

14. The *NEC*® permits a single 20-ampere circuit to supply all the receptacles in dwelling bathrooms. Does this *NEC*® section permit other loads to be connected to this circuit?

Answer: _____

_____

_____

_____

Reference: _____

15. Many large assembly and exhibition halls utilize cable trays for distribution of temporary wiring for an event. Is it permissible to use SO or SJT flexible cords in these cable trays for such temporary installations?

    Answer: _____

    _____

    _____

    _____

    Reference: _____

16. The *Code* requires that electric motors used on fire pumps be _____ for fire pump service.

    Answer: _____

    _____

    _____

    _____

    Reference: _____

17. Do the equipotential plane requirements of agriculture buildings apply to chicken houses?

    Answer: _____

    _____

    _____

    _____

    Reference: _____

18. If an equipment grounding conductor is not run with a 4-wire, 3-phase feeder to a second building, can the grounded conductor of the feeder circuit be grounded to the same contiguous metal water pipe as the main electrical service in the first building where that water line runs to the second building?

    Answer: _____

    _____

    _____

    _____

    Reference: _____

19. Many new dwellings have a conduit or tubing sleeve installed from the top of a panelboard up into the attic space so that future circuits can be gained from the panelboard without having to fish cables or cut sheetrock. Are such installations proper because future cables cannot be secured to the panelboard?

    Answer: _____

    _____

    _____

    _____

    Reference: _____

20. Are all 15- and 20-ampere receptacles for dwelling-unit unfinished accessory buildings required to be GFCI protected?

Answer: _____

_____

_____

_____

Reference: _____

2005 *CODE* CHANGES QUIZ 11

1. Where multiple-conductor power cables are installed in parallel, is it necessary for the equipment grounding conductor in each cable to be sized per the feeder or branch-circuit overcurrent device protecting the circuit according to Table 250.122?

Answer: _____

_____

_____

_____

Reference: _____

2. What is a power panelboard?

Answer: _____

_____

_____

_____

Reference: _____

3. Does the *Code* allow splicing of service-entrance conductors?

Answer: _____

_____

_____

_____

Reference: _____

4. Is it permissible to use NM cable with temporary wiring on a construction site?

   Answer: _____

   _____

   _____

   _____

   Reference: _____

5. Is it permissible to prewire electrical nonmetallic tubing before installing it in a concrete slab?

   Answer: _____

   _____

   _____

   _____

   Reference: _____

6. Is there a limit on how many individual branch-circuits are permitted to be run from one building to another on the same property under single management?

   Answer: _____

   _____

   _____

   _____

   Reference: _____

7. Why is MC cable allowed in assembly-type occupancies with over 100 occupants whereas Type AC cable is not allowed?

   Answer: _____

   _____

   _____

   _____

   Reference: _____

8. Do the requirements that state *receptacle outlets shall not be installed in a face-up position in the work surfaces or countertops* apply to wet bar sink locations?

   Answer: _____

   _____

   _____

   _____

   Reference: _____

9. What is the definition of a supervised industrial installation as opposed to a commercial establishment?

   Answer: _____

   _____

   _____

   _____

   Reference: _____

10. All 15- and 20-ampere, 125-volt receptacles in pediatric wards, rooms, or areas must be _____ or employ _____ covers.

    Answer: _____

    _____

    _____

    _____

    Reference: _____

11. Is it ever permissible to mount a small and lightweight luminaire (lighting fixture) to a device box, especially if the device box is a flush-mounted-type existing in a wall?

    Answer: _____

    _____

    _____

    _____

    Reference: _____

12. When trimming-in a job, how many inches (mm) of individual conductor must be left from the front of a 2 × 4 inch (50 × 100 mm) outlet box for the connection of or future replacement of toggle switches and receptacles?

    Answer: _____

    _____

    _____

    _____

    Reference: _____

13. Does the *NEC*® permit NM cable to be used for lease space wiring in a four-story building with 6000 square feet per floor that houses novelty shops?

    Answer: _____

    _____

    _____

    _____

    Reference: _____

14. Where in the *Code* is a "curb" around the floor opening required for busways passing vertically through more than one floor of all buildings?

Answer: _____

_____

_____

_____

Reference: _____

15. How many ground rods are required where the rod is to supplement a metallic water line electrode?

Answer: _____

_____

_____

_____

Reference: _____

16. Does Article 830—Network-Powered Broadband Communications Systems cover high-power communication systems?

Answer: _____

_____

_____

_____

Reference: _____

17. Is at least one general-care patient bed location receptacle outlet always required to be served by the normal system power supply, or can all general-care patient bed receptacles be served through emergency system connected panelboards?

Answer: _____

_____

_____

_____

Reference: _____

18. How many "main power feeders" may a single-family dwelling have and do the loads they serve make any difference? For example, a large single-family dwelling has a 600-ampere-rated service with two 100-ampere fused service switches, one 200-ampere fused switch, and a 400-ampere fused switch each feeding interior located panelboards. Are all "main power feeders"?

Answer: _____

_____

_____

_____

Reference: _____

19. Is GFCI protection for 120-volt receptacle outlets required on the third floor outdoor balcony of a dwelling building?

Answer: _____

_____

_____

_____

Reference: _____

20. Where in the *Code* are "handholes" covered?

Answer: _____

_____

_____

_____

Reference: _____

PRACTICE EXAM 1

> **Each lesson is designed purposely to require the student to apply the entire *NEC®* text, and not specific chapters or articles. It has been found that when studying for a timed, open-book examination, the student must gain proficiency in the Table of Contents, the Index, and the ability to move quickly from cover to cover to find the correct answer to each question in a timely fashion.**

1. Are all 125-volt, 15- and 20-ampere receptacles in the service area of a commercial garage required to be protected by a GFCI?

   Answer: _____

   _____

   _____

   _____

   Reference: _____

2. A luminaire (lighting fixture) is installed over a hydromassage bathtub. Is GFCI protection required for the luminaire (fixture)?

   Answer: _____

   _____

   _____

   _____

   Reference: _____

3. A nongrounding-type receptacle is to be replaced with a GFCI receptacle because a grounding means does not exist within the box. Can you supply downstream receptacles from the GFCI? Are these downstream receptacles required to be 2-wire, nongrounding type?

   Answer: _____

   _____

   _____

   _____

   Reference: _____

4. A kitchen range has a built-in 125-volt receptacle and is within 6 feet (1.8 m) of the sink. Does the *Code* require such a receptacle to be protected with a GFCI?

   Answer: _____

   _____

   _____

   _____

   Reference: _____

5. Does the *NEC*® prohibit the use of nonmetallic outlet boxes with metal raceways?

   Answer: _____

   _____

   _____

   _____

   Reference: _____

6. Does the *NEC*® permit luminaires (lighting fixtures) to be installed on trees?

   Answer: _____

   _____

   _____

   _____

   Reference: _____

7. Electrical nonmetallic tubing (ENT) can be used concealed within walls, ceilings, and floors where the walls, ceilings, and floors provide a thermal barrier of material that has at least _____.

   Answer: _____

   _____

   _____

   _____

   Reference: _____

8. Does the *Code* require any extra protection for electrical nonmetallic tubing (ENT) when it is installed through openings in metal studs?

   Answer: _____

   _____

   _____

   _____

   Reference: _____

9. What are the strapping requirements for ENT when installed in metal studs?

   Answer: _____

   _____

   _____

   _____

   Reference: _____

10. Does the *Code* require any extra protection for Type NM cable when it is installed through openings in metal studs?

    Answer: _____

    _____

    _____

    _____

    Reference: _____

11. When UF cable is used for interior wiring, are the conductors required to be rated at 90°C?

    Answer: _____

    _____

    _____

    _____

    Reference: _____

12. Must "hospital grade" receptacles be installed throughout hospitals?

    Answer: _____

    _____

    _____

    _____

    Reference: _____

13. When railings or lattice work is used for room dividers, does the *Code* require receptacles to be installed as if they were solid walls?

    Answer: _____

    _____

    _____

    _____

    Reference: _____

14. In an apartment project, multiwire branch-circuits are routed through outlet boxes to supply receptacles. Can the neutral conductor continuity be assured by two connections to the screw terminals of the receptacle?

    Answer: _____

    _____

    _____

    _____

    Reference: _____

15. As shown in the diagram above, a small appliance circuit receptacle is above the counter next to the refrigerator. The refrigerator is plugged into this receptacle that is 5 feet (1.5 m) from the kitchen sink. Is the receptacle required to have GFCI protection?

Answer: _____

_____

_____

_____

Reference: _____

16. Does the *Code* require outdoor receptacles to be installed on balconies of high-rise apartment buildings?

Answer: _____

_____

_____

_____

Reference: _____

17. The *Code* did require that switches and circuit breakers used as switches be installed so that the center of the grip of the operating handle when in its highest position was no more than 6 feet (1.8 m) above the floor or working platform. Is this still the case?

Answer: _____

_____

_____

_____

Reference: _____

18. Where can luminaires (lighting fixtures) be used as raceways for circuit conductors?

    Answer: _____

    _____

    _____

    _____

    Reference: _____

19. In what areas of a hospital are hospital-grade receptacles required?

    Answer: _____

    _____

    _____

    _____

    Reference: _____

20. Is it a requirement that receptacles over a wet bar be spaced according to cabinet or floor line requirements?

    Answer: _____

    _____

    _____

    _____

    Reference: _____

21. A weatherproof receptacle cover must be weatherproof when in use (attachment cap inserted) or shall have a _____ feature.

    Answer: _____

    _____

    _____

    _____

    Reference: _____

22. Would electrical nonmetallic tubing be an acceptable wiring method for a wet-niche luminaire (lighting fixture) or some portion of the circuit?

    Answer: _____

    _____

    _____

    _____

    Reference: _____

23. Can nonmetallic-sheathed cable be used to supply a recessed fluorescent luminaire (fixture)?

Answer: _____

_____

_____

_____

Reference: _____

24. Do conductors for festoon lighting have to be rubber covered?

Answer: _____

_____

_____

_____

Reference: _____

25. Can an electric discharge luminaire (fixture) such as a fluorescent strip be used as a raceway for circuit conductors?

Answer: _____

_____

_____

_____

Reference: _____

PRACTICE EXAM 2

1. Are all luminaires (fixtures) located in agriculture buildings to be listed as dusttight or watertight?

Answer: _____

_____

_____

_____

Reference: _____

2. When installing AC cable horizontally through metal studs, are insulators needed?

Answer: _____

_____

_____

_____

Reference: _____

3. What types of cable assemblies are acceptable for use as feeders to a floating building?

    Answer: _____

    _____

    _____

    _____

    Reference: _____

4. Would nonmetallic-sheathed cable be acceptable to wire a pool-associated motor?

    Answer: _____

    _____

    _____

    _____

    Reference: _____

5. Would liquidtight flexible conduit be acceptable for enclosing conductors from an adjacent weatherproof box to an approved swimming pool junction box?

    Answer: _____

    _____

    _____

    _____

    Reference: _____

6. Is the cord-and-plug connection of an electric water heater acceptable under the *NEC*®? If so, would a standard dryer cord be acceptable (240 volts with neutral)? (Assume that ampacities, and so on, are acceptable under the *NEC*®.)

    Answer: _____

    _____

    _____

    _____

    Reference: _____

7. Does the *Code* permit motor controllers and disconnects to be used as junction boxes or wireways?

    Answer: _____

    _____

    _____

    _____

    Reference: _____

8. In determining the locked-rotor kVA per horsepower of a dual voltage electric motor, which voltage must be considered?

Answer: _____

_____

_____

_____

Reference: _____

9. Would electrical nonmetallic tubing be acceptable to enclose the insulated conductors for a panelboard (not part of service equipment) feeding pool-associated equipment?

Answer: _____

_____

_____

_____

Reference: _____

10. Does the *Code* permit the installation of 15-ampere-rated receptacles on the 20-ampere small appliance branch-circuits required in a dwelling?

Answer: _____

_____

_____

_____

Reference: _____

11. Does the *Code* permit the installation of a 15-ampere-rated duplex receptacle on a 20-ampere separate circuit for a microwave oven with a nameplate reading of 13 amperes?

Answer: _____

_____

_____

_____

Reference: _____

12. What does the term "in free air" mean at the top of Table 310.17?

Answer: _____

_____

_____

_____

Reference: _____

13. Has the *National Electrical Code*® restricted rigid nonmetallic conduit (PVC) from being installed in cold weather areas?

Answer: _____

_____

_____

_____

Reference: _____

14. All temporary wiring 15-ampere and 20-ampere, 125-volt receptacles not a part of the permanent building structure need to have GFCI protection. Does this mean that if a permanently installed receptacle is installed and energized, no GFCI protection is required even though this outlet is to be used by construction personnel?

Answer: _____

_____

_____

_____

Reference: _____

15. In the *NEC*®, would a GFCI-protected receptacle be required on a wet bar in a dwelling?

Answer: _____

_____

_____

_____

Reference: _____

16. Must the garbage disposal receptacle located under the sink be GFCI protected because it is within 6 feet (1.8 m) of the sink?

Answer: _____

_____

_____

_____

Reference: _____

17. What is the maximum number of duplex receptacles permitted on a 2-wire, 15-ampere circuit in a dwelling?

Answer: _____

_____

_____

_____

Reference: _____

18. How many receptacles are required at the patient's bed location in a critical care facility?

Answer: _____

_____

_____

_____

Reference: _____

19. A ¼-horsepower circulating pump is to be located in a closet space housing a gas-fired water heater in a large single-family residence. Can the circulating pump be cord-and-plug connected?

Answer: _____

_____

_____

_____

Reference: _____

20. In lieu of switching a ceiling luminaire (lighting fixture), the *Code* permits switching a receptacle. Can the switched receptacle be counted as one of the receptacles required for the 12-foot (3.6 m) rule?

Answer: _____

_____

_____

_____

Reference: _____

21. A laundry room in a single-family residence has a receptacle outlet on one wall and a second receptacle on the opposite wall. Can these two outlets be fed by the same laundry branch outlet?

Answer: _____

_____

_____

_____

Reference: _____

22. Antique stores and others are selling hanging luminaires (fixtures) without a grounding conductor. Are these luminaires (fixtures) in violation of the *NEC®*?

Answer: _____

_____

_____

_____

Reference: _____

23. Can 6 feet (1.8 m) of flexible metal conduit be installed between rigid nonmetallic raceways used for service-entrance conductors?

Answer: _____

_____

_____

_____

Reference: _____

24. Many kitchen and bathroom sinks, tubs, and showers are provided with short sections of metal pipe (6 inches [150 mm] to 2 feet [600 mm]) for connection to nonmetallic pipe systems. Is it required to bond the several short sections of metal pipe to the service equipment?

Answer: _____

_____

_____

_____

Reference: _____

25. Is there a limit on the number of extension boxes that can be joined together?

Answer: _____

_____

_____

_____

Reference: _____

PRACTICE EXAM 3

1. A switch box was installed in a floor of a dwelling. A receptacle outlet with a standard plate was installed on the box. Does this standard outlet box and receptacle installed in the floor comply with the *Code*?

Answer: _____

_____

_____

_____

Reference: _____

2. A 4 inch × 2⅛ inch-square NM box is marked with the number of 14, 12, and 10 AWG conductors that can be installed in it. Does this mean that 8 AWG conductors are not permitted in the box?

Answer: _____

_____

_____

_____

Reference: _____

3. Are outdoor receptacle outlets required at each dwelling unit of a multifamily dwelling?

Answer: _____

_____

_____

_____

Reference: _____

4. Do portable lamps wired with flexible cord require a polarized attachment plug?

Answer: _____

_____

_____

_____

Reference: _____

5. When figuring receptacles for a bedroom in a residence, must you include the space behind the door?

Answer: _____

_____

_____

_____

Reference: _____

6. Are metal boxes required for splices in temporary wiring?

Answer: _____

_____

_____

_____

Reference: _____

7. What is the maximum number of 12 AWG conductors permitted in a 4 inch × 1½ inch-deep octagon box?

Answer: _____

_____

_____

_____

Reference: _____

8. When a junction box contains a combination of conductors, such as 14 AWG and 12 AWG wires, what table is used to determine the size of the box?

Answer: _____

_____

_____

_____

Reference: _____

9. Can you use threaded intermediate metal conduit in a Class 1, Division 1 location?

Answer: _____

_____

_____

_____

Reference: _____

10. Does the *Code* consider EMT as conduit?

Answer: _____

_____

_____

_____

Reference: _____

11. Can suspended fluorescent luminaires (fixtures) be connected together with unsupported EMT between them?

Answer: _____

_____

_____

_____

Reference: _____

12. Must lay-in fluorescent luminaires (lighting fixtures) be fastened to the grid T-bars of a suspended ceiling system?

    Answer: _____

    _____

    _____

    _____

    Reference: _____

13. Can luminaires (fixtures) marked "suitable for damp locations" be used in wet locations?

    Answer: _____

    _____

    _____

    _____

    Reference: _____

14. A UL-listed fluorescent luminaire (fixture) is equipped with a Class "P" ballast. Does the *Code* permit such a luminaire (fixture) to be surface mounted on a combustible low-density cellulose fiberboard ceiling?

    Answer: _____

    _____

    _____

    _____

    Reference: _____

15. Are all recessed incandescent luminaires (fixtures) required to have thermal protection?

    Answer: _____

    _____

    _____

    _____

    Reference: _____

16. Is it necessary to identify the high leg of a 240/120-volt system at the motor disconnect of a 3-phase motor?

    Answer: _____

    _____

    _____

    _____

    Reference: _____

17. Is a GFCI protection required for a receptacle located on a food preparation island that is less than 6 feet (1.8 m) from the countertop sink?

    Answer: _____

    _____

    _____

    _____

    Reference: _____

18. Can conductors that pass through a panelboard also be spliced in the panelboard?

    Answer: _____

    _____

    _____

    _____

    Reference: _____

19. Are the enclosures for mercury vapor luminaires (lighting fixtures) installed 8 feet (2.4 m) above grade on poles outdoors required to be grounded?

    Answer: _____

    _____

    _____

    _____

    Reference: _____

20. Would an exit luminaire (light) with an emergency pack capable of supplying more than 1½ hours of power have to be connected to an emergency circuit ahead of the main disconnect?

    Answer: _____

    _____

    _____

    _____

    Reference: _____

21. A grounded 20-ampere receptacle is installed in a residential garage to serve an air compressor. Would this receptacle require GFCI protection?

    Answer: _____

    _____

    _____

    _____

    Reference: _____

22. If additional receptacles are installed in a bathroom, not adjacent to the wash basin, are they required to be GFCI protected?

Answer: _____

_____

_____

_____

Reference: _____

23. Must a receptacle in a dwelling-unit kitchen installed for a luminaire (lighting fixture) be GFCI protected?

Answer: _____

_____

_____

_____

Reference: _____

24. When installing recessed luminaires (fixtures) (hi-hats) in the ceiling, does 314.20 apply to the luminaire (lighting fixture) housing?

Answer: _____

_____

_____

_____

Reference: _____

25. What are the support requirements for service-entrance cables?

Answer: _____

_____

_____

_____

Reference: _____

PRACTICE EXAM 4

1. An underground cable is routed from the service equipment in the building underground and up a pole in a commercial parking lot to supply an overhead luminaire (light). What is the minimum burial depth for the cable below the base of the pole?

Answer: _____

_____

_____

_____

Reference: _____

2. Does the *Code* permit 15-ampere duplex receptacles to be installed on a 20-ampere branch-circuit?

Answer: _____

_____

_____

_____

Reference: _____

3. We are designing a new hotel with 250 guest rooms. It is our intent to install permanent beds, desks, and dressers in each guest room, bolted to the wall. If receptacles are located in accordance with 210.52, the receptacles will be located behind the headboard of the bed and behind the dresser. Are the receptacles required behind these pieces of furniture? Would the number of required receptacles be less if the furniture was not bolted to the wall?

Answer: _____

_____

_____

_____

Reference: _____

4. Can a paddle fan be supported only by an outlet box if it weighs less than 35 pounds (16 kg)?

Answer: _____

_____

_____

_____

Reference: _____

5. Where can luminaires (lighting fixtures) be used as raceways for circuit conductors?

Answer: _____

_____

_____

_____

Reference: _____

6. Do recessed incandescent luminaires (lighting fixtures) require thermal protection?

Answer: _____

_____

_____

_____

Reference: _____

7. Are medium-base lampholders in industrial occupancies permitted on 277-volt luminaire (lighting) circuits?

Answer: _____

_____

_____

_____

Reference: _____

8. Can a battery-powered (unit equipment) emergency light be supplied directly from a circuit in a panelboard that supplies normal lumination (lighting) in the area that will be illuminated by the battery luminaire (light) under emergency conditions?

Answer: _____

_____

_____

_____

Reference: _____

9. A single transfer switch is permitted to serve one or more branches of the essential electrical system in a small hospital. What constitutes a small hospital?

Answer: _____

_____

_____

_____

Reference: _____

10. Would a 3-phase, 4-wire, 1200-ampere, 480/277-volt service switch with 900-ampere fuses be required to have ground-fault protection?

Answer: _____

_____

_____

_____

Reference: _____

11. Are luminaires (lighting fixtures) operating at less than 15 volts between conductors required to be protected by a GFCI over hot tubs or spas?

Answer: _____

_____

_____

_____

Reference: _____

12. Could you use electrical metallic tubing (EMT) in the earth below a concrete slab?

Answer: _____

_____

_____

_____

Reference: _____

13. An industrial building will be served by several services of different voltage ratings. Is it required that each system be separately grounded?

Answer: _____

_____

_____

_____

Reference: _____

14. In residential occupancies, locating equipment to comply with the *Code* becomes a problem. Can a service panel or distribution panelboard be located over a counter or an appliance (washing machine, dryer) that extends away from the wall?

Answer: _____

_____

_____

_____

Reference: _____

15. A surface-mounted fluorescent luminaire (fixture) with an integral receptacle is installed above the countertop in the kitchen of a dwelling and is not more than 5½ feet (1.7 m) above the floor. Is this luminaire (fixture) required to be supplied by a 20-ampere circuit as are other kitchen receptacles? Is the luminaire (fixture) receptacle required to be protected by GFCI?

Answer: _____

_____

_____

_____

Reference: _____

16. What are the requirements for grounding an agricultural building where livestock are housed?

Answer: _____

_____

_____

_____

Reference: _____

17. Can a fused disconnect switch or circuit breaker located on the meter pole of a farm installation be used as the service equipment for the residence?

Answer: _____

_____

_____

_____

Reference: _____

18. A Type UF cable feeds a 120-volt yard luminaire (light) location on residential property. The circuit is protected by a 15-ampere overcurrent protection device. It is not GFCI protected. What is the minimum burial depth permitted for the cable?

Answer: _____

_____

_____

_____

Reference: _____

19. Can a fixed storage-type water heater having a capacity of 120 gallons or less be cord-and-plug connected?

Answer: _____

_____

_____

_____

Reference: _____

20. A TV dish antenna is located in the side yard of a single-family dwelling. (a) What is the minimum depth required by the *NEC*® for the coaxial signal cable? (b) How are these antenna units required to be grounded?

Answer: _____

_____

_____

_____

Reference: _____

21. Does the *Code* permit low-voltage control cables to be supported from the conduit containing the power circuit conductors feeding an air-conditioning unit?

Answer: _____

_____

_____

_____

Reference: _____

22. Is it true that if the underground metallic water piping is not at least 10 feet (3 m) long, the underground piping system is not adequate as a grounding electrode?

    Answer: _____

    _____

    _____

    _____

    Reference: _____

23. Does the luminaire (lighting) outlet in the attic have to be switch controlled?

    Answer: _____

    _____

    _____

    _____

    Reference: _____

24. With regards to recessed luminaires (lighting fixtures) installed over showers, is it permissible to use a gasketed vapor-proof trim only to comply with the *Code* or must the entire luminaire (fixture) be rated for a wet location?

    Answer: _____

    _____

    _____

    _____

    Reference: _____

25. May overhead outside branch-circuit conductors of 10 AWG copper be used in an unsupported open span up to 50 feet (15 m)? Is 10 AWG listed copper multiconductor UF cable suitable for such an application?

    Answer: _____

    _____

    _____

    _____

    Reference: _____

PRACTICE EXAM 5

1. Is a nonmetallic cable tray permitted by the *NEC*®?

    Answer: _____

    _____

    _____

    _____

    Reference: _____

2. The *Code* requires clearance for luminaires (lighting fixtures) installed in clothes closets. (a) What are the clearance requirements for surface-mounted luminaire (fixtures) mounted on the wall above the door? (b) What are clearance requirements for surface-mounted luminaire (fixtures) mounted on the ceiling? (c) What are the clearance requirements for recessed incandescent or fluorescent luminaire (fixtures) ceiling mounted?

Answer: _____

_____

_____

_____

Reference: _____

3. Are 16 AWG luminaire (fixture) wires counted for the number of conductors in outlet boxes to determine the box fill?

Answer: _____

_____

_____

_____

Reference: _____

4. Are switchboards and control panels rated 1200 amperes or more, 600 volts, and over 6 feet (1.8 m) wide required to have one entrance at each end of the room?

Answer: _____

_____

_____

_____

Reference: _____

5. What size grounding-electrode conductor is required where a grounding conductor is routed to a driven rod and from the rod to a concrete-encased electrode?

Answer: _____

_____

_____

_____

Reference: _____

6. Can handle locks on circuit breakers be locked so that the power to loads such as emergency lighting, sump pumps, alarm warning circuits, and other types of equipment cannot be cut off by mistake? What section of the *NEC®* applies?

   Answer: _____

   _____

   _____

   _____

   Reference: _____

7. Can a service be mounted on a mobile home if the home is designed to be installed on a foundation?

   Answer: _____

   _____

   _____

   _____

   Reference: _____

8. Is an insulated equipment ground required for a pool panelboard feeder fed from the service equipment?

   Answer: _____

   _____

   _____

   _____

   Reference: _____

9. Would the *Code* allow the use of stainless steel ground rods?

   Answer: _____

   _____

   _____

   _____

   Reference: _____

10. The *Code* now allows flexible metal conduit for services. Does the *Code* permit liquidtight flexible conduit, metallic or nonmetallic?

    Answer: _____

    _____

    _____

    _____

    Reference: _____

11. The grounding conductor can be connected to the grounding electrode by listed connectors, clamps, or other listed means and _____.

    Answer: _____

    _____

    _____

    _____

    Reference: _____

12. Does the *Code* require a ground wire in all flexible raceways?

    Answer: _____

    _____

    _____

    _____

    Reference: _____

13. Is AC cable a permitted wiring method in places of assembly?

    Answer: _____

    _____

    _____

    _____

    Reference: _____

14. Where are arc-fault circuit-interrupter protection devices required by the *NEC*®?

    Answer: _____

    _____

    _____

    _____

    Reference: _____

15. Can liquidtight flexible nonmetallic conduit be used on residential work?

    Answer: _____

    _____

    _____

    _____

    Reference: _____

16. Are there any *Code* requirements against installing a bare neutral wire inside the service conduit?

    Answer: _____

    _____

    _____

    _____

    Reference: _____

17. Can the branch-circuit and Class II control wires for two 3-phase motors be installed in a common raceway of the proper size?

    Answer: _____

    _____

    _____

    _____

    Reference: _____

18. The *Code* explicitly requires bonding around concentric knockouts on service equipment, but does the *Code* require that equipment downstream from the service equipment be bonded in the same manner?

    Answer: _____

    _____

    _____

    _____

    Reference: _____

19. Is nonmetallic-sheathed cable permitted to run through cold air returns if it is sleeved with thin-wall conduit or greenfield in short lengths?

    Answer: _____

    _____

    _____

    _____

    Reference: _____

20. Is it necessary to pigtail the neutral when connecting receptacles to a multiwire circuit?

    Answer: _____

    _____

    _____

    _____

    Reference: _____

21. When a motor-control circuit extends beyond the control enclosure, what is the maximum overcurrent protection acceptable for 12 AWG copper control wires?

Answer: _____

_____

_____

_____

Reference: _____

22. Is an equipment bonding conductor required to bond a cover-mounted receptacle to a surface-mounted outlet box?

Answer: _____

_____

_____

_____

Reference: _____

23. When calculating branch-circuit and feeder loads, what are the normal system voltages that shall be used?

Answer: _____

_____

_____

_____

Reference: _____

24. Is it permissible to run four 12 AWG THW copper conductors in a cable or raceway where they encounter 100°F ambient temperatures to supply cord-and-plug-connected loads protected at 20 amperes?

Answer: _____

_____

_____

_____

Reference: _____

25. When flat conductor cable, Type FCC, is installed under carpet, what is the maximum size carpet squares permitted?

Answer: _____

_____

_____

_____

Reference: _____

PRACTICE EXAM 6

1. Are bonding jumpers required on metal feeder and branch-circuit raceways containing circuits of more than 250 volts-to-ground where oversize concentric or eccentric knockouts are encountered?

   Answer: _____

   _____

   _____

   _____

   Reference: _____

2. What is the maximum number of service disconnects allowed for a set of service-entrance conductors installed in an apartment per the *NEC*®?

   Answer: _____

   _____

   _____

   _____

   Reference: _____

3. (a) Is a GFCI required on or within an outdoor portable sign? (b) Would a GFCI-protected branch-circuit supplying the sign be acceptable?

   Answer: _____

   _____

   _____

   _____

   Reference: _____

4. Is a luminaire (lighting) outlet required in the crawl space of a manufactured building if equipment requiring service is installed in the space?

   Answer: _____

   _____

   _____

   _____

   Reference: _____

5. Can an office partition assembly be cord-and-plug connected?

   Answer: _____

   _____

   _____

   _____

   Reference: _____

Remote panelboard

Gasoline dispenser

Diesel fuel dispenser

6. In the diagram above, a diesel fuel dispenser is mounted adjacent to a gasoline dispenser on a service station island. A sealing fitting is installed in the conduit entering the gasoline dispenser. Is a sealing fitting required in the conduit entering the diesel fuel dispenser?

Answer: _____

_____

_____

_____

Reference: _____

7. In residential-type occupancies, does the wall space behind doors where the door opens against the wall count when spacing the receptacles in a room?

Answer: _____

_____

_____

_____

Reference: _____

8. A four-family apartment building is supplied by an overhead drop to a 300-kcmil THW aluminum riser. Four electric meters are grouped and have 100-ampere breakers below the meters to supply a panel in each apartment. 3 AWG THW copper feeders are run to each apartment unit. An individual water meter is provided for each apartment. Each water meter is supplied with nonmetallic pipe. The water system for each apartment is copper above grade. (a) What size grounding electrode conductor is required? (b) How would this conductor be attached to the metallic water piping in each apartment? (c) Is a supplementary driven electrode required?

Answer: _____

_____

_____

_____

Reference: _____

9. Is a 14 AWG copper conductor allowed in a kitchen of a dwelling unit if it serves a single fixed appliance such as a dishwasher?

Answer: _____

_____

_____

_____

Reference: _____

10. Can Type UF cable be used to supply a swimming pool motor?

Answer: _____

_____

_____

_____

Reference: _____

11. Are 16 AWG and 18 AWG luminaire (lighting fixture) wires counted for the number of conductors in outlet boxes for box fill?

Answer: _____

_____

_____

_____

Reference: _____

12. Is a receptacle outlet required for a wet bar and/or if an outlet is installed within 6 feet (1.8 m) of the wet bar sink, would it be required to be GFCI protected?

Answer: _____

_____

_____

_____

Reference: _____

13. When calculating the demand for a 3-kW and a 6-kW range in a dwelling, would you use Table 220.55, Column B for the 3-kW and Column C for the 6-kW range or would you combine them and use either B or C for the two ranges?

Answer: _____

_____

_____

_____

Reference: _____

14. When sizing the branch-circuit conductors to a 4.5-kW dryer, would you use 4.5 kW or 5 kW in making the calculations?

Answer: _____

_____

_____

_____

Reference: _____

15. What is the maximum weight of a ceiling fan that can be supported by an outlet box?

Answer: _____

_____

_____

_____

Reference: _____

16. Where does the *NEC*® list or calculate the number of conductors allowed in flexible metal conduit?

Answer: _____

_____

_____

_____

Reference: _____

17. Are ceiling grid support wires that rigidly support a lay-in ceiling acceptable for the sole support of junction boxes above the lay-in ceiling?

Answer: _____

_____

_____

_____

Reference: _____

18. Can electrical equipment approved for a Class I location be used in a Class II location?

Answer: _____

_____

_____

_____

Reference: _____

19. Can **THHN** insulated conductors be used outside for service-entrance conductors where exposed to the elements?

    Answer: _____

    _____

    _____

    _____

    Reference: _____

20. Who has the authority to determine acceptability of electrical equipment and materials?

    Answer: _____

    _____

    _____

    _____

    Reference: _____

21. Can more than one receptacle in a laundry room be supplied by the laundry branch-circuit?

    Answer: _____

    _____

    _____

    _____

    Reference: _____

22. Many air-conditioning units are being installed on roof tops of apartment buildings for different reasons. Is a receptacle adjacent to such equipment required for servicing this equipment?

    Answer: _____

    _____

    _____

    _____

    Reference: _____

23. *NEC®* 430.72 applies to motor-control circuits that are tapped from the load side of motor branch-circuit short-circuit, and ground-fault protective devices. Because Column B of Table 430.72(B) covers control circuits that do not extend beyond the motor control equipment enclosure and Column C covers those that do extend beyond, what does Column A cover?

    Answer: _____

    _____

    _____

    _____

    Reference: _____

24. A 240-volt, single-phase circuit with a rated load of 1.8 kW draws 9 amperes of current. The power factor is _____ percent.

    Answer: _____

    _____

    _____

    _____

    Reference: _____

25. Can you run multioutlet assembly from one room to another room through a sheetrock wall?

    Answer: _____

    _____

    _____

    _____

    Reference: _____

PRACTICE EXAM 7

1. Illuminated exit signs are required by local building codes, and battery packs will be used. Do these signs have to be installed on a luminaire (lighting) circuit in the area of the sign or could they be on a separate branch-circuit?

    Answer: _____

    _____

    _____

    _____

    Reference: _____

2. Can a metal cable tray be used as an equipment grounding conductor?

    Answer: _____

    _____

    _____

    _____

    Reference: _____

3. Can surface nonmetallic raceway be used outside in a wet location?

    Answer: _____

    _____

    _____

    _____

    Reference: _____

4. Switchboards and control panels rated 1200 amperes or more, 600 volts or less, and over 6 feet (1.8 m) wide are required to have one entrance at each end of the room. What conditions would permit only one entrance to this room?

Answer: _____

_____

_____

_____

Reference: _____

5. Is Type MC cable required to have a bushing installed like the bushings used on Type AC armored cable?

Answer: _____

_____

_____

_____

Reference: _____

6. Can electrical nonmetallic tubing be surface mounted in a warehouse building?

Answer: _____

_____

_____

_____

Reference: _____

7. How deep must a residential branch-circuit be buried if it passes under the driveway?

Answer: _____

_____

_____

_____

Reference: _____

8. Where more than one building or structure is on the same property and under single management, the *Code* requires each building to have a disconnecting means at each such building or structure of 600 volts or less. Under what conditions can the disconnecting means be located elsewhere on the premises?

Answer: _____

_____

_____

_____

Reference: _____

9. Are cable trays permitted in a grain elevator?

Answer: _____

_____

_____

_____

Reference: _____

10. If a luminaire (fixture) is listed for use as a raceway, can this raceway be used for conductors to supply a circuit or circuits beyond the luminaire (fixture)?

Answer: _____

_____

_____

_____

Reference: _____

11. What *Code* section prohibits bare neutral feeders?

Answer: _____

_____

_____

_____

Reference: _____

12. Would a large mercantile store be considered a place of assembly if it can hold more than 100 people?

Answer: _____

_____

_____

_____

Reference: _____

13. Can the neutral be reduced by 70 percent for data processing equipment over 200 amperes?

Answer: _____

_____

_____

_____

Reference: _____

14. What ampacity correction factor would you use for 12 AWG THW copper conductors used in an ambient temperature of 78°F? (Not more than three conductors in raceway, in free air, 240-volt circuit.)

Answer: _____

_____

_____

Reference: _____

15. How are box volumes calculated when the box contains different size conductors and a combination of clamps, devices, and studs?

Answer: _____

_____

_____

Reference: _____

16. What is the minimum service-drop clearance required over a residential yard where the drop conductors do not exceed 150 volts-to-ground?

Answer: _____

_____

_____

Reference: _____

17. An equipment room containing electrical, telephone, and air-handling equipment uses the space in the room for air-handling purposes. Is this room considered a plenum?

Answer: _____

_____

_____

Reference: _____

18. As applied to lumination (lighting) in a clothes closet, is the entire 12-inch (300 mm) or width-of-shelf area required to be unobstructed to the floor or could part of this area be used for storage?

Answer: _____

_____

_____

Reference: _____

19. Would an outdoor luminaire (lighting fixture) be considered as grounded by the physical connection to a grounded metal support pole or would an equipment grounding conductor connection directly to the luminaire (fixture) be required?

    Answer: _____

    _____

    _____

    _____

    Reference: _____

20. Would entrance-type luminaires (lighting fixtures) that are installed under roof overhangs, porch roofs, or canopies be required to be suitable for damp locations?

    Answer: _____

    _____

    _____

    _____

    Reference: _____

21. Table 430.147 through Table 430.150 list the horsepower and the full-load currents of motors. How could you determine the full-load current of a motor not listed, such as a 35-horsepower, 460-volt motor?

    Answer: _____

    _____

    _____

    _____

    Reference: _____

22. Where a single Type NM cable is pulled into a raceway, what percent may the cross section of the conduit be filled?

    Answer: _____

    _____

    _____

    _____

    Reference: _____

23. Can you use Table 310.15(B)(6) for a feeder to each apartment in a building regardless of whether the mains are in the apartment units or on the outside of the building?

    Answer: _____

    _____

    _____

    _____

    Reference: _____

24. Can swimming pool equipment such as a pump motor be calculated as a fixed appliance when applying 220.14(A) to a dwelling?

    Answer: _____

    _____

    _____

    _____

    Reference: _____

25. Is a GFCI receptacle required in a detached garage of a dwelling?

    Answer: _____

    _____

    _____

    _____

    Reference: _____

PRACTICE EXAM 8

1. Are all recessed luminaires (lighting fixtures) that are to be installed in a suspended ceiling required to be thermally protected?

    Answer: _____

    _____

    _____

    _____

    Reference: _____

2. How can you ground an agricultural building and comply with the *Code*?

    Answer: _____

    _____

    _____

    _____

    Reference: _____

3. Is Table 220.55 applicable to microwave ovens and convection cooking ovens?

    Answer: _____

    _____

    _____

    _____

    Reference: _____

4. Is an autotransformer recognized for use in a motor-control circuit?

Answer: _____

_____

_____

_____

Reference: _____

5. Can service equipment be mounted on a floating dwelling unit that is moved frequently?

Answer: _____

_____

_____

_____

Reference: _____

6. Is it permissible to use neon lumination (lighting) on or in a dwelling unit?

Answer: _____

_____

_____

_____

Reference: _____

7. In a large industrial plant, does the *Code* allow a branch-circuit over 50 amperes to feed several outlets such as to supply power to portable welders that move from place to place?

Answer: _____

_____

_____

_____

Reference: _____

8. Can nonmetallic raceways be used in healthcare facilities?

Answer: _____

_____

_____

_____

Reference: _____

9. Can a conductor be protected by an overcurrent device sized for its ampacity even though the allowable load current is only 50 percent of the assigned ampacity after derating?

   Answer: _____

   _____

   _____

   _____

   Reference: _____

10. Does the *Code* permit single-pole circuit breakers in lighting panelboards to be used as a switch for controlling a lighting circuit as an off and on switch?

    Answer: _____

    _____

    _____

    _____

    Reference: _____

11. When determining the required size box, items such as fittings, cable clamps, hickeys, where combinations of different size conductors are used, which conductor size do you deduct for the device and fittings?

    Answer: _____

    _____

    _____

    _____

    Reference: _____

12. Where rigid metal conduit is used as service raceways, are double locknuts permitted for the service equipment required continuity?

    Answer: _____

    _____

    _____

    _____

    Reference: _____

13. A receptacle outlet is to be added to an existing installation, and Type NM cable is to be "fished in" the partition wall. Does the *Code* allow a nonmetallic box without cable clamps to be used for such an installation?

    Answer: _____

    _____

    _____

    _____

    Reference: _____

14. What size general-use switch is required on an air-conditioning compressor, sealed (hermetic-type) when the nameplate states FLA of 100 amperes, 3-phase, 230 volts, LRA 580?

Answer: _____

_____

_____

_____

Reference: _____

15. A fluorescent luminaire (fixture) is supplied with trade size $^3/_8$ (12) flexible metal conduit 5 feet (1.5 m) long. Is an equipment grounding required to ground the luminaire (fixture)?

Answer: _____

_____

_____

_____

Reference: _____

16. What is the full-load current of an electric furnace rated 240 volts, single-phase, 10 kW when connected to a 208-volt circuit? (Show calculations.)

Answer: _____

_____

_____

_____

Reference: _____

17. Can four 400-watt wet-niche luminaires (lights) be connected to one GFCI for a swimming pool?

Answer: _____

_____

_____

_____

Reference: _____

18. How does one determine how many conductors conduit bodies (condulets) are approved for?

Answer: _____

_____

_____

_____

Reference: _____

19. How can a recessed luminaire (lighting fixture) approved for direct contact with thermal insulation be identified?

    Answer: _____

    _____

    _____

    _____

    Reference: _____

20. Does the 10-foot (3 m) tap rule apply only to conductors such as wires and not panelboard bus?

    Answer: _____

    _____

    _____

    _____

    Reference: _____

21. How are working clearances measured between enclosed electrical equipment facing each other across an aisle?

    Answer: _____

    _____

    _____

    _____

    Reference: _____

22. Can either the disposal, dishwasher, or trash compactor be installed on a 20-ampere kitchen small appliance circuit?

    Answer: _____

    _____

    _____

    _____

    Reference: _____

23. What is the required horizontal distance between service drops and swimming pools?

    Answer: _____

    _____

    _____

    _____

    Reference: _____

24. Can a bare 4 AWG copper grounding electrode conductor be buried below a concrete floor slab, and is there any depth requirement?

Answer: _____

_____

_____

_____

Reference: _____

25. Where the grounding electrode conductor connects to the water system on the house side of the water meter, is bonding required around valves with sweated connections? Also, what other equipment is required to be bonded around?

Answer: _____

_____

_____

_____

Reference: _____

PRACTICE EXAM 9

1. Can a floodlight and/or receptacle be on the GFCI circuit protecting an underwater luminaire (lighting fixture)?

Answer: _____

_____

_____

_____

Reference: _____

2. For a 600-ampere service where 350 kcmil copper conductors are parallel in two conduits and only 3-phase, 3-wire is needed, but the transformer is Y-connected and grounded: (a) Must the grounded conductor be brought into the service? (b) Can the grounded conductor be brought in through a single raceway or is it required to be installed in each of the parallel raceways? (c) What is the maximum size of the grounded conductor?

Answer: _____

_____

_____

_____

Reference: _____

3. What are the minimum dimensions of a junction box illustrated above with only a U-pull of two trade size 3 (78) conduits in one side?

Answer: _____

_____

_____

_____

Reference: _____

4. With two 42-circuit panelboards bolted together, can the 42 circuits from one panelboard feed through the gutter of the second panelboard if the 40 percent fill in that gutter is not exceeded?

Answer: _____

_____

_____

_____

Reference: _____

5. Can suspended fluorescent luminaires (fixtures) be connected together with unsupported EMT between them if the length of the EMT is less than 3 feet (900 mm)?

Answer: _____

_____

_____

_____

Reference: _____

322 • **Master Electrician's Review**

6. (a) Are 3-phase generators permitted to be connected to a three-pole transfer switch with a solid neutral or is a four-pole transfer switch that breaks the neutral required? (b) If either is OK, must a grounding electrode be provided at the generator in both cases?

Answer: _____

_____

_____

_____

Reference: _____

7. A 600-ampere underground service with four 4/0 AWG conductors per phase paralleled is installed in two rigid nonmetallic conduits. Must the phase conductors and neutral be grouped in a single raceway if the conduits are nonmetallic?

Answer: _____

_____

_____

_____

Reference: _____

8. The raceway for a 240-volt motor circuit includes both the power circuit and control circuit conductors. The power circuit conductors use 600-volt insulation. Must the control circuit conductors also have 600-volt insulation or would 300-volt insulation be acceptable?

Answer: _____

_____

_____

_____

Reference: _____

9. Does the *Code* either permit or prohibit a sprinkler head over a switchboard?

Answer: _____

_____

_____

_____

Reference: _____

10. A 2000-ampere, 3-phase, 480-volt ungrounded service is located 300 feet (91.4 m) from the water line within a building. Can the grounding electrode conductor from this service be connected to the building steel in the vicinity of the service, and then 300 feet (91.4 m) away bond the steel to the water line such that the building steel is being used as the grounding electrode conductor?

Answer: _____

_____

_____

_____

Reference: _____

11. Four 12 AWG THNN conductors in one conduit supply a continuous lighting load. What is the maximum permissible circuit breaker size and permissible load allowed on each conductor?

Answer: _____

_____

_____

_____

Reference: _____

12. A receptacle outlet is located in a soffit or roof overhang for connection of Christmas decorations or roof heating cable at a dwelling. When this receptacle is located out of reach from the grade, approximately 9 feet (2.75 m) above grade, is this receptacle required to be GFCI protected?

Answer: _____

_____

_____

_____

Reference: _____

13. Does the *NEC*® permit more than one Type NM cable under one device box cable clamp?

Answer: _____

_____

_____

_____

Reference: _____

14. A 400-ampere service uses two 200-ampere panelboards, each with a main breaker. Additional load is added so that one more 100-ampere service disconnect is required. Must the service-entrance conductors be increased in size? (The calculated load is not increased.)

    Answer: _____

    _____

    _____

    _____

    Reference: _____

15. Can an attachment plug be used as a disconnecting means for a 3-horsepower motor?

    Answer: _____

    _____

    _____

    _____

    Reference: _____

16. Where in the *Code* does it permit using 4 AWG copper THW conductors for service conductors supplying a 100-ampere service?

    Answer: _____

    _____

    _____

    _____

    Reference: _____

17. In three-way and four-way switch circuitry in a metal raceway, is the neutral required to be in the same raceway with the travelers (switch legs)?

    Answer: _____

    _____

    _____

    _____

    Reference: _____

18. Single-conductor UF cable comes underground from a submersible pump through a building wall to the pump controller. Are these conductors permitted to be run inside the building?

    Answer: _____

    _____

    _____

    _____

    Reference: _____

19. In commercial garages, can regular receptacles and plates be used over 18 inches (450 mm) above the floor? What are the wiring restrictions in a garage that has a minimum of four air changes per hour?

    Answer: _____

    _____

    _____

    _____

    Reference: _____

20. A single-family dwelling has two electric dryer receptacle outlets wired with two 10 AWG black conductors to each one and one 10 AWG white conductor neutral. Each outlet has a separate 30-ampere circuit breaker. Is one 10 AWG neutral conductor permitted to be used with this installation?

    Answer: _____

    _____

    _____

    _____

    Reference: _____

21. Can trade size ⅜ (12) flexible metal conduit be used on machinery and boilers for limit switches, flow switches, can switches, and solenoids?

    Answer: _____

    _____

    _____

    _____

    Reference: _____

22. Article 100—Definitions states that the definition of a building is *a structure that stands alone or that is cut off from adjoining structures by fire walls with all openings therein protected by approved fire doors.* Could this definition apply to a building with fire-rated floors and all elevator shafts and stairways protected by approved fire rating and fire doors? In other words, could each floor be considered a separate building?

    Answer: _____

    _____

    _____

    _____

    Reference: _____

23. Can unit equipment used for emergency lighting be directly connected on the branch-circuit rather than use a plug-in receptacle?

Answer: _____

_____

_____

_____

Reference: _____

24. *NEC®* 517.20 specifies GFCI receptacles for wet locations that by definition in Article 100 include outside areas exposed to the weather. Are receptacles required to be GFCI protected when outdoors at healthcare facilities?

Answer: _____

_____

_____

_____

Reference: _____

25. Does a gas furnace in a dwelling require a disconnect switch on the furnace? May the furnace be directly connected onto a basement lighting circuit?

Answer: _____

_____

_____

_____

Reference: _____

PRACTICE EXAM 10

1. Are open tube fluorescent luminaires (fixtures) acceptable in a basement garage under an apartment building?

Answer: _____

_____

_____

_____

Reference: _____

2. Hot tubs are popular. Are there size and depth requirements to classify them as a pool or are they classified as a pool regardless of size for wiring of circulation pumps, heaters, and so forth?

   Answer: _____

   _____

   _____

   Reference: _____

3. Rigid metal conduit runs in a commercial garage floor and extends up through an 18-inch (450 mm) hazardous area with no couplings or fittings in that 18-inch (450 mm) area. Is a seal-off fitting required in this conduit?

   Answer: _____

   _____

   _____

   Reference: _____

4. Three single-phase, 480-volt transformers rated at 104 amperes each are connected in a wye bank for a 3-phase system. What is the maximum overcurrent protection permitted for the primary when the secondary is also protected? Would the primary protection be different if connected in a delta bank?

   Answer: _____

   _____

   _____

   Reference: _____

5. How is the load for a section of multioutlet assembly (plug-mold) determined?

   Answer: _____

   _____

   _____

   Reference: _____

6. A motor is connected to a branch-circuit breaker in a panel that is out of sight of the motor location. The breaker is used as the controller and also the disconnecting means for the motor. If this breaker is of the "lock off" type, would this meet *Code* requirements?

   Answer: _____

   _____

   _____

   Reference: _____

7. A 12 AWG remote-control circuit is protected by 80-ampere fuses in the motor disconnect (magnetic contact). Is this control circuit properly protected?

Answer: _____

_____

_____

_____

Reference: _____

8. Bus duct is reduced in size from 1000 amperes to 400-ampere bus duct and runs 40 feet (12 m) to where it terminates in a distribution panel. Is overcurrent protection required?

Answer: _____

_____

_____

_____

Reference: _____

9. Can UF cable be used from the meter, down the pole, then underground 20 feet (6 m) to the disconnect for a pasture pump?

Answer: _____

_____

_____

_____

Reference: _____

10. On a 480/277-volt, 3-phase, 4-wire system where the neutral is not used, must it be grounded at the service panel?

Answer: _____

_____

_____

_____

Reference: _____

11. If available on the premises, the *Code* requires the bonding of all grounding electrodes together to form the grounding electrode system. Are these requirements the same for a separately derived system?

Answer: _____

_____

_____

_____

Reference: _____

12. Define tap conductors. Is it permissible to make a tap from a tap?

Answer: _____

_____

_____

_____

Reference: _____

13. Can two tandem circuit breakers mounted adjacent to each other in a panelboard with adjacent handles tied together be used in place of two double-pole breakers to supply two 220-volt electric baseboard heaters?

Answer: _____

_____

_____

_____

Reference: _____

14. What minimum size copper grounding electrode conductor can you use to ground a 400-ampere lighting service when two 3/0 AWG copper conductors in parallel are used for the phase conductors?

Answer: _____

_____

_____

_____

Reference: _____

15. Can *aluminum* rigid metal conduit be used in hazardous locations such as in sewer plants?

Answer: _____

_____

_____

_____

Reference: _____

16. A two-lamp exit luminaire (light) is supplied from both a battery-operated unit of equipment and a normal source. Must the normal supply to the exit luminaire (light) be from a recognized emergency source such as ahead of the service disconnect?

Answer: _____

_____

_____

_____

Reference: _____

17. A service is changed and moved to a different location. The old electric range run is Type SE cable that is too short to reach the new service, so a metal junction box is installed and a short piece of new Type SE cable is installed to reach the new service. This is in a residential basement on a 7-foot (2 m) ceiling. Does this metal junction box require grounding, and if so, does it require a separate grounding conductor to the service equipment (not the neutral of the 3-wire Type SE cable)?

Answer: _____

_____

_____

_____

Reference: _____

18. Can service conductors be run directly to a fire pump controller?

Answer: _____

_____

_____

_____

Reference: _____

19. A trucking firm has added a second service. They have 240-volt service to the old part of the building and 208-volt service to the new part. Common computer equipment is connected to both services with shielded TWINAX (signal cable) between devices. The two services are not tied to the same ground. This causes ground loop on signal shield. What is the best way per the *NEC*® to achieve equal potential bonding on network?

Answer: _____

_____

_____

_____

Reference: _____

20. A luminaire (lighting fixture) is installed over a hydromassage bathtub. Is GFCI protection required for the luminaire (fixture)?

Answer: _____

_____

_____

_____

Reference: _____

21. Frequently the *NEC*® requires GFCI protection of receptacles. Is there any requirement that limits the number of receptacles that can be protected by a single GFCI circuit breaker or a feed-through receptacle?

    Answer: _____

    _____

    _____

    _____

    Reference: _____

22. Can an electrical discharge luminaire (fixture) such as a fluorescent strip be used as a branch-circuit junction box or termination point for circuit conductors?

    Answer: _____

    _____

    _____

    _____

    Reference: _____

23. Does the *Code* allow electrical nonmetallic tubing (ENT) to be installed above a suspended ceiling provided that a thermal barrier of material has at least a 15-minute rating of fire-rated assemblies?

    Answer: _____

    _____

    _____

    _____

    Reference: _____

24. Does the *Code* permit 8 AWG conductors to be installed in a $4 \times 2\frac{1}{8}$ inch-square nonmetallic box that is marked with the number of 14 AWG, 12 AWG, and 10 AWG conductors that can be installed in it?

    Answer: _____

    _____

    _____

    _____

    Reference: _____

25. Must lay-in fluorescent luminaires (lighting fixtures) be fastened or can they lay in grid T-bars of a suspended ceiling system?

    Answer: _____

    _____

    _____

    _____

    Reference: _____

PRACTICE EXAM 11

1. When Type UF cable is used for interior wiring as permitted, are the conductors required to be rated at 90°C?

   Answer: _____

   _____

   _____

   _____

   Reference: _____

2. Are swimming pool underwater luminaires (lighting fixtures) operating at less than 15 volts between conductors required to be protected by a GFCI device or circuit breaker?

   Answer: _____

   _____

   _____

   _____

   Reference: _____

3. Can handle locks on circuit breakers be locked so that the power to loads such as emergency lighting, sump pumps, alarm warning circuits, and other types of equipment cannot be cut off by mistake? If so, what sections of the *NEC®* would apply?

   Answer: _____

   _____

   _____

   _____

   Reference: _____

4. Can nonmetallic-sheathed cable be run parallel to framing members through cold air returns?

   Answer: _____

   _____

   _____

   _____

   Reference: _____

5. Is a GFCI properly protected circuit supplying an outdoor portable sign an acceptable method for installation in accordance with the *NEC®*?

   Answer: _____

   _____

   _____

   _____

   Reference: _____

Grounding
electrode

6. *Refer to the diagram above.* Where should the grounding electrode conductor terminate in a multiple disconnect service?

Answer: _____

_____

_____

_____

Reference: _____

8" × 8" (200 mm × 200 mm) wireway

Switch C
200 ampere

Switch A
200 ampere

Switch B
200 ampere

1/0 AWG copper

Water pipe

Refer to the diagram above for Questions 7 and 8.

7. A 600-ampere service is built with parallel sets of 350-kcmil CU THWN service conductors. Three 200-ampere fused service disconnects are supplied with 4/0 AWG THWN service taps within an 8" × 8" (200 mm × 200 mm) wireway. All service bonding is continuous from the CT cabinet through all service enclosures. Does the *Code* permit only one grounding electrode conductor connection at switch "C" if sized for 700-kcmil service conductors?

Answer: _____

_____

_____

_____

Reference: _____

8. With reference to Question 7, how should the trough above the three service disconnects be sized? Is the trough to be sized as a wireway or a junction box?

   Answer: _____

   _____

   _____

   _____

   Reference: _____

9. An EMT and steel box system was installed throughout an office/warehouse building. An additional equipment grounding conductor was pulled with all receptacle branch-circuits and terminated at all receptacle grounding screws. Must the equipment grounding conductor also be attached to each metal box?

   Answer: _____

   _____

   _____

   _____

   Reference: _____

10. Is an insulated equipment grounding conductor required in nonmetallic sheathed cable that is supplying a swimming pool pump motor installed in the garage of a single-family dwelling?

    Answer: _____

    _____

    _____

    _____

    Reference: _____

11. A 225-ampere rated power panel has main lugs only and powers a calculated load of 180 amperes. It is supplied by a 500-kcmil CU feeder with 400-ampere overcurrent protection. Does this installation comply with the *Code*?

    Answer: _____

    _____

    _____

    _____

    Reference: _____

12. If a small dental office uses limited amounts of inhalation anesthetics, does *NEC*® 517.31 require standby emergency power to supply emergency lighting and the anesthetic equipment?

    Answer: _____

    _____

    _____

    _____

    Reference: _____

13. Does the *Code* allow any derating of loads in a commercial laundry? The normal load of a washing machine is 4 amperes; the spin cycle draws 9 amperes. Because it is unlikely that all machines will be spinning at the same time, what load figure should be used for load calculations?

    Answer: _____

    _____

    _____

    _____

    Reference: _____

14. Is the examining room of a dental office considered to be a general-care patient care area? Is Type NM cable a permitted wiring method in this area?

    Answer: _____

    _____

    _____

    _____

    Reference: _____

15. Do the metal enclosures for individual channel-type neon letters (either front-lit or back-lit) require grounding?

    Answer: _____

    _____

    _____

    _____

    Reference: _____

16. Is it permissible to ground the secondary of a separately derived system back to the neutral terminal of a service switchboard instead of going directly to the water line when there is no effectively grounded structural steel nearby?

    Answer: _____

    _____

    _____

    _____

    Reference: _____

17. When conductors are paralleled, does the ampacity double (assuming the same type, size, length)?

    Answer: _____

    _____

    _____

    _____

    Reference: _____

18. Suppose you use three 300-kcmil conductors in a raceway in free air with 75°C temperature rating of 285 amperes for each conductor. Instead of this, can you use six 1/0 AWG conductors with 75°C rating and two conductors in parallel per phase, all in the same conduit at 300 amperes ampacity, or must you derate because you have more than three conductors? (Assume that you are using Type THWN conductors in a trade size 2½ (63) conduit with device terminations rated for 75°C.)

Answer: _____

_____

_____

_____

Reference: _____

19. Are medium-base HID lampholders permitted to be installed on a 277-volt circuit?

Answer: _____

_____

_____

_____

Reference: _____

20. Is GFCI protection required for receptacles located on decks where the receptacles are located less than 6 feet, 6 inches (2 m) above grade and access to the deck is by steps or stepping onto a low deck? (The receptacle cannot be reached while standing on the ground.)

Answer: _____

_____

_____

_____

Reference: _____

21. Is the bathroom GFCI-protected circuit permitted to supply luminaire (lighting fixture) outlets in addition to the bathroom receptacles?

Answer: _____

_____

_____

_____

Reference: _____

22. Is it acceptable to connect a garbage disposal and dishwasher on the same branch-circuit?

    Answer: _____

    _____

    _____

    _____

    Reference: _____

23. Are receptacles required in a four-car garage at an apartment building if the garage is detached and provided with electric power and lights?

    Answer: _____

    _____

    _____

    _____

    Reference: _____

24. Does an outdoor entrance luminaire (lighting fixture) with integral photocell satisfy the *Code* without a wall switch control? What about a motion detector used to activate the luminaire (lighting fixture)?

    Answer: _____

    _____

    _____

    _____

    Reference: _____

25. Does the *Code* require feeder conductors to have an ampacity of 125 percent of the continuous load plus the noncontinuous load, or is it only necessary for the feeder overcurrent device to have this rating? If only the feeder overcurrent device requires this rating, can the feeder conductors be sized only for total continuous plus noncontinuous load?

    Answer: _____

    _____

    _____

    _____

    Reference: _____

PRACTICE EXAM 12

1. In calculating feeder loads for electric space heating, is it required to figure 125 percent of space heating loads or is the 125 percent required only for branch-circuit loads?

   Answer: _____

   _____

   _____

   _____

   Reference: _____

2. For household range load calculations, is it acceptable to combine ovens and countertops and treat them as one appliance for feeder calculations?

   Answer: _____

   _____

   _____

   _____

   Reference: _____

3. Because the disconnecting means for separate buildings on one property are required to be suitable for use as service equipment, are the number of disconnects at each building's entrance limited to six?

   Answer: _____

   _____

   _____

   _____

   Reference: _____

4. The *Code* requires multiple services to be grounded to the same electrode. Must the grounding connections be made to the same point on the electrode or can they be grounded to the same electrode (water line or structural steel) at different locations?

   Answer: _____

   _____

   _____

   _____

   Reference: _____

5. Does Article 514 apply to underground wiring (such as to a sign) that is at least 30 feet (9 m) from any hazardous location in a service station?

   Answer: _____

   _____

   _____

   _____

   Reference: _____

6. What is the required minimum clearance between a thermally protected Type IC recessed incandescent luminaire (fixture) and wood framing?

Answer: _____

_____

_____

_____

Reference: _____

7. Does the *Code* permit the metal coaxial sheath of CATV cable to be grounded to a separate driven rod or is it required to be bonded to the building's power system ground or service equipment enclosure?

Answer: _____

_____

_____

_____

Reference: _____

8. An automotive repair garage is classified as Class I, Division 2 up 18 inches (450 mm) above the floor. An attached sales and office area is not classified. Is Type NM-B cable permitted to be used in the sales and office area?

Answer: _____

_____

_____

_____

Reference: _____

9. Do Type AC and MC cables installed on 277-volt circuits require bonding if connected into concentric knockouts?

Answer: _____

_____

_____

_____

Reference: _____

10. A swimming pool is installed at a single-family dwelling. Can 2-wire Type NM-B cable with a bare grounding conductor be used to connect the pump motor located in the basement?

Answer: _____

_____

_____

_____

Reference: _____

11. Ceiling lighting consists of lay-in fluorescent luminaires (fixtures). Is it acceptable to use trade size ⅜ (12) flexible metal conduit from luminaire to luminaire (fixture to fixture) in lengths less than 6 feet (1.8 m) provided an equipment grounding conductor is installed in the conduit?

    Answer: _____

    _____

    _____

    _____

    Reference: _____

12. Can a 12-volt dry-niche luminaire (lighting fixture) mounted on the outside of a permanent aboveground swimming pool be supplied by flexible cord connected to a receptacle located 11 feet (3.35 m) from the pool?

    Answer: _____

    _____

    _____

    _____

    Reference: _____

13. Can a timer switch for a spa be located within 5 feet (1.5 m) of the spa? What if it is located on the unit and not on the wall?

    Answer: _____

    _____

    _____

    _____

    Reference: _____

14. Is there any requirement relative to the location of the wall switch for an outdoor entrance luminaire (light)? As an example, must the switch for an entrance luminaire (light) at a sunroom door be located at the entrance door from outside or can it be located 10 feet (3 m) away at another door leading to the sunroom?

    Answer: _____

    _____

    _____

    _____

    Reference: _____

15. When calculating the demand for a 3-kW and a 6-kW range, would you use Table 220.55, Column B for the 3-kW and Column C for the 6-kW or would you combine them and use either B or C for the two ranges?

    Answer: _____

    _____

    _____

    _____

    Reference: _____

16. Can more than one receptacle in a laundry room be supplied by the laundry branch-circuit?

    Answer: _____

    _____

    _____

    _____

    Reference: _____

17. For a large bathroom with provisions for a washing machine, is GFCI protection required for an inaccessible receptacle located behind and dedicated to the washing machine?

    Answer: _____

    _____

    _____

    _____

    Reference: _____

18. Is there an acceptable method whereby control wiring can be run in the same conduit with power conductors to a central air-conditioning unit?

    Answer: _____

    _____

    _____

    _____

    Reference: _____

19. Does the *Code* require grounding the metal sheath of CATV cable entering a dwelling?

    Answer: _____

    _____

    _____

    _____

    Reference: _____

20. The *Code* requires bonding a metallic conduit enclosing a grounding electrode conductor. How is the bonding jumper sized when the conduit enters a concentric knockout?

    Answer: _____

    _____

    _____

    _____

    Reference: _____

21. Does the *Code* require a GFCI-protected receptacle when installed within 6 feet (1.8 m) of a dwelling wet bar or laundry sink?

    Answer: _____

    _____

    _____

    _____

    Reference: _____

22. A 500-kcmil Type THW CU feeder protected with 400-ampere overcurrent device is routed through a junction box. What is the minimum size tap conductor required to supply a 30-ampere load using the 25-foot (7.5 m) tap rule? What is the minimum size using the 10-foot (3 m) tap rule?

    Answer: _____

    _____

    _____

    _____

    Reference: _____

23. A large recreation room in a dwelling has three sets of sliding doors opening onto a covered porch. Is a switch required at each door to control porch lumination (lighting)?

    Answer: _____

    _____

    _____

    _____

    Reference: _____

24. Existing feeder conduits are being repulled using conductors with 90°C Type THHN insulation. What ampacity values are allowed for these conductors that are connected to a circuit breaker or panel main lug terminals?

    Answer: _____

    _____

    _____

    _____

    Reference: _____

25. Branch-circuits serving patient care areas must be installed in metal raceways. Which areas in nursing homes and residential custodial care facilities are designated as patient care areas?

    Answer: _____

    _____

    _____

    _____

    Reference: _____

FINAL EXAMINATION

Examination Instructions:

For a positive evaluation of your knowledge and preparation awareness, you must:

1. locate yourself in a quiet atmosphere (room by yourself).

2. have with you at least two sharp No. 2 pencils, the 2005 *NEC®*, and a handheld calculator.

3. time yourself (three hours) with no interruptions. Do not spend more than 3½ minutes per question.

4. after the test is complete, grade yourself honestly and concentrate your studies on the sections of the *NEC®* in which you missed the questions.

Caution: Do not just look up the correct answers, because the questions in this examination are only an exercise and not actual test questions. Therefore, it is important that you be able to quickly find answers from throughout the *NEC®*.

1. According to the *National Electrical Code®*, open conductors for communication equipment on a building shall be separated at least _____ feet (_____ m) from *lightning* conductors.
   A. 2 (600 mm)
   B. 4 (1.2)
   C. 6 (1.8)
   D. 8 (2.4)

   Answer: _____   Reference: _____

2. A megohmmeter is an instrument used for
   A. polarizing a circuit.
   B. measuring high resistances.
   C. shunting a generation system.
   D. determining amperes.

   Answer: _____   Reference: _____

3. The total opposition to alternating current in a circuit that includes resistance, inductance, and capacitance is called
   A. reactance.
   B. resistance.
   C. reluctance.
   D. impedance.

   Answer: _____   Reference: _____

4. A 240-volt, single-phase circuit has a resistive load of 8500 watts. The net calculated current to supply this load is _____ amperes.
   A. 35
   B. 39
   C. 44
   D. 71

   Answer: _____   Reference: _____

Switch S1 (on position)

120V

Lamp
L1

Refer to the diagram above for Questions 5 and 6.

5. In the diagram above, switch S1 is in the "on" position, but luminaire (light) L1 does not come on. Voltage across L1 is measured to be 120 volts. Voltage across S1 is measured to be 0 volt. The luminaire (light) does not come on because
   A. the luminaire (light) is open (burned out).
   B. the luminaire (light) and switch are shorted.
   C. the luminaire (light) is good but the switch does not make contact.
   D. there is a break in the wire of the circuit.

   Answer: _____   Reference: _____

6. In the diagram above, with a 9-ampere current in the circuit, the power factor is _____ percent.
   A. 71
   B. 83
   C. 93
   D. 108

   Answer: _____   Reference: _____

480V 3-phase

Balanced 3-phase load

7. In the diagram above, three balanced resistive 100-ampere loads are connected to a 480-volt, 3-phase, 3-wire circuit. The total power of this circuit is _____ kilowatts.
   A. 48
   B. 72
   C. 83
   D. 144

   Answer: _____   Reference: _____

8. A 100-horsepower induction motor is loaded to 30 horsepower. To improve the power factor of this motor, the amount of load should be
   A. increased.
   B. left unchanged because load has no effect.
   C. decreased.
   D. removed to give 0 load.

   Answer: _____   Reference: _____

9. An electrical installation requires a total of 200 feet (61 m) of conductor with a maximum line resistance of 0.5 ohm. The minimum size conductor must have a resistance of no more than _____ ohm(s) per 1000 feet (305 m).
   A. 0.5
   B. 1.0
   C. 2.5
   D. 5.0

   Answer: _____   Reference: _____

10. To get the maximum total resistance using three resistors,
    A. all three resistors should be connected in series.
    B. all three resistors should be connected in parallel.
    C. two resistors should be connected in parallel then one connected in series.
    D. two resistors should be connected in series then one connected in parallel.

    Answer: _____   Reference: _____

11. Four resistance heaters are connected in parallel. Their resistances are heater 1, 20 ohms; heater 2, 30 ohms; heater 3, 60 ohms; and heater 4, 10 ohms. The total resistance of the parallel circuit is _____ ohms.
    A. 5
    B. 12
    C. 50
    D. 120

    Answer: _____   Reference: _____

12. For circuits supplying luminaire (lighting) units having ballasts, transformers, or autotransformers, the calculated load shall be based on the _____ of the luminaire (lighting) units.
    A. total ampere rating
    B. size of the conductors
    C. total wattage of the lamps
    D. voltage rating

    Answer: _____   Reference: _____

13. A building on a blueprint is 16 inches × 10 inches. If the drawing scale is ¼ inch = 1 foot, what is the area of the building in square feet?
    A. 160 square feet
    B. 640 square feet
    C. 2560 square feet
    D. 5120 square feet

    Answer: _____   Reference: _____

14. In the diagram above, each of the supply-side EMT conduits contains three 500-kcmil copper THW service conductors in parallel. As shown, a separate bonding jumper is installed from each conduit to the grounded bus terminal. Each bonding jumper must be at least _____ AWG copper.
    A. 1/0
    B. 2/0
    C. 3/0
    D. 4/0

    Answer: _____   Reference: _____

15. A group of conveyors is operated by individual motors in a manufacturing facility. The total ampacity required to run the group is 58 amperes at 240 volts, single-phase. There are 250 feet of conduit between the main service panel and the subpanel. The voltage drop in the feeder conductors can be a maximum of 3 percent.

    Each conductor in the feeder circuit to supply these conveyors must be at least _____ AWG copper THW. (Use resistance values in the *National Electrical Code®*, Chapter 9, Table 8.)
    A. 2
    B. 3
    C. 4
    D. 6

    Answer: _____   Reference: _____

16. A single-phase, 3-wire service has two ungrounded conductors of 2/0 AWG copper THWN. The neutral conductor is 1 AWG copper THWN. The conduit size needed for the service-entrance conductors is at least trade size _____ (_____).
    A. 1½ (41)
    B. 2 (53)
    C. 2½ (63)
    D. 3½ (91)

    Answer: _____   Reference: _____

17. Two 1 AWG, one 1/0 AWG, and one 2/0 AWG copper conductors, Type THHN are in a 10-foot (3.05-m) conduit. The size of IMC conduit must be at least _____ inches (_____ mm).
    A. 1¼ (35)
    B. 1½ (41)
    C. 2 (53)
    D. 2½ (63)

    Answer: _____   Reference: _____

18. A one-family dwelling contains the following:

    • 1200 square feet of floor space
    • Service: 120/240 volt, single-phase
    • Heat (separate control): two 500-watt, 240-volt baseboard heaters; three 1000-watt, 240-volt baseboard heaters; one 2000-watt, 240-volt baseboard heater
    • Range: one 11,000-watt, 240-volt
    • Dryer: one 6000-watt, 240-volt
    • Water: one 4000-watt, 240-volt water heater

    Using the optional calculation method, each ungrounded conductor in the service for this dwelling has a total net calculated load of _____ amperes.
    A. 146
    B. 120
    C. 99
    D. 84

    Answer: _____   Reference: _____

19. A metallic cold water piping system is available within the structure being supplied by an electrical service with 1/0 AWG THW copper service-entrance conductors. The copper grounding-electrode conductors run to the metal water pipe shall have a minimum size of _____ AWG.
    A. 8
    B. 6
    C. 4
    D. 2

    Answer: _____   Reference: _____

20. Ground-fault protection is required to be installed on a 2000-ampere, solidly grounded, wye service. What is the maximum setting of the ground fault protection?
    A. 800 amperes
    B. 1000 amperes
    C. 1200 amperes
    D. 1600 amperes

    Answer: _____ Reference: _____

21. Overhead service conductors shall be installed so that the minimum clearance from a window opening is _____ feet (_____ mm/m).
    A. 2 (600 mm)
    B. 3 (900 mm)
    C. 6 (1.8 m)
    D. 7 (2.4 m)

    Answer: _____ Reference: _____

22. The 240-volt, single-phase feeder shown in the diagram above supplies a branch-circuit panel that has an improperly balanced load of 57 amperes on one ungrounded conductor and 45 amperes on the other ungrounded conductor. What is the load on the grounded (neutral) conductor?
    A. no load
    B. 12 amperes
    C. 57 amperes
    D. 102 amperes

    Answer: _____ Reference: _____

23. The maximum number of power or lighting conductors that can be installed in a raceway before the derating factors must be applied is
   A. 1.
   B. 3.
   C. 5.
   D. 7.

   Answer: _____   Reference: _____

24. A set of blueprints for an electrical installation calls for a trade size ¾ (21) conduit to be installed as a raceway to enclose a 3-wire circuit consisting of 8 AWG, Type TW conductors. The load to be served is 36 amperes. After the conduit is installed, it is discovered that 10 feet (3 m) of the conduit pass through an area where the ambient temperature is 115° Fahrenheit (115°F). Which of the following actions is an acceptable action that can be taken to correct the condition?
   I.   8 AWG THW conductors can be substituted for the 8 AWG TW conductors.
   II. 6 AWG TW conductors can be substituted for the 8 AWG TW conductors.
   A. I only
   B. II only
   C. Both I and II
   D. Neither I nor II

   Answer: _____   Reference: _____

25. A circuit consisting of three 6 AWG THW insulated copper conductors is run through an area of a building where the temperature is normally 120° Fahrenheit (120°F). Which of the following is the maximum allowable load current for each conductor?
   A. 48.75 amperes
   B. 56.25 amperes
   C. 60.00 amperes
   D. 65.50 amperes

   Answer: _____   Reference: _____

26. A 230-volt, single-phase, 100-ampere circuit is installed in a nonmetallic raceway. Which of the following conditions must apply to the equipment grounding conductor installed with the circuit?
   I.   It must be counted when determining conductor fill of the raceway.
   II. It must be the same size as that of the circuit conductor.
   A. I only
   B. II only
   C. both I and II
   D. neither I nor II

   Answer: _____   Reference: _____

27. Three 4 AWG THWN and four 1/0 AWG THW conductors are to be installed in a single run of IMC conduit. The minimum trade size of conduit permitted is
  A. 1 (27).
  B. 2 (53).
  C. 3 (78).
  D. 4 (103).

  Answer: _____   Reference: _____

28. Eight 4 AWG THWN conductors are to be installed in a single run of rigid nonmetallic conduit. The minimum size of conduit permitted is
  A. 1 (27).
  B. 1½ (41).
  C. 2 (53).
  D. 2½ (63).

  Answer: _____   Reference: _____

29. A municipality has adopted the *National Electrical Code*® without amendments. A conflict occurs about the interpretation of a section of the adopted electrical code. Which of the following is responsible for making the interpretation of the *Code*?
  A. The engineer overseeing the construction
  B. The electrical contractor performing the work
  C. The chief electrical inspector (authority having jurisdiction)
  D. The International Association of Electrical Inspectors

  Answer: _____   Reference: _____

30. A service disconnecting means can be installed at which of the following locations?
  I. Outside a building at a readily accessible point nearest the point of entrance of the service-entrance conductors.
  II. Inside a building at a readily accessible point nearest the point of entrance of the service-entrance conductors.
  A. I only
  B. II only
  C. either I or II
  D. neither I nor II

  Answer: _____   Reference: _____

Conductors crossing and connecting

Figure 1   Figure 2

31. Figure 1 and Figure 2 shown above are standard electrical symbols used on blueprints representing crossing conductors. Which symbol represents two conductors crossing but not connecting?
    A. Figure 1 only
    B. Figure 2 only
    C. Figure 1 and Figure 2
    D. Neither Figure 1 nor Figure 2

    Answer: _____   Reference: _____

32. The following figure is a standard symbol used on blueprints. The symbol represents which of the following?
    A. an air circuit breaker
    B. a lightning arrester
    C. a fuse
    D. a thermal element

    Answer: _____   Reference: _____

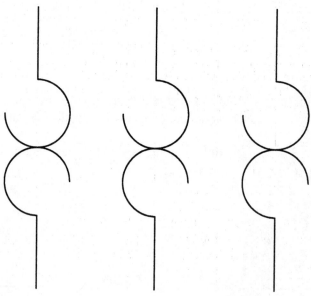

33. An electrical contractor installs the wiring in a new 1400-square-foot single-family dwelling unit before a permit is obtained from the city electrical inspector. The electrical inspector stops the contractor from continuing the job. The city where the structure is located has adopted an electrical code that contains the following provisions:

Permits for New Construction

1. One- and two-family dwelling: On all new single- and two-family dwelling construction, the electrical fee shall be as follows: $.02/square feet under roof.
2. All other

Penalty for Failure to Obtain Permit

In case it shall be discovered that any electrical work has been installed or put into use for which no permit has been issued, the violator shall pay a fee equal to three times the permit fee that shall have been paid for work done in violation thereof and no additional permits shall be granted until all fees have been paid.

Before the electrical contractor can complete the work, a permit fee must be paid. Which of the following is the minimum permit fee required?
A. $2.80
B. $28.00
C. $56.00
D. $84.00

Answer: _____ Reference: _____

34. A service disconnect is supplied by conductors having 38,500 amperes RMS fault current available at the supply terminals of the disconnect. Which of the following statements is correct?
A. Only the overcurrent device has to have an interrupting rating at or above the 38,500 amperes.
B. The overcurrent device and the panel must be rated for the maximum fault current available.
C. The overcurrent device must be rated at 200,000 amperes RMS.
D. Neither the panel nor the overcurrent device has to be rated with fault-current interrupting rating.

Answer: _____ Reference: _____

35. A motor is protected against short-circuit and ground-fault by an adjustable instantaneous trip circuit breaker that is part of a combination controller having motor overload and short-circuit and ground-fault protection in each conductor. The setting of the instantaneous trip breaker shall be permitted to exceed the motor full-load current by not more than _____ percent.
A. 250
B. 1300
C. 700
D. 1000

Answer: _____ Reference: _____

36. Thermal overload relays are used for the protection of polyphase induction motors. Their primary purpose is to protect the motor in case of
A. reversal of phases in the supply.
B. low line voltage.
C. short-circuit between phases.
D. sustained overload.

Answer: _____ Reference: _____

37. A motor controller and motor branch-circuit disconnecting means a 2300-volt motor shall have a continuous ampere rating of not less than the
   A. trip setting of the short-circuit protective device rating.
   B. trip setting of the overload protection device.
   C. trip setting of the fault-current protection device.
   D. locked-rotor rating of the motor.

   Answer: _____ Reference: _____

38. The disconnecting means for both a 2300-volt motor and the controller shall be
   A. permitted within the same enclosure as the controller.
   B. located separately from the controller enclosure.
   C. located at the service equipment.
   D. permitted within the watt-hour meter enclosure.

   Answer: _____ Reference: _____

39. Speed-limiting devices shall be provided with which of the following?
   A. polyphase squirrel-cage motors
   B. synchronous motors
   C. compound motors
   D. series motors

   Answer: _____ Reference: _____

40. Which of the following appliances can be grounded to the grounded (neutral) conductor in an existing dwelling?
   A. electric water heater
   B. kitchen disposal
   C. dishwasher
   D. electric dryer

   Answer: _____ Reference: _____

41. A metal luminaire (lighting fixture) shall be grounded if located
   A. 10 feet (3 m) vertically or 6 feet (1.8 m) horizontally from a kitchen sink.
   B. 8 feet (2.4 m) vertically or 5 feet (1.5 m) horizontally from a kitchen sink.
   C. 6 feet (1.8 m) vertically or 3 feet (900 mm) horizontally from a kitchen sink.
   D. 8 feet (2.4 m) vertically or 3 feet (900 mm) horizontally from a kitchen sink.

   Answer: _____ Reference: _____

42. When used, a driven ground rod shall be installed so that the soil will be in contact with a length of the rod not less than
   A. 4 feet (1.2 m).
   B. 6 feet (1.8 m).
   C. 8 feet (2.4 m).
   D. 10 feet (3 m).

   Answer: _____ Reference: _____

43. Portable and stationary electrically heated appliances having metal frames are connected to an electrical system operating at 277 volts to ground. Which of the following statements about the metal frames of each appliance is (are) correct?
    A. It must be grounded.
    B. It must be grounded only if supplied by a portable cord.
    C. It can be permanently insulated from ground by special permission.
    D. It is connected by the manufacturer to the grounded conductor thermal.

    Answer: _____   Reference: _____

44. Which of the following most accurately describes the condition of a motor known as "locked-rotor?"
    A. When the electrician places a lock on the motor controller to keep the motor from being energized
    B. Mechanical brakes used on the motor shaft to stop the motor during shutdown
    C. An electronic control device used to lock the speed of the motor at that specified by the manufacturer
    D. When the circuits of a motor are energized but the rotor is not turning

    Answer: _____   Reference: _____

45. The following diagram represents overhead conductors between two buildings on an industrial site. The voltage is 240/480 volts ac. The conductors pass over a driveway leading to a loading dock at one of the buildings. What is the minimum vertical clearance permitted between the overhead conductors and the driveway?

Minimum clearance above industrial driveway

To warehouse loading dock

    A. 10 feet (3 m)
    B. 12 feet (3.7 m)
    C. 15 feet (4.5 m)
    D. 18 feet (5.5 m)

    Answer: _____   Reference: _____

46. When a ground-fault protection for equipment is installed within service equipment, it shall be performance tested
    A. at the factory before shipment.
    B. before being installed on-site.
    C. when first installed on-site.
    D. after the electrical system has been used for one week.

    Answer: _____   Reference: _____

47. Three continuous-duty motors with full-load current ratings of 5.6 amperes, 4.5 amperes, and 4.5 amperes, respectively, are to be installed on a single branch-circuit. The circuit conductors shall have a minimum ampacity of _____ amperes.
    A. 16
    B. 20
    C. 24
    D. 30

    Answer: _____   Reference: _____

48. Which of the following statements about the number of electrical services to a building is (are) correct?
    I. A service supplied by a wind-powered generator can be installed on a building in addition to the service supplied by the local electrical utility.
    II. A service supplied by a solar photovoltaic system can be installed on a building in addition to the service supplied by the local electrical utility.
    A. I only
    B. II only
    C. Both I and II
    D. Neither I nor II

    Answer: _____   Reference: _____

49. Which of the following statements about the termination and bonding of conductors at service equipment is (are) true?
    I. The grounded service conductor and the equipment grounding conductor shall be bonded together within the service equipment.
    II. The grounding electrode conductor and the system grounded conductor shall be bonded together within the service equipment.
    A. I only
    B. II only
    C. Both I and II
    D. Neither I nor II

    Answer: _____   Reference: _____

50. When a universal series ac motor is caused to run without being connected to a load, the motor will
    A. run at a constant speed for an indefinite time.
    B. vary in running speed from about 80 percent to 125 percent of the normal rated speed.
    C. decrease in running speed until a locked-rotor condition occurs.
    D. increase in running speed to a dangerous level that can damage the motor.

    Answer: _____   Reference: _____

51. The following diagram represents an electrical service with a fused switch as the main disconnect. What is the maximum height the center of the grip of the operating handle of the switch is permitted to be located above the ground?

A. 5 feet (1.5 m)
B. 6 feet, 7 inches (2 m)
C. 7 feet (2.1 m)
D. ½ foot (150 mm)

Answer: _____   Reference: _____

52. When fuses are used for motor overload protection for a 3-wire, 3-phase motor, the fuse shall protect
    A. each ungrounded conductor.
    B. only two ungrounded conductors.
    C. only one ungrounded conductor.
    D. the equipment grounding conductor.

Answer: _____   Reference: _____

53. How is the size of an electrical conductor supplying a circuit determined?
    A. voltage
    B. amperage
    C. the length
    D. all of the above

    Answer: _____   Reference: _____

54. Ohm's law is
    A. measurement of the I²R losses.
    B. the relationship between voltage, current, and power.
    C. an equation for determining power.
    D. a relationship between voltage, current, and resistance.

    Answer: _____   Reference: _____

55. Electrical pressure is measured in
    A. volts.
    B. amperes.
    C. coulombs.
    D. watts.

    Answer: _____   Reference: _____

56. The resistance of a 1500-watt, 120-volt resistance heater element is
    A. 14.4 ohms.
    B. 9.6 ohms.
    C. 11.2 ohms.
    D. 12.5 ohms.

    Answer: _____   Reference: _____

57. Total ampacity of stranded or solid conductors is the same if they have the same
    A. diameter.
    B. circumference.
    C. cross-sectional area.
    D. insulation.

    Answer: _____   Reference: _____

58. Two 6-volt batteries connected in parallel will give
    A. longer service than one battery.
    B. 12 volts.
    C. higher currents.
    D. lower currents.

    Answer: _____   Reference: _____

59. A branch-circuit that supplies a number of outlets for lighting and appliances is called a _____ branch-circuit.
    A. general-purpose
    B. utility
    C. multipurpose
    D. none of the above

    Answer: _____   Reference: _____

60. An electric heater will produce less heat on low voltage because
    A. its total watt output decreases.
    B. the current will decrease.
    C. the resistance does not change.
    D. all of the above.

    Answer: _____   Reference: _____

61. A 10-ohm resistance carrying 10 amperes of current uses _____ watts of power.
    A. 100
    B. 200
    C. 500
    D. 1000

    Answer: _____   Reference: _____

62. A _____ stores energy in much the same manner as a spring stores mechanical energy.
    A. coil
    B. capacitor
    C. resistor
    D. none of the above

    Answer: _____   Reference: _____

63. More heat is created when current flows through which of the following?
    A. 10-ohm capacitor
    B. 10-ohm inductance coil
    C. 10-ohm resistor
    D. all of the above are equal

    Answer: _____   Reference: _____

64. _____ means that it is constructed or protected so that exposure to the weather will not interfere with its successful operation.
    A. Weatherproof
    B. Weathertight
    C. Weather resistant
    D. all of the above

    Answer: _____   Reference: _____

65. The service conductor clearance from windows can be less than 3 feet if it is
    A. run along the left side of the window.
    B. run above the top level of the window.
    C. run below the bottom level of the window.
    D. none of the above

    Answer: _____   Reference: _____

66. The *NEC*® states that electrical equipment shall be installed
    A. not exceeding the provisions of the *Code*.
    B. not less than the *Code* permits.
    C. according to the *Code* and local code amendments.
    D. none of the above

    Answer: _____   Reference: _____

67. The equivalent resistance of three resistors of 8 ohms, 8 ohms, and 4 ohms that are connected in parallel is
    A. 2 ohms.
    B. 4 ohms.
    C. 6 ohms.
    D. 8 ohms.

    Answer: _____   Reference: _____

68. Because fuses are rated by both amperage and voltage, a fuse will operate correctly on
    A. ac only.
    B. ac or dc.
    C. dc only.
    D. any voltage.

    Answer: _____   Reference: _____

69. A 10-horsepower, 480-volt, 3-phase motor operating at 80 percent efficiency has an input current of _____ amperes. (1 hp = 746 watts)
    A. 11.2
    B. 21.5
    C. 33.3
    D. 42

    Answer: _____   Reference: _____

70. A branch-circuit supplying a detached garage requires a total of 380 feet (116 m) of circuit conductor with a maximum conductor resistance of 0.8 ohm. The minimum size conductor must have a resistance of no more than _____ ohms per 1000 feet (305 m).
    A. 1.6
    B. 2.1
    C. 3.4
    D. 4.8

    Answer: _____   Reference: _____

71. An electric resistance heater is rated for 2400 watts at 240 volts. What power is consumed when the heater is operated at 120 volts?
    A. 600 watts
    B. 1200 watts
    C. 2400 watts
    D. 4800 watts

    Answer: _____   Reference: _____

72. A 230-volt, single-phase circuit has 10 kilowatts of load and 50 amperes of current. The power factor is _____ percent.
    A. 78
    B. 87
    C. 94
    D. 115

    Answer: _____   Reference: _____

73. A 1½-horsepower single-phase motor that has an efficiency of 80 percent, operates at 230 volts, and has an input current of _____ amperes. (1 hp = 746 watts)
    A. 1.3
    B. 3.2
    C. 6.1
    D. 6.8

    Answer: _____   Reference: _____

74. A 30-horsepower, wound-rotor induction motor with no code letter is to be installed with 460-volt, 3-phase, alternating current. Disregarding all the exceptions, the nontime-delay fuse for short-circuit protection of the motor branch-circuit must be rated at a maximum _____ amperes.
    A. 40
    B. 50
    C. 60
    D. 80

    Answer: _____   Reference: _____

75. The following 480-volt, 3-phase, 3-wire, intermittent use equipment is in a commercial kitchen: two 5000-watt water heaters; four 3000-watt fryers; and two 6000-watt ovens. Each ungrounded conductor in the feeder circuit for this kitchen equipment must be sized to carry a minimum calculated load of _____ amperes.
    A. 15
    B. 27
    C. 41
    D. 46

    Answer: _____   Reference: _____

Appendix 1: Symbols

(Courtesy of American Iron and Steel Institute)

Electrical Wiring Symbols

Selected from American National Standard Graphic for
Electrical Wiring and Layout Diagrams Used in Architecture and Building Construction

ANSI Y32.9-1972

1. Lighting Outlets

Ceiling　　　　　　　　　*Wall*

1.1 Surface or pendant incandescent, mercury-vapor, or similar lamp fixture

1.2 Recessed incandescent, mercury-vapor, or similar lamp fixture

1.3 Surface of pendant individual fluorescent fixture

1.4 Recessed individual fluorescent fixture

1.5 Surface or pendant continuous-row fluorescent fixture

1.6 Recessed continuous-row fluorescent fixture

1.8 Surface or pendant exit light

1.9 Recessed exit light

1.10 Blanked outlet

1.11 Junction box

1.12 Outlet controlled by low-voltage switching when relay is installed in outlet box

2. Receptacle Outlets

Grounded　　　　　　　　*Ungrounded*

2.1 Single receptacle outlet

2.2 Duplex receptacle outlet

2.3 Triplex receptacle outlet

2.4 Quadrex receptacle outlet

2.5 Duplex receptacle outlet – split wired

2.6 Triplex receptacle outlet – split wired

2.7 Single special-purpose receptacle outlet – split wired

NOTE 2.7A: Use numeral or letter as a subscript alongside the symbol, keyed to explanation in the drawing list of symbols, to indicate type of receptacle or use

2.8 Duplex special-purpose receptacle outlet
See note 2.7A

2.9 Range outlet (typical)
See note 2.7A

2.10 Special-purpose connection or provision for connection
Use subscript letters to indicate function
(SW – dishwasher; CD – clothes dryer, etc).

DW   UNG
            DW

2.12 Clock hanger receptacle

ⓒ   ⓒ UNG

2.13 Fan hanger receptacle

Ⓕ   Ⓕ UNG

2.14 Floor single receptacle outlet

⊟   ⊟ UNG

2.15 Floor duplex receptacle outlet

⊟   ⊟ UNG

2.16 Floor special-purpose outlet
See note 2.7A

          UNG

2.17 Floor telephone outlet – public

◀

2.18 Floor telephone outlet – private

◁

2.19 Underfloor duct and junction box for triple, double, or single duct system (as indicated by the number of parallel lines)

2.20 Cellular floor header duct

3. Switch Outlets

3.1 Single-pole switch

S

3.2 Double-pole switch

S2

3.3 Three-way switch

S3

3.4 Four-way switch

S4

3.5 Key-operated switch

SK

3.6 Switch and pilot lamp

SP

3.7 Switch for low-voltage switching system

SL

3.8 Master switch for low-voltage switching system

SLM

3.9 Switch and single receptacle

─⊖ S

3.10 Switch and double receptacle

═⊖ S

3.11 Door switch

SD

3.12 Time switch

ST

3.13 Circuit breaker switch

SCB

3.14 Momentary contact switch or pushbutton
for other than signaling system

SMC

3.15 Ceiling pull switch

(S)

5.13 Radio outlet

R

5.14 Television outlet

TV

6. Panelboards, switchboards, and related equipment

6.1 Flush-mounted panelboard and cabinet
NOTE 6.1A: Identify by notation or schedule

6.2 Surface-mounted panelboard and cabinet
See note 6.1A

6.3 Switchboard, power control center, unit
substations (should be drawn to scale)
See note 6.1A

6.4 Flush-mounted terminal cabinet
See note 6.1A

NOTE 6.4A: In small-scale drawings the TC may be
indicated alongside the symbol

TC

6.5 Surface-mounted terminal cabinet
See note 6.1A and 6.4A

TC

6.6 Pull box
Identify in relation to wiring system section and size

6.7 Motor or other power controller
See note 6.1A

MC

6.8 Externally operated disconnection switch
See note 6.1A

6.9 Combination controller and disconnection
means
See note 6.1A

7. Bus Ducts and Wireways

7.1 Trolley duct
See note 6.1A

T		T		T

7.2 Busway (service, feeder, or plug-in)
See note 6.1A

B		B		B

7.3 Cable through, ladder, or channel
See note 6.1A

BP		BP		BP

7.4 Wireway
See note 6.1A

W		W		W

9. Circuiting

Wiring method identification by notation on
drawing or in specifications.

9.1 Wiring concealed in ceiling or wall

NOTE 9.1A: Use heavyweight line to identify
service and feed runs.

9.2 Wiring concealed in floor
See note 9.1A

Appendix 2: Basic Electrical Formulas

Basic Electrical Formulas
(Courtesy of American Iron and Steel Institute)

DC Circuit Characteristics

Ohm's Law:

$$E = IR \quad I = \frac{E}{R} \quad R = \frac{E}{I}$$

E = voltage impressed on circuit (volts)

I = current flowing in circuit (amperes)

R = circuit resistance (ohms)

Resistances in Series:

$$R_t = R_1 + R_2 + R_3 + \ldots$$

R_T = total resistance (ohms)

R_1, R_2 etc. = individual resistances (ohms)

Resistances in Parallel:

$$R_t = \frac{1}{\frac{1}{R_1} + \frac{1}{R_2} + \frac{1}{R_3} + \ldots}$$

Formulas for the conversion of electrical and mechanical power:

$$HP = \frac{watts}{746} (watts \times .00134)$$

$$HP = \frac{kilowatts}{.746} (kilowatts \times 1.34)$$

Kilowatts = HP × .746

Watts = HP × 746

HP = Horsepower

In direct-current circuits, electrical power is equal to the product of the voltage and current:

$$P = EI = I_2 R = \frac{E_2}{R}$$

P = power (watts)

E = voltage (volts)

I = current (amperes)

R = resistance (ohms)

Solving the basic formula for I, E, and R gives

$$I = \frac{P}{E} = \sqrt{\frac{P}{R}}; \quad E = \frac{P}{I} = \sqrt{RP}; R = \frac{E2}{P} = \frac{P}{T^2}$$

Energy

Energy is the capacity for doing work. Electrical energy is expressed in kilowatt-hours (kWhr), one kilowatt-hour representing the energy expended by a power source of 1 kW over a period of 1 hour.

Efficiency

Efficiency of a machine, motor or other device is the ratio of the energy output (useful energy delivered by the machine) to the energy input (energy delivered to the machine), usually expressed as a percentage:

$$\text{Efficiency} = \frac{output}{input} \times 100\%$$

$$\text{or Output} = \text{Input} \times \frac{efficiency}{100\%}$$

Torque

Torque may be described as a force tending to cause a body to rotate. It is expressed in pound-feet or pounds of force acting at a certain radius:

Torque (pound-feet) = force tending to produce rotation (pounds) × distance from center of rotation to point at which force is applied (feet).

Relations between torque and horsepower:

$$\text{Torque} = \frac{33,000 \times HP}{6.28 \times rpm}$$

$$HP = \frac{6.28 \times rpm \ time \ torque}{33,000}$$

rpm = speed of rotating part (revolutions per minute)

AC Circuit Characteristics

The instantaneous values of an alternating current or voltage vary from zero to maximum value each half cycle. In the practical formula that follows, the "effective value" of current and voltage is used, defined as follows:

Effective value = 0.707 × maximum instantaneous value

Inductances in Series and Parallel:

The resulting circuit inductance of several inductances in series or parallel is determined exactly as the sum of resistances in series or parallel as described under dc circuit characteristics.

Impedance:

Impedance is the total opposition to the flow of alternating current. It is a function of resistance, capacitive reactance and inductive reactance. The following formulae relate these circuit properties:

$$X_L = 2\pi Hz L X_c = \frac{1}{2\pi Hz C} Z = \sqrt{R^2 + (X_L - X_C^2)}$$

X_L = inductive reactance (ohms)

X_c = capacitive reactance (ohms)

Z = impedance (ohms)

Hz - (Hertz) cycles per second

C = capacitance (farads)

L - inductance (henrys)

R - resistance (ohms)

$\pi = 3.14$

In circuits where one or more of the properties L, C, or R is absent, the impedance formula is simplified as follows:

Resistance only: $Z = R$

Inductance only: $Z = X_L$

Capacitance only: $Z = XC$

Resistance and Inductance only: $Z = \sqrt{R^2 + X_L^2}$

Resistance and Capacitance only: $Z = \sqrt{R^2 + X_C^2}$

Inductance and Capacitance only: $Z = \sqrt{X_L - X_C^2}$

Ohm's law for AC circuits:

$E = 1 \times Z$ $I = \dfrac{E}{Z}$ $Z = \dfrac{E}{I}$

Capacitances in Parallel:

$C_t = C_1 + C_2\ C_3 + ...$

C_t = total capacitance (farads)

$C_1 C_2 C_3 ...$ = individual cpacitances (farads)

Capacitances in series:

$C_t = \dfrac{1}{\dfrac{1}{C_1} + \dfrac{1}{C_2} + \dfrac{1}{C_3} + ...}$

Phase Angle

An alternating current through an inductance lags the voltage across the inductance by an angle calculated as follows:

Tangent of angle of lag = $\dfrac{X_L}{R}$

An alternating current through a capacitance leads the voltage across the capacitance by an angle calculated as follows:

Tangent of angle of lead = $\dfrac{X_C}{R}$

The resultant angle by which a current leads or lags the voltage in an entire circuit is called the phase angle and is calculated as follows:

Cosine of phase angle = $\dfrac{R \text{ of circuit}}{Z \text{ of circuit}}$

Power Factor

Power factor of a circuit or system is the ratio of actual power (watts) to apparent power (volt-amperes), and is equal to the cosine of the phase angle of the circuit:

$PF = \dfrac{\text{actual power}}{\text{apparent power}} = \dfrac{\text{watts}}{\text{volts} \times \text{amperes}} = \dfrac{kW}{kVA} = \dfrac{R}{Z}$

KW = kilowatts

kVA = kilowatt-amperes = volt-amperes + 1,000

PF = power factor (expressed as decimal or percent)

Single-Phase Circuits

$kVA = \dfrac{EI}{1,000} = \dfrac{kW}{PF}$ $kW = kVA \times PF$

$I = \dfrac{P}{E \times PF}$ $E = \dfrac{P}{I \times PF}$ $PF = \dfrac{P}{E \times I}$

$P = E \times I \times PF$

P = power (watts)

Two-Phase Circuits

$I = \dfrac{P}{2 \times E \times PF}$ $E = \dfrac{P}{2 \times I \times PF}$ $PF = \dfrac{P}{E \times I}$

$kVA = \dfrac{2 \times E \times I}{1000} = \dfrac{kW}{PF}$ $kW = kVA \times PF$

$P = 2 \times E \times 1 \times PF$

E = phase voltage (volts

Three-Phase Circuits, Balanced Star or Wye

$I_N = O$ $I = I_P$ $E = \sqrt{3} E_P = 1.73 E_P$

$E_P = \dfrac{E}{\sqrt{3}} = \dfrac{E}{1.73} = 0.577E$

I_N = current in neutral (amperes)

I = line current per phase (amperes)

I_P = current in each phase winding (amperes)

E = voltage, phase to phase (volts)

E_P = voltage, phase to neutral (volts)

Three-Phase Circuits, Balanced Delta

$I = 1.732 \times I_P$ $I_P = \dfrac{1}{\sqrt{3}} = 0.577 \times I$

$E = E_P$

Power:

Balanced 3-Wire, 3-Phase Circuit, Delta or Wye

For unit power factor (PF = 1.0):

$P = 1.732 \times E \times I$

$I = \dfrac{P}{\sqrt{3}}$ $E = 0.577\dfrac{P}{E}$ $E = \dfrac{P}{\sqrt{3}} \times I = 0.577\dfrac{P}{I}$

P = total power (watts)

For any load:

$P = 1.732 \times E \times I \times PF$ $VA = 1.732 \times E \times I$

$E = \dfrac{P}{PF \times 1.73 \times I} = 0.577 \times \dfrac{P}{PF \times I}$

$I = \dfrac{P}{PF \times 1.73 \times E} = 0.577 \times \dfrac{P}{I} \times E$

$PF = \dfrac{P}{1.73 \times I \times E} = \dfrac{0.577 \times P}{I \times E}$

VA = apparent power (volt-amperes)

P = actual power (watts)

E = line voltage (volts)

I = line current (amperes)

Power Loss:

Any ac or dc Circuit

$P = I_2 R I = \sqrt{\dfrac{P}{R}} R = \dfrac{P}{I^2}$

P = power heat loss in circuit (watts)

I = effective current in conductor (amperes)

R = conductor resistance (ohms)

Load Calculations

Branch Circuits—Lighting & Appliance 2-Wire:

$$I = \frac{\text{total connected load (watts)}}{\text{line voltage (volts)}}$$

I = current load on conductor (amperes)

3-Wire:

Apply same formula as for 2–wire branch circuit, considering each line to neutral separately. Use line-to-neutral voltage; result gives current in line conductors.

USEFUL FORMULAS

TO FIND	SINGLE PHASE	THREE PHASE	DIRECT CURRENT
AMPERES when kVA is known	$\frac{\text{kVA} \times 1000}{E}$	$\frac{\text{kVA} \times 1000}{E \times 1.73}$	not applicable
AMPERES when horsepower is known	$\frac{\text{HP} \times 746}{E \times \% \text{ eff.} \times \text{pf}}$	$\frac{\text{HP} \times 746}{E \times 1.73 \times \% \text{ eff.} \times \text{pf}}$	$\frac{\text{HP} \times 746}{E \times \% \text{ eff.}}$
AMPERES when kilowatts are known	$\frac{\text{kW} \times 1000}{E \times \text{pf}}$	$\frac{\text{kW} \times 1000}{E \times 1.73 \times \text{pf}}$	$\frac{\text{kW} \times 1000}{E}$
HORSEPOWER	$\frac{I \times E \times \% \text{ eff.} \times \text{pf}}{746}$	$\frac{I \times E \; 1.73 \times \% \text{ eff.} \times \text{pf}}{746}$	$\frac{I \times E \times \% \text{ eff.}}{746}$
KILOVOLT AMPERES	$\frac{I \times E}{1000}$	$\frac{I \times E \times 1.73}{1000}$	not applicable
KILOWATTS	$\frac{I \times E \times \text{pf}}{1000}$	$\frac{I \times E \times 1.73 \times \text{pf}}{1000}$	$\frac{I \times E}{1000}$
WATTS	$E \times I \times \text{pf}$	$E \times I \times 1.73 \times \text{pf}$	$E \times I$

$$\text{ENERGY EFFICIENCY} = \frac{\text{Load Horsepower} \times 746}{\text{Load Input kVA} \times 1000}$$

$$\text{POWER FACTOR (pf)} = \frac{\text{Power Consumed}}{\text{Apparent Power}} = \frac{W}{VA} = \frac{kW}{kVA} = \cos\varnothing$$

I = Amperes	E = Volts	kW = Kilowatts	kVA = Kilovolt-amperes
HP = Horsepower	% eff. = Percent Efficiency		pf = Power Factor
	e.g., 90% eff. is 0.90		e.g., 95% pf is 0.95

EQUATIONS BASED ON OHM'S LAW:

P = POWER, IN WATTS
I = CURRENT, IN AMPERES
R = RESISTANCE, IN OHMS
E = ELECTROMOTIVE FORCE, IN VOLTS

Appendix 3: Answer Key

SAMPLE EXAM QUESTION REVIEW

1. A. Ref.—Article 350.10
2. B. Ref.—352.26
3. C. Ref.—360.20(B)
4. C. Ref.—314.28(A)(2)
5. B. Ref.—300.6(C)
6. B. Ref.—368.30
7. C. Ref.—Chapter 9, Table 1
8. B. Ref.—400.13 and 240.5(A)
9. D. Ref.—310.14
10. D. Ref.—300.4(F)
11. D. Ref.—334.30
12. A. Ref.—Table 310.15(B)(6)
13. D. Ref.—300.5(D)
14. C. Ref.—378.22 and 378.12
15. D. Ref.—Table 310.15(B)(6)
16. D. Ref.—280.25
17. C. Ref.—408.15
18. B. Ref.—230.54(C) Exception
19. A. Ref.—230.31(B) Exception
20. B. Ref.—230.9(A) Exception
21. C. Ref.—230.24(A)
22. D. Ref.—240.83(C)
23. D. Ref.—110.26(D)
24. B. Ref.—230.95
25. B. Ref.—Knowledge
26. D. Ref.—General knowledge; Appendix 2
27. B. Ref.—Table 430.37
28. D. Ref.—Knowledge
29. D. Ref.—430.24 and Table 430.150, Table 310.16
30. B. Ref.—430.22 and Table 430.150, Table 310.16
31. C. Ref.—430.22, Table 430.150, Table 310.16
32. A. Ref.—430.81(B)
33. B. Ref.—404.8(B)
34. C. Ref.—450.43(A)
35. A. Ref.—300.7(A)

36. B. Ref.—410.66(A)
37. C. Ref.—410.76(B)
38. A. Ref.—700.12(B)(2)
39. C. Ref.—520.25(A) and (C)
40. D. Ref.—501.15(A)
41. A. Ref.—500.5(C)
42. D. Ref.—511.1(4)(A)
43. B. Ref.—517.18(B)
44. C. Ref.—675.15, 250.106 (FPN) 2
45. A. Ref.—440.61, 440.3, and 250.114(3)(A)
46. B. Ref.—250.66(A)
47. D. Ref.—250.52(A)
48. C. Ref.—Knowledge; Appendix 2
49. A. Ref.—210.52, 210.11(C)
50. B. Ref.—422.13 and 422.10
51. C. Ref.—422.11(E)
52. C. Ref.—Chapter 3. P = EI (Appendix 2)
53. A. Ref.—Chapter 3; Appendix 2
54. B. Ref.—230.95(A)
55. C. Ref.—Chapter 9, Tables 1, 4, 5
56. B. Ref.—430.24, Table 310.16
57. A. Ref.—430.62(A) and Table 430.152
58. C. Ref.—424.3(B)
59. B. Ref.—Chapter 3, this text
60. C. Ref.—424.3(B)
61. D. Ref.—90.2(B)
62. C. Ref.—General knowledge
63. C. Ref.—Chapter 2 text
64. B. Ref.—392.8(C)
65. B. Ref.—312.2(A)
66. C. Ref.—344.30(A)
67. B. Ref.—Table 314.16(A)
68. D. Ref.—Table 430.91
69. D. Ref.—410.73(F)
70. A. Ref.—450.47
71. B. Ref.—450.13(2)
72. C. Ref.—460.2
73. C. Ref.—440.32, Table 310.16
74. D. Ref.—Table 430.52
75. D. Ref.—450.21(A) and (B)

QUESTION REVIEW CHAPTER 2

1. No. However, there is an exception: 356.12 for electrical signs. Ref.—356.12 Exception

2. Governmental bodies exercising legal jurisdiction over the electrical installation, insurance inspectors, or others with the authority and responsibility for governing the electrical installation. Ref.—90.4

3. Ships, railway rolling stock, aircraft, or automotive vehicles. Ref.—90.2(B). Four installations are not covered, for example, see 90.2(B)(1)–(5).

4. Fine print notes (FPN). Ref.—90.5(C)

5. 110.3(B)

6. When adopted by the regulatory authority over the intended use. Ref.—Pages 70-i, which states that the *Code* is advisory as far as NFDA and ANSI are concerned but is offered for use in law for regulatory purposes.

7. No. It is the intent that all premise wiring or other wiring of the utility-owned meter read equipment on the load side of service point of buildings, structures, or other premises not owned or leased by the utility are covered. It also is the intent that buildings used by the utility for purposes other than listed in 90.2(B)(5), such as office buildings, warehouses, machine shops, or recreational buildings, that are not an integral part of the generating plant, substation, or control center are also covered. Ref.—90.2(B)(5)

8. No, there has been a change in this edition of the *NEC®* from the wording of the previous edition.

9. (A) 2005, (B) September 2004.

10. 1881, The National Fire Engineers met in Richmond, Virginia, which resulted in the first document that led to the *National Electrical Code®*. Various meetings were held after that; then in 1897 some various allied interests of the church, electrical, and architectural concerns met, and in 1911 NFPA was set up as the sponsor.

11. Yes. Ref.—Article 590, 590.4(C)

12. Chapter 8. Ref.—90.3

13. No, it is not a training manual for untrained persons. Ref.—90.1(C)

14. No, only underground mines and rolling stock for surface mines. All surface mining other than the rolling stock are covered by the *NEC®*. Ref.—90.2(B)(2)

15. Nowhere in the *Code* does it state that it is the minimum allowed. Only when adopted by a local jurisdiction would it be considered the minimum of that jurisdiction. When an installation is installed for the *National Electrical Code®*, it provides a safeguarding of persons and property, and the provisions are considered necessary for safety and compliance with the *NEC®* will result in an installation essentially free from hazards. However, it may not be efficient or convenient or adequate in some installations. Ref.—90.1(A) and (B)

16. Definitions, Article 100, to continue for three hours or more

17. 1999. *The National Electrical Code®* is adopted every three years; there being a 1999, a 2002, and a 2005. Therefore, not adopting the two editions, they would be using the 1999 *NEC®*.

18. Article 725 covers Class 2 wiring.

19. Article 517—Health Care Facilities, 517.2, Definitions

20. 250.52(A)(2). The water pipe must be used if available, and supplemented with another grounding electrode of one of the types listed in 250.52 (A)(1) through 250.52(A)(7).

21. Article 422 lists specific requirements for specific appliances. Ref.—422.16(B)

22. Yes. Ref.—Article 770

23. Phase converters are covered in *NEC®* Article 455. Both static and rotary types are covered.

24. Article 426 covers fixed outdoor electrical deicing- and snow-melting equipment.

25. Article 680 for swimming pools, fountains, and similar installations. Fine print note to 680.1 clearly states that a fountain covers reflective pools, ornamental pools, and display pools. However, it does not cover a drinking fountain. An argument could be made that the new Article 682 covers this installation; however, this is not clear.

CHAPTER 3—PRACTICE EXAM

1. C
2. D
3. D
4. C
5. C

6. A

7. D

8. B

9. C

10. B

11. C

12. A

13. B

14. C

15. A

QUESTION REVIEW CHAPTER 3

(Reduce fractions to their lowest terms.)

1. ½

2. ¼

3. ¹⁄₁₀

4. ⅖

5. ¾

6. ¾

7. ½

8. ⅗

9. ⅗

10. ⅕

11. ¼

12. ¼

(Change mixed numbers to improper fractions.)

1. ³⁄₂

2. ¹³⁄₄

3. ⁴³⁄₈

4. ¹⁷⁄₄

5. ⁵⁷⁄₈

6. ¹³⁄₆

7. ²³⁄₄

8. ²⁵⁄₃

9. ⁴⁶⁄₇

10. ²⁹⁄₈

(Change improper fractions to mixed numbers.)

1. 4½

2. 2⅖

3. 12⅘

4. 3¼

5. 9⅔

6. 15½

7. 1⅔

8. 4⅓

(Multiply whole numbers and fractions. Answer in proper reduced form.)

1. 4

2. ⅜

3. ¹⁄₄₀

4. ⁹⁄₂ = 4½

5. ²¹⁄₄ = 5¼

6. ⁵⁄₂ = 2½

7. 9

8. ¹⁵⁄₇ = 2¹⁄₇

(Convert fractions to decimal equivalents.)

1. .25

2. .625

3. .75

4. .866

5. .375

(Resolve to whole numbers.)

1. 16

2. 81

3. 2500

4. 1

5. 1000

(Find the square root.)

1. 5

2. 9

3. 7

4. 56

5. 101

6. 29

7. 78

(Solve the following problems.)

1. .05 or 5%. Ref.—Ohm's law; Appendix 2

2. 100%. Ref.—Chapter 3

3. 5 amperes. Ref.—Chapter 3; Ohm's law; Appendix 2

4. 1.33 ohms. Ref.—Chapter 3

5. 14.4 ohms. Ref.—Chapter 3

6. 6 volts. Ref.—Chapter 3

7. 33.34 ohms

8. 83.34 ohms

9. 23.81 ohms

10. 107.15 ohms

11. 37 volts

12. 56 volts

13. 27 volts

14. 1.12 amperes

QUESTION REVIEW CHAPTER 4

1. Yes. Ref.—90.2(A). A floating restaurant must be wired as a typical commercial restaurant.

2. Chapter 8. Article 250 does not apply in Chapter 8 unless specifically referenced.

3. Article 406

4. False. The rules for calculating branch-circuits are referenced in many articles in the *NEC®*.

5. Article 516. Luminaires (fixtures) installed in a booth must be type Class I, Division 1, generally.

6. Article 422 for appliances, specifically 422.2. Electrical heating is covered in Article 424 and heat pumps in Article 440.

7. Chapter 9, Tables 1 through 8 and all Notes

8. Lighting and power in mine shafts are not covered by the *NEC®*, Article 90, 90.2(B)

9. 300.50(A)

10. Article 225, which covers outside branch-circuits and feeders, 225.31

11. Chapter 9, Table 10. Expansion fittings are required by 300.7, and 352.44, Tables 352.44(A) and 352.44(B).

12. Places of assembly, Article 518

13. *NEC®* Article 250, 250.34

14. Article 680, 680.61(B), Part II

15. Article 90, 90.9(C)(2)

16. These requirements appear in the building codes or the NFPA Life Safety *Code*; they are not found in the *NEC®*.

17. Article 760

18. Yes. Ref.—590.6(A)

19. Article 502—(a) Class 2, (b) Division 1. Article 500 for the definitions and Article 502(10)(A) and (B) for the specific requirements.

20. It is a violation to put over 1000-volt wiring in a residence. Neon signs requirements are found in Article 600. However, Article 410 requirements limit installation under 1000 volts in dwellings. Ref.—410.75 and 410.80(B)

21. Article 500 defines this area. Article 501—Volatile Liquids would be the article required in which to make the installation. Article 511 would not apply, see 511.1 scope.

22. Yes, Article 220, 220.88

23. Article 300, 300.4(D)

24. Article 702

25. Article 705, which covers interconnected electrical power production sources, otherwise known as co-generation

QUESTION REVIEW CHAPTER 5

1. 4 feet (1.2 m). Ref.—110.26, Table 110.26(A). Some inspectors may permit condition 2.

2. Two. Ref.—110.26(C)

3. Yes. Ref.—110.26(D)

4. 6½ feet (2 m). Ref.—110.26(E)

5. One. Where the required working space is doubled, only one door is required. Ref.—110.26(C)

6. No. The required disconnecting means and supplementary overcurrent device must have a working clearance in accordance with 110.26. Clear working clearance of 30 inches (750 mm) wide and 40 inches (1 m) deep is required from the ground level up 6½ feet (2 m).

7. 30 inches (750 mm). Ref.—110.26(A)(2)

8. Ungrounded. Ref.—240.20(A)

9. 60 percent. Ref.—Chapter 9, Table Note (4)

10. Clothes closet. Ref.—240.24(D)

11. 1¼ inches (32 mm). Ref.—300.4(A)(1)

12. 600 volts. Ref.—362.12(6)

13. Yes. Ref.—230.43(16)

14. Yes. It shall be bonded to the electrode at each end and all intervening raceway boxes and enclosures between the service equipment and the grounding-electrode conductor. Ref.—250.92(A)(3)

15. Overcurrent device. Ref.—250.122

16. Two. Ref.—210.52(E)

17. No. Ref.—680.74, except manufacturer's instructions may require 110.3(B)

18. Yes. Ref.—210.52(A)

19. No. Ref.—210.52(A) for site-built dwellings; 550.13(D) Exception 4 for mobile homes

20. 210.63. Ref.—Article 210

21. Article 312

22. Article 410, 410.4(D)

23. No. 310.4 states that parallel conductors shall not be permitted unless they are size 1/0 AWG or larger.

24. 2/0 AWG. Table 310.15(B)(6); single phase dwellings

25. Chapter 9, Table 8

QUESTION REVIEW CHAPTER 6

1. 18.04 amperes—$P = E \times 1.73/I$ or $I = P/E \times 1.73$

2. 10, 80 percent loading. Ref.—220.11

3. 1120 square feet. Ref.—Knowledge

4. 2. Ref.—Table 220.12, 210.11(C)(2)

5. 5—2 kitchen appliance circuits, 1 laundry circuit, 1 heating circuit (Ref.—422.12), and 1 bathroom receptacle (210.11(C))

6. 2. Ref.—210.11(C)(1)

7. Yes. Ref.—210.11(C)(2)

8. 125 percent of nameplate. Ref.—422.13 and 422.10

9. None. The *NEC*® does not require power to the structure. Ref.—210.52(G)

10. 2. Ref.—210.52(E)

11. Four minimum, one for each of the two small appliance branch-circuits, one for the outdoor receptacles, and one for the bathroom receptacles. Ref.—210.8(A), also see 210.52 and all subsections

12. Yes. Illumination is required at these locations. Ref.—210.70 (A)(2)(b)

13. No. Ref.—210.70

14. Yes. Ref.—210.63

15. 20 ampere. Ref.—210.11(C)(2)

16. No. Ref.—422.12, Definition of Individual Circuit (Article 10)

17. No. Ref.—410.8

18. Yes. Ref.—210.52(B)(2) Exception 2

19. Yes. Ref.—Definition of direct grade access, 210.8(3) and 210.52(E)

20. Yes, two receptacles at direct grade access are required. Ref.—210.52(E) and 210.8

21. 230.79(C)

    (a) 220.12(J) and Table 220.12; $28 \times 40 = 1120$ sq. ft.; 3 VA $\times$ 1120 = 3360 volt-amperes

    (b) 220.52; 3000 volt-amperes

    (c) 220.14(A); 4500 volt-amperes. Ref.—422.13

    (d) 220.52(B); 1500 volt-amperes

    (e) 220.54; 5000 volt-amperes

    (f) 0. Electricity not required; however, if supplied, the load would have to be included in calculation. Ref.—210.52(G)

    (g) 220.14(A); size not stated, therefore, a 20-ampere branch-circuit is recommended.

    (h) 0. 220.12; 210.52(B) Example 2 could be general lighting or on appliance load

    (i) 0. None. Read question again.

    (j) 0. Gas range igniter can be fed from appliance circuit, 210.52(B)(2) Exception 2

    (k) 0. None. This load is included in general lighting calculations, 220.3.

22. 125 percent. Ref.—430.22

23. 1 volt-ampere per sq. ft. or $100 \times 200 = 20,000$ VA Ref.—Table 220.12

24. Two small appliance circuits minimum required. Ref.—210.11(C)(1)

25. Size; Formulas in Appendix 2

QUESTION REVIEW CHAPTER 7

1. 6. Ref.—230.208

2. 94 percent. Ref.—Chapter 3

3. A disconnecting means is required at each building. Ref.—Article 225, Part II

4. Yes, you are limited to not more than six disconnects on this secondary side. Ref.—450.3(B)—conductors must be protected in accordance with 240.4.

5. No; however, 230.54(E) requires that the conductors be separated where of a different potential.

6. No, 240.24(E) prohibits branch-circuit overcurrent devices from being located within a residential bathroom. Ref.—240.24(E), 230.70(A)(2)

7. Yes. SE cable can be used in the same manner as Type NM cable. In this application, however, it is SE cable, and must comply with Articles 300 and 334, Part II. Ref.—338.10(B)(4)

8. Usually no. Yes, if the window is not designed to be opened, or where the conductors run above the top level of the window. Where windows are designed for opening and closing, then a minimum of 3 feet (900 mm) from the windows, doors, porches, fire escapes, or similar locations is required. Ref.—230.9

9. (a) Yes. Ref.—250.64(D)

   (b) They shall be sized in accordance with 250.66 for the largest conductor entering each respective enclosure.

10. Requirement to open all of the conductors is a disconnect requirement and not an overcurrent requirement. Ref.—240.20(B) gives the requirements for circuit breakers, 240.40 gives the requirements for fuses. 210.4 is the requirement for multiwire branch-circuits. Disconnecting, not overcurrent, is the purpose.

11. A separately derived system is defined in Article 100, and the grounding installation requirements are found in 250.30. Ref.—Article 100 and 250.30

12. 450.3—not more than six disconnects are permitted on the secondary side of the transformer.

13. No. Identification of the hi-leg is only required where the neutral is present. Ref.—110.15, 230.56, and 408.3(E)

14. No. The grounding-electrode conductor must be continuous, and the connection shall be made solidly and used for no other purpose. Ref.—250.64(C)

15. Less than .25 (¼) inch (6.4 mm). Ref.—352.44 Exception

16. 90.5—mandatory rules of the *Code* are characterized by the word "shall." Explanatory material is in the form of fine print notes (FPN).

17. Grounded effectively is defined in Article 100. The authority having jurisdiction will make that determination.

18. Article 334, nonmetallic-sheathed cable. Romex is a trade name.

19. Article 360 covers the requirements of flexible nonmetallic tubing. It is manufactured in trade sizes ½ (16) and ¾ (21) only. *Note:* This is a flexible metallic tubing that is liquidtight without a nonmetallic jacket. Rarely used in construction installation.

20. Article 404

21. Chapter 9, Table 1, Footnote 9 states that multiconductor cable of two or more conductors shall be treated as a single conductor cable, and cables that have an elliptical cross-section, the cross-section layering calculation shall be based on using the major diameter for the ellipse as the circle diameter.

22. Article 511 does not apply. 511.1 and 511.2 clearly exempt these parking structures. Therefore, Chapters 1 through 4 apply, and the wiring method would be taken from those four chapters.

23. Article 388

24. 180 volt-amperes each. Article 220, 220.14(I) covers the requirements.

25. It is not permitted. 240.24(E) prohibits the installation of overcurrent devices in bathrooms of residences.

QUESTION REVIEW CHAPTER 8

1. Yes. Ref.—240.16(A). However, handle ties are required.

2. No maximum length needed, provided they are outdoors. Ref.—240.21(B)(5)

3. No. Ref.—240.60(A). Voltage to ground can exceed 300 volts on a wye connection and, therefore, would not be permitted.

4. It is not a limiting factor. However, the six disconnect rule applies. Ref.—230.90(A) Exception 3

5. 100 percent. Ref.—424.3 and 220.51

6. No. Ref.—501.10(B)(2), 350.60, 501.30(B), and must comply with 250.102

7. Yes. Both the overcurrent device and conductors would have to meet these requirements. Ref.—215.2(A)(1)

8. There is no limit. Ref.—Table 220.12 and 220.14(J). *Note:* Many local jurisdictions set limits on 15- and 20-ampere branch-circuits in all installations. However, the *NEC®* does not specify a limit on dwelling branch-circuits for general-use electrical outlets.

9. No. Ref.—220.14(I), also 240.4 (D) limits 12 AWG conductors to a 20-ampere branch-circuit. 16 × 180 VA = 2880 ÷ 120 V = 24 amperes.

10. No, you must use 310.15, see Note at the top of Annex B. The Annex is not a part of the *NEC*® requirements; it serves only for use as reference.

11. (A) No. Ref.—220.60

   (B) Yes. Ref.—424.3(B)

12. No, only if installed in cable with 12 inches (300 mm) of cover. Otherwise, can be installed at the greater depth or in raceways. Ref.—300.5 and Table 300.5

13. Yes. Ref.—240.16(A)

14. No. Ref.—215.2. Footnote: Branch-circuits are required to be calculated at 125 percent for water heaters—422.13.

15. No, No, No. Ref.—240.21(B)

16. Yes. Ref.—210.19 and 110.3(B)

17. No. Ref.—210.6—12 amperes or less must be 120 volts; over that can be 240 volts. 1400 VA ÷ 240 V = 5.833 amperes

18. 180 VA. Ref.—220.14(I)

19. Yes, it complies with the tap rules in 240.21(B)(1) for the 10-foot tap rules.

20. Yes, it complies with the 25-foot tap rule. Ref.—240.21(B)(2)

21. No, it does not comply with any of the tap rules in 240.21.

22. True. Ref.—368.17(B)

23. B. Ref.—240.21(B)(4)

24. Yes. Ref.—Table 250.122, 250.122(B)

25. A fuse must be installed. Ref.—110.3(B). The manufacturer's instructions must be followed.

QUESTION REVIEW CHAPTER 9

1. Soldered. Ref.—250.8

2. 14. Ref.—600.7(D)

3. Not required to be accessible. Ref.—250.68(A)

4. White or gray. Ref.—200.6(A)

5. 4/0. Ref.—250.24

   3-500 kcmil = 1500 kcmil or 3 each 500 kcmil = 1500 kcmil × 12% = 187.5 kcmil

   1500 × 12.5% = 187.5 kcmil

   Chapter 9, Table 8—3/0 AWG 167.8 kcmil too small; therefore, use 4/0 211.6 kcmil

6. 4. Ref.—Table 250.102(D), 250.122

7. Yes, bonding is not required. Ref.—250.94(3)

8. They are required to be run in each nonmetallic raceway and are required to be sized in accordance with the sum of the paralleled phase conductors. Ref.—215.6 and 250.122 and Table 250.122

9. Yes. Ref.—250.112(L)

10. No, the language is specific. Ref.—250.30

11. Yes, it must be bonded. No, it is not sufficient to bond at one end. It must be bonded at both ends and at all enclosures in which it passes through. Ref.—250.64(E)

12. No, a driven rod is not acceptable. An equipment grounding conductor must be carried with the circuit. Ref.—551.76(A)

13. Yes, the requirements for separately derived transformers are found in 250.30. These requirements do not amend the requirements of 250.50, but they are the specific requirements. Therefore, a supplementary grounding electrode is required.

14. 8 AWG copper or equivalent. Ref.—551.56(C)

15. No. Ref.—250.104. *Note:* The interior gas piping system is required to be grounded for safety. It can be grounded with the circuit conductor equipment ground most likely in which it would come in contact. For example, a central heating electric furnace may have 12 AWG conductors supplying the fan motor. The 12 AWG equipment grounding conductor run with those circuit conductors could be the conductor used to bond the gas piping system in this case. **Warning:** The underground gas piping system may not be grounded. Electrolysis may occur, creating dangerous underground gas leaks.

16. Yes, a 1 AWG copper or a 2/0 AWG aluminum or copper-clad aluminum wire is required in each conduit. Ref.—Table 250.122 and 250.122, which state that where run in parallel in multiple raceways or cables as permitted, the equipment grounding conductors also shall be run in parallel, each equipment grounding conductor shall be sized on the basis of the ampere rating of the overcurrent device protecting the circuit conductor in the raceway in accordance with Table 250.122.

17. No. Ref.—110.14(B)

18. 8 AWG. Ref.—Table 250.122. This is a minimum size conductor and may have to be larger if the fault current available exceeds that for which the 8 AWG conductor will carry.

19. Yes. The flexible metal conduit cannot exceed 6 feet (1.8 m) for the combined two pieces of flexible metal conduit. Ref.—250.118, 348.60

20. The maximum rating of the grounding impedance, no smaller than 8 AWG copper or 6 AWG aluminum. Ref.—250.24, 250.36

21. By a ground-fault circuit interrupter. Ref.—Part VII, Article 680.71

22. Yes, all backfed circuit breakers must have a retainer kit. Ref.—408.16(F)

23. Maybe. Ref.—250.50 does not permit the water line to be used as a grounding-electrode conductor. All connections must be made within the first 5 feet (1.52 m). However, the other types can be connected at any point and, therefore, the answer is *maybe*.

24. 250.80 Exception for Services and 250.86 Exception for Feeders or Other Conductors permit these metal 90s where maintained at a burial depth in accordance with Table 300.5 need not be grounded. Otherwise, they must be bonded and at most practical places where they emerge from the ground, such as turning up into a concrete slab or turning up under a panelboard, or turning to go up the side of a pole or other structure. Bonding is required and provides a greater degree of safety.

25. Main grounding-electrode conductor would be based on the four 750-kcmil service conductors, and they must be sized at 2/0 AWG copper or 4/0 AWG aluminum. Switch A, the tap conductor from the grounding-electrode conductor to Switch A, would be sized only on the conductor feeding that switch, 3 AWG, and the size of the conductor, based on Table 250.66, would be 8 AWG copper or 6 AWG aluminum. Switch B would be based on the same Table 250.66, and would be 4 AWG copper or 2 AWG aluminum. Switch C would be 8 AWG copper or 6 AWG aluminum, and Switch D would be the 2/0 AWG conductor for the main going in. However, if you chose to tap that conductor to Switch D and run the main grounding electrode to one of the others, then based on the four 1/0 AWGs, that conductor would have to be a 6 AWG copper or 4 AWG aluminum. Ref.—Table 250.66

QUESTION REVIEW CHAPTER 10

1. Within 12 inches (300 mm) of the outlet or box, and every 4½ feet (1.4 m). Ref.—338.4 and 334.30

2. Maximum number of conductors shall be permitted using the provisions of 314.16(B). Conduit bodies shall be supported in a rigid and secure manner. Ref.—314.16(C)

3. Yes. Ref.—348.20(A)(1) and 350.20(A) Exception

4. Yes. Ref.—342.30(A)

5. Yes. Ref.—362.30(B)—Metal studs are the required support when they do not exceed 3 feet (900 mm) apart.

6. Type MC cable must be supported at intervals not exceeding 6 feet (1.8 m), and it shall be secured within 12 inches (300 mm) from every outlet box, junction box, cabinet, or fitting, except where fished. Ref.—330.30

7. NM cable shall be supported at intervals not exceeding 4½ feet (1.4 m) and within 12 inches (300 mm) of every cabinet, box, or fitting. Two conductor cables shall not be stapled on edge. Where type NM cable is run through holes in wood studs or rafters, it shall be considered supported. Ref.—334.30

8. No. Ref.—378.12

9. Aboveground conductors must be installed in a rigid metal conduit, intermediate metal conduit, electrical metallic tubing, rigid nonmetallic conduit in cable tray or busway or cable bus or in other identified raceways, or as open runs of metal clad cable suitable for the use. Open runs of Type MV cable, bare conductors, or bare bus bars also are permitted as they apply. Ref.—300.37

10. Yes, provided that carpet squares not exceeding 36 inches square (914 mm²) are used in a wiring system in accordance with Article 324 Type FCC cable and associated accessories are used. Ref.—Article 324.10(H)

11. Metallic wireway shall be established in accordance with 376.22 and shall not exceed 30 current-carrying conductors at any cross-sectional area. The sum of cross-sectional areas of all conductors at any cross-section of the wireway shall not exceed 20 percent. The derating factors in Table 310.15(B)(2) *do not* apply. For nonmetallic wireway, 378.22 applies. The cross-sectional area containing conductors at any cross-section of the nonmetallic wireway shall not exceed 20 percent of the interior cross-sectional area. The derating factors in Table 310.15(B)(2) *do* apply to the current-carrying conductors at the 20 percent specified. ***Note:*** This is a critical difference. Caution

should be used when reading questions as they apply to a specific wiring method, metallic or nonmetallic, because the conditions and rules are very different.

12. 1¼ inches (32 mm) from the nearest edge of the stud. Ref.—300.4(D)

13. No, the electrical metallic tubing must be securely fastened in place at least every 10 feet (3 m) and within 3 feet (900 mm) of each outlet box, junction box, device box, cabinet, conduit body, or other tubing termination. Support within 3 feet (900 mm) of each coupling is not required. Ref.— 358.30

14. There is no difference in the permitted uses or installation requirements in the two products. Article 342 and Article 344 are essentially the same. Ref.—Article 342 and Article 344

15. 1/0 AWG. Ref.—310.4

16. Yes, in any building exceeding three floors, non-metallic tubing must be concealed within a wall, floor, or ceiling where the material has at least a 15-minute finished rating as identified in the listings of the fire-rated assemblies. *Exception: Where a fire sprinkler system(s) is installed in accordance with NFPA 13-1999,* Standard for the Installation of Sprinkler Systems, *on all floors, ENT is permitted to be used within walls, floors, and ceilings, exposed or concealed, in buildings exceeding three floors above grade* [reprinted with permission from NFPA 70-2005]. Ref.—362.10(2)

17. No, they must be removed. They are not permitted for a period to exceed 90 days. Ref.—590.3(B)

18. No, lengths not exceeding 4 feet (1.2 m) can be used to connect physically adjustable equipment and devices that are permitted in the duct or plenum chamber. Ref.—300.22(B)

19. ¼ inch (6 mm). Ref.—352.44

20. Yes, even though 300.3 requires that all conductors of the same circuit, including the neutral and equipment grounding conductors, must be run within the same raceway, cable trench, cable, or cord; Exception 2 to that section permits that with column-type panelboards; the auxiliary gutter and pull box may contain the neutral terminations. Ref.—300.3(B)(4)

21. No. Generally, each cable shall be secured to the panelboard individually. Ref.—312.5(C). The Exception would allow this under conditions.

22. No, a branch-circuit or a lighting panelboard is one that 10 percent of its overcurrent devices are rated 30 amperes or less for which neutral connections are provided. A lighting and branch-circuit panelboard, as required by the *Code,* shall have not more than two circuit breakers or two sets of fuses for disconnecting. Ref.—408.14 and 408.16—A disconnect is required ahead of the lighting and branch-circuit panelboard although the load is relatively small in this cabin.

23. It is prohibited by the *Code.* The reference you are looking for is 404.8(B). The switches must be arranged so that the voltage does not exceed 300 volts, or permanently installed barriers between the adjacent switches must be installed. You are correct; it is a dangerous situation.

24. Yes, knob-and-tube wiring is still an acceptable wiring method and is covered in Article 394. Extensions are permitted in hollow spaces, walls, and ceilings. However, they are not permitted in the hollow spaces of walls, ceilings, and attics where the spaces are insulated by loose, rolled, or formed-in-place insulated material that covers the conductors. You can make the installation, but you will have to route the conductors along a running board above the insulation. Ref.—Article 394, 394.12(5)

25. 1/0 AWG or larger. Ref.—392.3(B)(1)

QUESTION REVIEW CHAPTER 11

1. 430.6(A). Ref.—430.6(A)

2. 100 ampere. Ref.—430.6, Tables 430.250 and 430.52 (40 × 250% = 100)

3. 175 percent. Ref.—Tables 430.250 and 430.52

4. 8 AWG THW aluminum. Ref.—Table 430.248, 430.24 and 430.25

5. Same. Ref.—240.21

6. 20. Ref.—240.4, 240.6, and Table 310.16

7. 40. Ref.—240.4, 240.6, and Table 310.16

8. 100. Ref.—240.4, 240.6, and Table 310.16

9. 200. Ref.—240.4, 240.6, and Table 310.16

10. 125 percent. Ref.—450.3(B)

11. 175 percent. Ref.—Table 430.52

12. $V_D = 22 \times 2 \times 150 \times 1.24/1000 = 8.184$ volts

13. 3.5 percent

14. 8 AWG

15. 125 percent. Ref.—Article 430.32(A)(1)

16. 6. Ref.—Table 430.91

17. Indoors. Ref.—410.73(E) and (F)

18. Fire protection or equipment cooling. Ref.—450.47

19. Accessible. Ref.—450.13(A)

20. 3. Ref.—460.2

21. 8. Ref.—440.32 and Table 310.16

22. 300 percent. Ref.—Table 430.52

23. 112.5 kVA. Ref.—450.21(B)

24. 600 watts. Ref.—Ohm's law (*Clue:* Resistance does not change); Appendix 2

25. 87 percent. Ref.—Chapter 3; Appendix 2

QUESTION REVIEW CHAPTER 12

1. Nonconductive, conductive, composite. Ref.—770.5

2. True. Ref.—511.7

3. Yes, where the raceway emerges from the ground, both at the office and at the dispenser. Ref.—514.9(A) and (B)

4. False. Ref.—700.1

5. Class 1, Division 2. Ref.—511.3(B)(2)

6. Inadvertent parallel interconnection. Ref.—700.6

7. ⅝ (18 mm). Ref.—501.15(C)(3)

8. 12 inches (300 mm). Ref.—490.24

9. Two duplex or four single receptacles. Ref.—517.18(B)

10. 6. Ref.—220.14, 430.24, Tables 430.248 and 310.16

11. 15 feet. (4.5 m) Ref.—600.10(D)(2)

12. Length and design. Ref.—600.41(A)

13. True. Ref.—240.54(B)

14. False. Ref.—240.53

15. True. Ref.—240.53

16. False. Ref.—240.53

17. False. Ref.—240.53

18. False. Ref.—240.53

19. True. Ref.—240.53

20. True. Ref.—240.53; Bussmann SPD Protection Handbook

21. (a) Electrician or electrical contractor

    (b) Field marking is required to state "Caution—Series Rated System, ___ Amperes Available.

        Identified Replacement Components Required." Ref.—110.22

    (c) Manufacturer

22. 6 amperes. Ref.—$I = \dfrac{HP \times 746}{E \times PF \times eff\%}$

    Note: When power factor is not noted, it must be calculated at 100 percent.

23. 60. Ref.—Tables 430.250 and 430.52

24. 27. Ref.—Ohm's law, Chapter 3, Table 220.20; Appendix 2

25. 18. Ref.—Ohm's law; Appendix 2

QUESTION REVIEW CHAPTER 13

1. Yes. Ref.—810.20(C), 810.21

2. No. Chapter 8 stands alone. There are no burial depth requirements for satellite dishes installed anywhere there is no safety hazard, and, therefore, the only consideration is for the customer's protection of his or her own equipment, *NEC*® 800.10. *Note:* Also look at 90.3. Chapter 8 covers communication systems and is independent of the other chapters, except where they are specifically referenced. Chapter 8 stands alone, and other chapters cannot be applied unless specifically referenced.

3. Yes. Ref.—800.33

4. 6 feet (1.8 m). Ref.—800.13

5. It is an intrinsically safe circuit covered in Article 504 of the *National Electrical Code*®.

6. Yes, motor control centers are covered in Part VIII of Article 430. *Note:* The requirements in Article 408 for accessibility are also applicable as referenced in 430.1, (FPN 1).

7. Definition of Article 100 defines a separately derived system. 250.30 covers the grounding requirements for this separately derived transformer.

8. Article 406 covers these isolated ground receptacles intended for the reduction of electrical noise. See also 250.96(B). Orange triangle now required.

9. Article 530 covers film vaults. *Note:* Scope 530.1 covers these areas in a portion of a building.

10. When a term is used only once in the *NEC*®, the definition appears where it is being used. The definition of a bathroom area appears in Article 100—Definitions.

11. Yes. Ref.—Article 650

12. Yes, Article 110, 110.26 for installations of less than 600 volts, 110.32 for installations over 600 volts.

13. Yes. 110.12(C) clearly states that paint, plaster, and other abrasives may damage the bus bars, wiring terminals, insulators, and other surfaces in panelboards. Therefore, the internal parts of this equipment must be covered and protected.

14. Inspector was correct. 550.32 requires the service equipment to be located outside and not more than 30 feet (9 m) from the exterior wall of the mobile home. An exception permits the installation of the service equipment on the mobile home when installed by the manufacturer. Electrical contractors, however, cannot make this installation.

15. Optional standby system is covered in Article 702 and is defined in 702.2.

16. Color coding is not specifically required; however, 210.4(D) requires the identification of ungrounded conductors on multiwire branch-circuits under certain conditions. This identification can be by color coding, marking tape tagging, or other equally effective means.

17. No. The footnotes to Table 310.16 are mandatory language, and the note limits 14 AWG conductors to a 15-ampere overcurrent device. Ref.—240(4)(D)

18. 225.7(B) covers outdoor branch-circuits for lighting equipment and there is no maximum number. See 215.4(A) for feeders.

19. Yes. Article 547—Agriculture Buildings covers this type of installation.

20. Yes. Article 324.10(H) covers this type of installation. However, the carpet must be laid in squares not to exceed 36 inches (914 mm) square.

21. Article 280—Surge Arrestors covers that type of installation.

22. Article 590 permits a class less than would be required for permanent installation. However, there are time constraints. *Note:* 590.4(A) requires the service on that temporary pole to comply with Article 230.

23. 86 percent. Ref.—Chapter 3, Duff and Herman, *Alternating Current Fundamentals, 4E* by Thomson Delmar Learning; Appendix 2

24. 60 hp. Formula is $EFF = \dfrac{W}{\sqrt{3} \times EXI}$

Ref.—Loper, *Direct Current Fundamentals, 4E* by Thomson Delmar Learning; Appendix 2

25. 0.105. Ref.—Loper, *Direct Current Fundamentals, 4E* by Thomson Delmar Learning; Appendix 2

PRACTICE EXAMINATION

1. C. Ref.—Article 100—Definition
2. A. Ref.—310.11(C)
3. E. Ref.—725.23
4. D. Ref.—312.8
5. B. Ref.—240.51(A) and 240.53(A)
6. D. Ref.—760.52(A)
7. C. Ref.—Chapter 9, Note 4 to Table 1
8. D. Ref.—240.80
9. C. Ref.—720.5
10. D. Ref.—Table 402.5
11. A. Ref.—Chapter 9, Table 8, Col 6
12. B. Ref.—760.54, 55, and 58
13. B. Ref.—90.5(C)
14. D. Ref.—344.30(B) and Table 344.30(B)(2)
15. D. Ref.—Article 100, see definition "switch"
16. D. Ref.—725.11(B)
17. B. Ref.—250.12
18. C. Ref.—Article 100—Definition
19. D. Ref.—110.27
20. B. Ref.—110.6
21. B. Ref.—90.9
22. B. Ref.—Article 100—Definition
23. B. Ref.—725.54 and 725.55
24. B. Ref.—Article 100—Definition
25. A. Ref.—200.6
26. B. Ref.—110.19
27. A. Ref.—424.39
28. C. Ref.—660.5
29. A. Ref.—225.14(C)
30. C or the height of the equipment, whichever is greater. Ref.—110.26(E)
31. A. Ref.—220.14(H)
32. A. Ref.—810.11 Exception
33. B. Ref.—230.79(C)
34. A. Ref.—430.89
35. D. Ref.—225.4

36. D. Ref.—110.53
37. D. Ref.—225.36 Exception
38. A. Ref.—210.23(A)(2)
39. D. Ref.—366.23, copper (assumed) is 1000 A per square inch. $4 \times \frac{1}{2} = 2$ square inches, $1000 \times 2 = 2000$ amperes maximum
40. D. Ref.—590.7
41. D. Ref.—430.32(D)(1)
42. D. Ref.—550.16(A)
43. B. Ref.—600.9
44. D. Ref.—820.40(A)(1, 3, 6)
45. E. Ref.—110.32
46. D. Ref.—398.30(E)
47. A. Ref.—Article 100—Definition
48. A. Ref.—225.6(A)(1)
49. B. Ref.—210.52(B)(2)
50. B. Ref.—210.52(C)(1) through (3)

2005 *CODE* CHANGES QUIZ 1

1. Article 682—Natural and Artificially Made Bodies of Water. Ref.—Table of Contents
2. Article 353—High Density Polyethylene Conduit: Type HDPE Conduit. Ref.—Table of Contents
3. Article 590—Temporary Installations. Ref.—Index
4. Bathroom. Ref.—Article 100—Definitions
5. Unclassified locations. Ref.—500.2
6. No. Ref.—Article 80 was relocated as a new addendum G.
7. The system bonding jumper is a new term to describe the jumper used to bond the grounded conductor, grounding electrode conductor, and equipment grounds together Ref.—250.30(A)(1)
8. Grounding electrode conductor. Ref.—Article 100—Definitions
9. Handhole enclosure. Ref.—Article 100—Definitions
10. Commercial or Institutional Kitchen. Ref.—210.8(B)
11. Ground fault protection Ref.—210.(B) (4)
12. Chapter 9, Table 2. Ref.—Same
13. 90.2(B)(4). Ref.—830.1 FPN No. 2
14. The fill restrictions do not apply; therefore, there are no maximum requirements. Ref.—820.48 Exception

15. False. Ref.—513.12. The GFCI requirements only apply to receptacles used for diagnostic equipment, electrical hand tools, or portable lighting equipment.

2005 *CODE* CHANGES QUIZ 2

1. Metal raceways. Ref.—645.5(D)(6)
2. Enclosed in raceways. Ref.—610.11(A) and (B)
3. Rigid metal conduit; intermediate metal conduit. Ref.—550.15(H)
4. 230; 525.12(A) and (B). Ref.—Same
5. 506.15(A) (1) through (5) Ref.—Same
6. Resistors and reactors that are part of other apparatus are not covered by Article 470 but are generally covered by the listing for the apparatus. Ref.—470.1 Exception
7. 115. Ref.—430.110
8. Bathtubs; shower stalls. Ref.—406.8(C).
9. 410.15(B)(6); 300.19. Ref.—Same
10. 400.4 and Table 400.4. Ref.—Same
11. 600 volts. Ref.—398.12
12. No. Ref.—Article 374, Section 374.17
13. False. Ref.—392.3(E). It is not limited but can be installed in accordance with 392.3.
14. Corrosive vapors; physical damage. Ref.—368.12(A)
15. 100 feet (30.5 m) of trade size 1 (27) intermediate metal conduit (IMC). Ref.—Article 342, Sections 342.20, 300.1(C), and 342.130

2005 *CODE* CHANGES QUIZ 3

1. Above ground. Ref.—326.12(3)
2. Tamper resistant type. Ref.—517–18.
3. Earth. Ref.—250.4(A)(5)
4. NFPA 496. Ref.—500.2 (FPN) under the definition of "purged and pressurized."
5. 225.6. Ref.—300.3 *Exception: Individual conductors shall be permitted where installed as separate overhead conductors in accordance with 225.6* [reprinted with permission from NFPA 70-2005].
6. Branch-circuit panelboard. Ref.—200.6(D). They must be marked with one of the prescribed methods in 200.7.
7. 8. Ref.—400.7(C)

8. Bathtub; shower space; face-up; above. Ref.—550.15(F)

9. ½ (16); one. Ref.—600.32(A)(2) and (3)

10. 630. Ref.—Same

11. The disconnecting switch must be in-site. Ref.—430.102

12. 250.30(A)(5) 250.28(A) through (D) size in accordance with Table 250.66

13. No, they must be approved. Ref.—300.5(K); definition "approved" in Article 100

14. The *Code* requires listed or marked and listed. Ref.—310.8(D), 2005 adding sleeve to I.D. listed for "sunlight resistant"

15. No, the *Code* limits are 120 volts between conductors. Ref.—210.6(A)(1)

16. Yes. Ref.—210.8(A)(3)

17. 103 is the metric designator for trade size 4 conduit and tubing. Ref.—300.1(C)

18. No, EMT is an acceptable equipment grounding conductor. Ref.—250.118

19. Article 409, Part II. Ref.—Same

20. Ungrounded. Ref.—285.4

2005 *CODE* CHANGES QUIZ 4

1. Yes. Ref.—517.13(B) Exception 2

2. Nothing. See Definition, Article 100; luminaire is an international term

3. Yes. Ref.—*NEC*® Annex G

4. Yes. 90.1(D) was added, and the term "International" appears on the cover.

5. Signaling and communications was added to eliminate any confusion that the *NEC*® does include requirements for these systems and that they are covered.

6. Yes. *NEC*® Annex A and referenced in 90.7, FPN 3

7. No, the *Code* now uses a dual measuring system; either can be used as applicable. Ref.—90.9(B)

8. Article 406 was added to contain these requirements, which were formerly found in Article 410 for the 2002 *NEC*®.

9. Yes, the reference to types of fuel was deleted, and the definition now covers all self-propelled vehicles.

10. No. Each disconnect must be marked to indicate its purpose, unless the purpose is evident.

11. Yes. 110.26(C)(2) requires panic hardware on doors to electrical equipment spaces that contain equipment rated 1200 amperes or more.

12. 410.14(A) and 410.14(B)

13. Lighting fixtures. Article 100, definition of luminaire

14. Yes. Ref.—406.4(E)

15. Yes. Ref.—210.12(B)

16. No. Ref.—210.52(C)(5)

17. Yes. 424.44(G) requires GFCI protection in these locations.

18. Yes. Ref.—200.6

19. No. 210.6(E) only permits where conditions of maintenance and supervision ensure that only qualified persons service the installation.

20. Yes. 230.22 Exception permits the grounded conductor of a multiconductor cable to be bare.

2005 *CODE* CHANGES QUIZ 5

1. Yes. Ref.—230.33 in accordance with 110.14, 300.5(E), 300.13, and 300.15

2. No. 230.44 permits cable tray as a support system. Ref.—392.2 and 392.3

3. Yes. Ref.—110.14(B)

4. No. 230.70(A)(3) permits a shunt trip device, but the disconnecting means is still required 230.66, 230.70, 230.70(A)

5. Yes. Ref.—550.16(C)(3)

6. Yes. Ref.—430.83(A)(3)

7. No. The service disconnecting means shall be installed at a readily accessible location either outside of a building or structure or inside nearest the point of entrance of the service conductors.

8. Article 647, 647.1

9. 230.82(5)

10. The overcurrent protective device(s) shall be selected or set to carry indefinitely the sum of the locked-rotor current of the fire pump motor(s) and the pressure maintenance pump motor(s) and the full-load current of the associated fire pump accessory equipment when connected to this power supply. Ref.—695.4(B)(1)

11. Yes. Ref.—210.8(B)(3)

12. 240.24(B). However, Exception 2 of that section permits them to be accessible to only authorized management personnel.

13. Yes. No circuit breakers used as switches in 120-volt and 277-volt fluorescent lighting circuits shall be listed and shall be marked SWD or HID. Circuit breakers used as switches in high-intensity discharge lighting circuits shall be listed and shall be marked as HID. Ref.—240.81 and 240.83, 404.11

14. Yes, unless they are in metal raceways. Ref.— 645.5 (D)(6), 725.2, 760.2, 760.3(A), 770.2, and 770.3(A)

15. Article 590, 590.3(B)

16. Yes. Ref.—250.53(D)(2)

17. Yes. 250.54 but the earth shall not be the sole equipment grounding conductor.

18. A surge arrester is a protective device for limiting surge voltages by discharging or bypassing surge current, and it also prevents continued flow of follow current while remaining capable of repeating these functions. A Transient Voltage Surge Suppressor (TVSS) is a protective device for limiting transient voltages by diverting or limiting surge current; it also prevents continued flow of follow current while remaining capable of repeating these functions. Ref.—280.2 and 285.2

19. No. Ref.— 314.20, first paragraph

20. 50 pounds (23 kg). Ref.—314.27(B)

2005 *CODE* CHANGES QUIZ 6

1. 6 feet (1.8 m). Ref.—320.30(B)(3)

2. Yes. Ref.—300.7(A)

3. No. Ref.—300.11(C)

4. No. Ref.—300.22(A), (B), and (C)

5. Article 590

6. It is a device intended to provide protection from the effects of arc faults by recognizing characteristics unique to arcing and by functioning to de-energize the circuit when an arc fault is detected. Ref.—210.12

7. No. If the runs comply exactly with Exception 5 to 310.15(B)(2), then they would not apply.

8. No. Ref.—396.12

9. No. It is allowed to be used where the support is no more than 6 feet (1.8 m) from the last point of support for connections within an accessible ceiling to luminaire(s) (lighting fixture[s]) or equipment.

10. Yes. 511.1 would cover the repair. Article 625 does not cover the vehicle, only the supply related to vehicle charging.

11. Yes. 342.30(B)(3), where threaded and threadless couplings are used.

12. Yes, if the threadless fittings are listed for the purpose. Ref.—344.42(A)

13. No. Ref.—322.10(3)

14. Yes.

15. Yes. Ref.—404.9(B)

16. Yes. Ref.—110.16

17. No. Ref.—334.12(A)(1)

18. No, however it may be a violation of the serving utility. Ref.—230.28

19. No. 514.11(A), (B) and (C), and 514.13 both are required.

20. No. Ref.—110.26(F)

2005 *CODE* CHANGES QUIZ 7

1. Yes. All branch-circuits that supply 125-volt, single-phase, 15- and 20-ampere outlets installed in dwelling unit bedrooms shall be protected by an arc-fault circuit interrupter listed to provide protection of the *entire branch-circuit*.

2. It would be considered as part of the lighting load. Ref.—220.52(A) Exception

3. It is not one of the permitted methods in 230.43 and is not permitted in wet locations. Ref.—320(3)

4. Yes. Ref.—250.104(B), last sentence.

5. Yes. Ref.—334.30(B)(2) except as prohibited by 334.12(A)(1)

6. No, only steel IMC is permitted by Article 342. Ref.—342.1

7. No, unless identified as sunlight resistant. Ref.—362.12(9)

8. Yes, all raceway articles permit cables unless specifically prohibited by the respective cable article. Ref.—342.22 through 362.22

9. Yes, 4 quarter bends (360 degrees total). Ref.—350.26

10. Yes. Ref.—300.50(B)

11. Listed and marked. Ref.—410.31

12. NUCC is nonmetallic underground conduit with conductors as covered by Article 354. Yes, it is required to be listed (Ref.—354.6). Yes, the conductors and raceway are required to be listed (Ref.—354.100(B) and 354.100(C)).

13. No. Ref.—362.12(10)

14. In accordance with the installation instructions Ref.—392.6(C)

15. The highest level that water can reach before it spills out. Ref.—680.2

16. EMT is a permitted equipment grounding conductor. Ref.—250.118(4)

17. These tables are provided to assist the user. The amount of temperature change would be provided by the user. Ref.—352.44 and the Tables 352.44 (A) and (B)

18. Yes. Ref.—378.6

19. Yes. Ref.—230.41(A)

20. 647.4(D). *Note:* FPNs are not enforceable there. This is the only location in the *Code* where voltage drop is mandated.

2005 *CODE* CHANGES QUIZ 8

1. No. Ref.—356.12(4)

2. 250.102, 356.60

3. No. Article 376.

4. 115 percent. Ref.—445.13

5. LCDI or AFCI. Ref.—440.65

6. No. Ref.—430.125(A) Exception permits alarm devices where the operation of the motor is vital to the safeguard of persons.

7. 30. Ref.—426.32

8. Yes. Ref.—404.9(B)

9. Yes. Ref.—404.8(A) Exception 2

10. No. It is identified for one conductor. Ref.—408.41

11. Intermediate metal conduit (Type IMC) and rigid metal conduit (Type RMC)

12. Yes. Ref.—424.3(A)

13. Battery or generator. Ref.—517.45(A)

14. Division 1. Ref.—500.5(C)(9)

15. 50 volts. Ref.—480.4

16. Purpose; use. Ref.—408.4

17. No. Ref.—430.245(B); 310.3

18. Thermal protection. Ref.—410.73(E)(4)

19. No. Ref.—440.14

20. Temperatures. Ref.—422.16(B)(3)

2005 *CODE* CHANGES QUIZ 9

1. No. Ref.—Article 100, definition of service; it must be supplied from the serving utility.

2. Article 110 Part IV. Ref.—Table of Contents

3. 6 feet (1.8 m); lower. Ref.—110.26(F)

4. 110.54(A) and 110.54(B). Ref.—Same

5. Three continuous white stripes. Ref.—200.6

6. Yes. Ref.—210.8(A)(2)

7. An arc-fault circuit interrupter *is a device intended to provide protection from the effects of arc faults by recognizing characteristics unique to arcing and by functioning to de-energize the circuit when an arc-fault is detected.* [Reprinted with permission from NFPA 70-2005.] Ref.—210.12(A)

8. No, that would be a feeder by definition. Ref.—Article 100—Definitions

9. False. Ref.—110.26(A)(3). Equipment located above or below other electrical equipment is permitted to extend up to 6 inches (150 mm) beyond the front of other electrical equipment.

10. 110. Ref.—110.26(F)(1)(A)

11. 110, Part III. Ref.—Same

12. Three continuous white stripes on other than green insulation along its entire length. Ref.—200.6

13. GFCI (ground-fault circuit interrupter protection for personnel). Ref.—210.8(A)(2)

14. Arc-fault circuit interrupters (AFCI). Ref.—210.12(B)

15. No. Ref.—210.52(B)(3). Some dwellings have more than one kitchen, and this new requirement was added to reduce overloading.

16. 3; 900. Ref.—210.52(D)

17. Two; readily accessible. Ref.—210.60(B)

18. 150 volt-amperes. Ref.—220.43(B)

19. 12; (3.7 m). Ref.—225.18(2)

20. 100 amperes. Ref.—225.39(C)

2005 *CODE* CHANGES QUIZ 10

1. 230.40 Exception 4. Ref.—Same

2. 100 amperes. Ref.—230.79(C)

3. No, 240.4(D). Ref.—240.4(D)

4. 240.2(E). Ref.—240.2 Definitive

5. Unlimited length. Ref.—240.21(B)(5)

6. 240.2. Ref.—Same

7. They are required to be GFCI protected. Ref.—525.18(A)

8. First paragraph of 240.21, which states: *No conductor supplied under the provisions of Sec. 240.21 (A) through (G) shall supply another conductor except through an overcurrent device.* [Reprinted with permission from NFPA 70-2005.] Ref.—240.21, second sentence, first paragraph

9. No, the exception to 220.52 allows this load to be excluded from the calculation. Ref.—210.52 (B)(1) Exception 2 and 220.52(A) Exception

10. Yes, they would have to be accessible; however, only two would have to be readily accessible. Ref.—210-60(A) and (B)

11. No, they would still be accessible as defined by the *Code*. Ref.—Article 100, definition of accessible and readily accessible

12. Where the generator is a separately derived system. Ref.—250.20(D) and FPNs 1 and 2

13. There is no set distance other than it must be within 36 inches (900 mm) of the outside edge of each basin and it must be located on the wall immediately adjacent. Ref.—210.52(D)

14. Yes, an exception states that if the circuit supplies only a single bathroom, then it is permitted to supply other equipment within that single bathroom. Ref.—210.11(C)(3) and Exception

15. Yes, in places of assembly Article 518. Ref.—518.3(B) Exception

16. Listed. Ref.—695.10

17. No. Ref.—547.10(A)

18. No, an equipment grounding conductor in accordance with 250.118 is required, and the grounded (neutral) cannot be bonded to the grounding-electrode system where there are common metallic paths between the two buildings. Ref.—250.32(B) (1) and (2)

19. Yes, if it meets the conditions of 373.5(C) Exception. Ref.—Same

20. Yes, if they have a floor located at or below grade and are not intended as habitable, and are limited to storage areas, work areas, or similar use. Ref.—210.8(A)(2)

2005 *CODE* CHANGES QUIZ 11

1. Yes, or ground-fault protection is installed in accordance with 250.122(F)(2). Ref.—Same

2. A power panelboard is one having 10 percent or fewer of its overcurrent devices protecting lighting and appliance branch-circuits. Ref.—408.14(B)

3. Yes, in accordance with 110.14, 300.5(E), 300.13, and 300.15. Ref.—230.46

4. Yes, for feeders or branch-circuits. Ref.—590.4 (B) and (C)

5. Yes, in trade size ½ (16) through 1 (27) as a listed manufactured prewired assembly. Ref.—331.3(8)

6. Yes, only one is permitted generally. Additional feeders or branch-circuits are permitted in accordance with 225.32(A) through (E). Ref.—Same

7. Type AC cable with an insulated equipment grounding conductor sized to Table 250.122 is now allowed. Ref.—518.4(A)

8. Yes, they are not permitted to be installed in a face-up position in the countertops or work spaces. Ref.—406.4(E); 210.8(A)(7) and 210.52(C)(5)

9. The industrial portion of a facility meeting all the conditions of 240.2 Ref.—Same

10. Listed tamper resistant; listed tamper-resistant. Ref.—517.18(C)

11. Yes, if the luminaire (lighting fixture) does not exceed 6 pounds (3 kg). Ref.—314.27(A) Exception

12. At least 3 inches (75 mm) of conductor outside the opening. Ref.—300.14

13. No, the building codes may also prohibit combustible wiring methods. Ref.—334.12(A)(5)

14. It is not required for all buildings; industrial buildings are exempted. However, where the busway vertical riser penetrates two or more floors, a 4-inch (100 mm) minimum curb must be installed to retain liquids. Ref.—368.10(B)(2)

15. Two minimum where a made electrode is used and they have not been tested for compliance with 250.56. Ref.—250.52(A)(2)

16. No, only medium- and low-power systems are covered by Article 830. Ref.—830.4 and Table 830.4

17. Where a general-care patient bed location is served from two transfer switches on the emergency system, they are not required to have circuits from the normal system. Ref.—517.18(A) Exception 3

18. Yes, they could all be main power feeders if feeding lighting and branch-circuit panelboards. Ref.—310.15(B)(6)
19. Yes. Ref.—210.8(A)(3)
20. Article 314. Ref.—314.29

PRACTICE EXAM 1

1. Yes. Ref.—511.12
2. No. Ref.—680.72, however, 410.4(A) and (D) limit the use of pendant-type fixtures over this area.
3. The answer is Yes. Ref.—406.3(D)(3) and no, Ref.—406.3(D) grounding-type receptacles may be used as replacements, but they must be marked "GFCI Protected"
4. No. Ref.—210.8(A)(6)—only those serving the countertops. *Note:* Only very old ranges have receptacles mounted on them.
5. No. Ref.—314.3 Exceptions 1 and 2
6. Yes. Ref.—410.16(H)
7. 15-minute finish rating. Ref.—362.10 and 12
8. No. Ref.—300.4(B)(2)
9. Where run through the framing members, ENT shall be considered supported. Ref.—362.30(B)
10. Yes. Ref.—300.4(B) and 334.17
11. Yes. Ref.—340.10(4) and 334.112
12. No. Ref.—517.10 Exceptions 1 through 3
13. Yes. Ref.—210.52(A)
14. No. Ref.—300.13(B)
15. Yes. Ref.—210.52(C) and 210.8(A)(6) *Note:* The GFCI requirement would not apply if the receptacle was located behind the refrigerator.
16. No. Ref.—210.52(A) and (E)
17. No. The requirement is now 6 feet 7 inches (2 m). Ref.—404.8(A)
18. See referenced exceptions and general rule. Ref.—410.31
19. General- and critical-care patient bed location. Ref.—517.18(B)
20. No. Ref.—210.52(A)—shall be in addition to those required if over 5½ feet (1.7 m) above the floor. See 210.8(A)(7)
21. Self-closing. Ref.—406.8(B)
22. No. Ref.—680.23(B)

23. Yes, provided it is installed and supported in accordance with Article 334. Ref.—334.6, 10, 12, and 30
24. No, they may be rubber or thermoplastic types. Ref.—225.4
25. Yes. Ref.—410.31

PRACTICE EXAM 2

1. No. Ref.—Only as environmental conditions require—547.1 and 547.8(A), (B), (C)
2. No. Ref.—320.30, 300.4
3. Extra-hard usage portable power cables listed for wet locations and sunlight resistance. Ref.—553.7(A) and (B)
4. Yes. Ref.—680.42(C) and 680.21(A)(1)—NM cable has a covered or bare equipment ground conductor, not insulated, but is now permitted with the dwelling occupancy.
5. Yes. Ref.—680.23(F)(1)
6. No, water heaters are not listed with an attachment cord. Ref.—400.7 and 8(1); 110.3(B)
7. No, generally. Ref.—312.8
8. Voltage with the highest locked rotor kVA per horsepower. Ref.—430.7(B)(3)
9. Yes. Ref.—680.25(A)
10. Yes. Ref.—210.21 and Tables 210.21(B)(2) and (B)(3)
11. Yes. Ref.—210.23(A)
12. Not enclosed, buried, or in raceway. Ref.—Table 310.17
13. No. Ref.—352.10 (FPN) is not mandatory, only explanatory
14. No, GFCI protection is still required. Ref.—590.6(A)
15. Yes. Ref.—210.8(A)(7)
16. No. Ref.—210.8(A)(7)
17. No maximum. Ref.—Table 220.14(J)
18. 6. Ref.—517.19(B)
19. Yes. Ref.—430.81(B) and 430.42(C)
20. Yes. Ref.—210.70(A) Exception 1
21. Yes, one or more supplied by a 20-ampere current. Ref.—210.52(F)
22. Yes. Ref.—Article 250 and Article 410 Part V

23. Yes, flexible metal conduit must be bonded to grounded conductor. Ref.—230.43(15)

24. No, only interior metal water piping systems. Ref.—250.104(A)

25. No, generally. Ref.—314.22, 314.16(A)

PRACTICE EXAM 3

1. No. Ref.—314.27(C)

2. No. Ref.—314.16(A) and (B)

3. No. Ref.—210.52(E) only single- and two-family dwellings

4. No, polarized or grounding type. Ref.—410.42(A)

5. Yes. Ref.—210.52(A)
   Note: Due to the arrangement of furniture, the receptacle located behind the door may be the only receptacle accessible for frequent use such as vacuuming.

6. No. Ref.—590.4(G)

7. 6, where integrated clamps, and the like, are not used. Ref.—Table 314.16(A)

8. Table 314.16(B). Ref.—314.16(A)(2) as applicable

9. Yes. Ref.—501.10(A)

10. No, however, it is a raceway. Ref.—Article 358

11. Yes, up to 5 feet (1.5 m). Ref.—358.30(A) Exception 1

12. Yes. Ref.—410.16(C)

13. No. Ref.—Definitions, Article 100, and 410.4

14. No, must be spaced 1½ inches (38 mm) from the surface. Ref.—410.76(B)

15. Yes, with exceptions. Ref.—410.65(C)

16. No, only where the grounded (neutral) conductor is present. Ref.—110.15, 230.56, and 408.3(E)

17. Yes. Ref.—210.8(A)(6). All kitchen countertop receptacles are required to be GFCI protected.

18. Generally no. Ref.—312.8 and Exceptions

19. Yes. Ref.—250.112(J); Article 410, Part V

20. No. Ref.—700.12(E)

21. No. Ref.—210.8(A)(2) Exceptions 1 and 2

22. Yes "all." Ref.—210.8(A)(1)

23. No; however, it cannot be counted as one of the countertop receptacles as required by 210.52(C). Ref.—210.52(B) Exception 3

24. No. Ref.—314.1 and 410.64

25. Within 12 inches (300 mm) of service head, goose neck, and at intervals not exceeding 30 inches (750 mm). Ref.—338.10, 230.51(A)

PRACTICE EXAM 4

1. 24 inches (600 mm). Ref.—Table 300.5

2. Yes. Ref.—Table 210.21(B)(2)

3. No, you are not required to locate required receptacles behind furniture. However, the number required by 210.52(A) must be installed convenient to furniture arrangement. Ref.—210.60(B) and Exceptions

4. Yes, but only if the box is listed for fan support. Ref.—314.27(D) and 422.18

5. Ref.—410.31 and 410.32

6. Yes, generally. Ref.—410.65(C) Exceptions 1 and 2

7. Yes. Ref.—210.6(C)

8. Yes, it is required. Ref.—700.12(E)

9. 150 kVA or less. Ref.—Diagram 517.41(3)

10. Yes. Ref.—230.95

11. Yes, 15-volt luminaires (lighting fixtures). Ref.—680.43(B)(1)

12. Yes. Ref.—358.10(B)

13. No, it is not permitted. Ref.—250.58

14. No. Ref.—110.26

15. Yes. Yes. Ref.—210.8(A)

16. See 547.9 and 250.32.

17. No. Ref.—Part V, Article 230, specifically 230.70. *Note:* If service is properly established at the pole, the answer to this question is Yes.

18. 24 inches (600 mm). Ref.—Table 300.5

19. No, unless the cord and plug are part of listed appliance. Ref.—110.3(B), 400.8(1)

20. (a) No minimal burial depth requirements. Ref.—Article 810
    (b) As per 810.21 and 810.15

21. Yes, provided they control the power conductors in the conduit. Ref.—300.11(B)

22. Yes. Underground water piping system must be at least 10 feet (3 m) and be supplemented by an electrode for it to qualify as an electrode. Ref.—250.52(A)(1)

23. Yes. Ref.—210.70

24. Entire fixture. Ref.—410.4

25. Yes. No. Ref.—225.6, 340.10, and 340.12(11)

PRACTICE EXAM 5

1. Yes. Ref.—392.3(E)

2. Ref.—410.8(D)

   (a) 12-inch (300 mm) incandescent, 6-inch (150 mm) fluorescent

   (b) 12-inch (300 mm) incandescent, 6-inch (150 mm) fluorescent

   (c) 6-inch (150 mm) incandescent, 6-inch (150 mm) fluorescent

3. No, generally four or less 14 AWG fixture wires are not required to be counted. Ref.—314.16(B)(1)

4. Yes, generally. Ref.—110.26(C)

5. 4 AWG copper. Ref.—250.66(B)

6. Yes. Ref.—700.12(E) Exception

7. Yes, if all conditions are met. Ref.—550.32(B)(1)–(7)

8. Yes. Ref.—680.25(B)

9. Yes, when smaller than ⅝ inch (15.87 mm), must be listed. Ref.—250.52(C)(2)

10. Yes, both are acceptable. Ref.—230.43

11. Exothermic welds. Ref.—250.70

12. No. Ref.—250.118

13. Yes. Ref.—518.4(A)

14. All branch-circuits that supply 125 volt, 15- and 20-ampere outlets installed in dwelling unit bedroom. Ref.—210.12(B)

15. Yes. Ref.—356.10 and 356.12

16. Yes, but bare conductors are permitted. Ref.—230.41 and Exception

17. No. Ref.—300.3(C) and 725.54

18. No, only circuits over 250 volts to ground. Ref.—250.97. See the Exception for listed boxes with a specified type of concentric knockouts that are acceptable.

19. Yes. Exceptions would permit this practice in some instances without raceway protection. Ref.—300.22(C)

20. Yes. Ref.—300.13(B)

21. 60 amperes. Ref.—430.72(B) Exception 2—Table 430.72(B) Column C

22. Yes. Ref.—406.10, 250.146

23. *Code* uses nominal voltages. Ref.—See Article 100—Definition or 220.5

24. 12 AWG THW conductor in cable or raceway. Ref.—Table 310.16 = 25 amperes

    Four conductors in a raceway Table 310.15(B)(2)—four to six conductors in a raceway or cable shall be derated to 80% = 25 × 80% = 20 amperes the allowable ampacity.

    Ambient temperature correction factors shown at the bottom of Table 310.16 require an additional deduction of .82 — 20 × .82 = 16.4 amperes. 210.21—outlet devices shall have an ampere rating not less than the load to be served. Table 210.21(B)(2) permits a 20-ampere branch-circuit for a maximum of 16 amperes. 16.4 exceeds this requirement. The conductor will carry the maximum allowable load. *Note:* 240.4(D) limits the overcurrent protection on a 12 AWG conductor to 20 amperes.

25. 36" × 36" (900 mm × 900 mm). Ref.—324.10(H)

PRACTICE EXAM 6

1. Yes. Ref.—250.97, 250.94

2. Six. Ref.—230.71(A), 230.40 Exception 1

3. (a) Yes. Ref.—600.10(C)(2)

   (b) No. Ref.—600.10(C)(2)—There is no exception

4. Yes. Ref.—545.2, 210.70 (a dwelling), 210.70(C) (other types of buildings)

5. Yes, if they meet all the conditions. Ref.—605.8

6. Yes. Ref.—501.15(A), Article 500—Gasoline is a Group D, Class I hazard. Diesel is not. However, the sealing requirements are to prevent the passage of gases, vapors, and flames.

7. Yes. Ref.—210.52(A)

8. (a) 2 AWG copper or 1/0 AWG aluminum. Ref.—250.24 base on 300-kcmil Service-Drop Table 250.66

   (b) Bonded to the panelboard supplying each apartment size to Table 250.122 base and an overcurrent device 100 amperes 8 AWG copper or 6 AWG aluminum. Ref.—250.104

   (c) No. Ref.—250.52—A supplementary grounding electrode is only required where the underground water pipe is available. This installation is suggested by a nonmetallic water system.

9. Yes. Ref.—210.23(A)

10. No, generally. Yes, for interior portion of a one-family dwelling. Ref.—680.21(A)(4)

11. No, where an equipment grounding conductor and not over four fixture wires smaller than 14 AWG enter the box from the fixture. Ref.—314.16(A) Exception

12. Receptacle is not required specifically. 210.52(A) would require a receptacle near the wet bar. If located within 6 feet (1.8 m), a GFCI is required. Ref.—210.52(A), 210.8(A)(7)

13. Read heading for Table 220.55. Note 3 permits add nameplates of the two ranges. Use Column C. Ref.—Table 220.55 and Note 3

14. 4.5 kW. Ref.—220.14(B). 220.54 is used for feeder load calculations.

15. 35 pounds (16 kg). Ref.—422.18

16. Table 348.22 for trade size ⅜ (12). For ½ (16) through 4 (53) use Table 1, Chapter 9. Ref.—348.22

17. No. Ref.—314.23(B), 314.23(A)–(G). *Note:* 300.11(A) the general method requirements give limited permission for branch-circuits. However, junction boxes are covered by Article 314.

18. No, explosionproof not required or acceptable. Must be approved for Class II locations. Ref.—502.5

19. No. THHN is not acceptable in wet locations. Ref.—Table 310.13 and 310.8

20. Authority having jurisdiction. Ref.—90.4, also see definition in Article 100, "Approved"

21. Yes. Ref.—210.52(F)

22. Yes. Ref.—210.63

23. Column A is the basic rule to be applied when an exception does not apply. Ref.—430.72(B)

24. 83. Ref.—Duff and Herman, *Alternating Current Fundamentals, 4E*, Thomson Delmar Learning, and Chapter 3 of this book

25. Qualified yes "Metal Multioutlet Assembly." Ref.—Article 380, 380.3

PRACTICE EXAM 7

1. Lighting circuit serving the area. Ref.—700.12(E)

2. Yes, if they are classified for purpose and meet all conditions of 392.7(B). Ref.—392.7(A) and (B)

3. No. Ref.—388.10(1)

4. If the required workspace is unobstructed or doubles. Ref.—110.26(C)(2)(a) and (b)

5. No, the conductors in MC cable are wrapped in a nylon wrapping. AC cable conductors are individually wrapped in kraft paper. Ref.—320.40, Armored Cable; 330.40, Type MC Cable

6. Generally, no, if exposed to physical damage. Ref.—362.10 and 362.12. *Note:* If these conditions do not exist, then ENT could be installed.

7. 18 inches (450 mm). Ref.—Table 300.5

8. Large capacity multibuilding industrial installations under single management. Ref.—225.32 Exception 1

9. No for Class 2 Div. 1, Yes for Class 2 Div. 2. Ref.—502.10(A) and (B)

10. No. Ref.—410.31

11. Grounded (neutral) conductor must be insulated generally. Ref.—310.12; 250.140 and 142. The rules for an insulated neutral or grounded conductor for a feeder, although not clearly written, the service neutral is often bare and 250.140 permits dryers and ranges originating in the service panel to be Type SE, which has a bare grounded conductor.

12. No. Classifying structures are the responsibility of the building inspector or the fire marshal generally. They rely on the adopted building code or the NFPA 101 Life Safety Code. Ref.—Article 518

13. No. Ref.—220.61, also Table 310.15(B)(4)

14. None. Ref.—Table 310.16, Ambient Temperature Correction factors at bottom of table.

15. Table 314.16(B)

Example: An outlet box contains a 120-volt, 20-ampere single receptacle and a 120-volt, 30 ampere receptacle. The 20-ampere receptacle is fed with a 12.2 W/GRD NM cable that extends from the outlet on to additional outlets located elsewhere. The 30-ampere receptacle is fed by an individual branch-circuit with 10.2 W/GRD NM cable. What size box is needed?

Answer: 314.16(A) and Table 314.16(B)

12.2 W/GRD Feed	2.12 @ 2.25 cu. in.	4.5 cu. in.
12.2 extending on	2.12 @ 2.25 cu. in.	4.5 cu. in.
10.2 W/GRD Feed	2.10 @ 2.5 cu. in.	5.0 cu. in.

30 A Receptacle	2.10	5.0 cu. in.
20 A Receptacle	2.12 @ 2.5 cu. in.	4.5 cu. in.
GRD	–10	2.5
Clamp	–10	2.5
		28.5

A box with 28.5 cubic inches is required.

16. 10 feet (3 m). Ref.—230.24(B)

17. No, it is considered by the *NEC*® as "Other Space Used for Environmental Air." Ref.—300.22(C) and Exceptions, definition of "Plenum" in Article 100

18. No, this area must be unobstructed. Ref.—410.8(A)

19. Equipment grounding conductor required. Ref.—250.134, 250.136

20. Yes. Ref.—410.4(A)(1)

21. By interpolation. Ref.—430.6(A)(1)

22. 53 percent. Ref.—Chapter 9, Table 1 and Note 6

23. Yes. Ref.—Table 310.15(B)(6)

24. No, motors are computed in accordance with Article 430. Ref.—220.14, 680.3

25. Yes, if the garage is supplied with electricity. Ref.—210.8 (A)(2), 210.52(G)

PRACTICE EXAM 8

1. Yes, generally. Ref.—410.73(E)

2. As per 547.9 in compliance with Article 250. Ref.—250.32(E) (FPN) and 547.9

3. Yes, if known, and they occupy a dedicated space with the individual branch-circuit. Ref.—Table 220.55 "Heading." If they are plugging into one of the "small appliance" outlets, then they would not have to be added to the load calculations.

4. Yes, for controlled starting. Ref.—430.82(B)

5. No. Ref.—553.2 and 553.4

6. No. Ref.—Article 600, 210.6(A)

7. Yes. Ref.—210.21

8. Yes, limited application. Ref.—Articles 517.12, 13, and 30

9. No, generally. Ref.—240.3 and 310.15

10. Yes. Must be marked SWD. Ref.—240.83(D), 404.11

11. There shall be two deductions for the largest conductor connected to each device, and one deduction of the largest conductor entering the box for each of the following: the clamps, the hickeys, all of equipment grounding conductors, etc. Ref.—314.16(B)(1)

12. No. Ref.—250.94

13. No, the cable can be fished; however, Article 314 does not permit the fished NM cable without a box clamp. Ref.—334.30(B)(1), 314.17(C) Exception

14. 150 amperes will meet all conditions required for a general-use switch. Ref.—440.12, 430.109(D)(3)

15. No. Ref.—348.60 and 348.20(A)(2)

16. I P/E I = 10,000/240 I = 41.667
   R = E/I R = 240/41.667 R = 5.76
   P = IE P = 208 × 36.1 P = 7.51 kW
   I = E/R I = 208/5.76 I = 36.1 amperes
   36.1 amperes—Chapter 3, Ohm's law
   I = P/E I = 10,000/240 I = 41.667
   R = E/I R = 240/41.667 R = 5.76
   I = E/R I = 208/5.76 I = 36.1 amperes

17. Yes. Ref.—680.5, 680.22(A)(5), and 680.23(A)(3)

18. Chapter 9, Table 1 unless durably and legibly marked by manufacturer in cubic inches. Ref.—314.16(C)

19. By the markings on the luminaire (fixture) (IC) by the listing agency. Ref.—410.65(C) Exceptions 1 and 2

20. Conductors only. Ref.—240.21

21. Table 110.26(A)(1), Condition 3. Ref.—110.26(A)(1)

22. Yes, if cord-and-plug connected but not required or advisable. Generally, these appliances are wired to the requirements in Article 422. Ref.—210.52(B)(1) and (2)

23. 10 feet (3 m). Ref.—680.8(A)

24. Yes, no depth requirement but is required to be protected from severe physical damage. Ref.—250.64 and 250.120

25. Water meter and all insulating joints. Bonding around valves and sweated joints is not generally required but the water meter, filters, and insulating joints would be required to be bonded around. Ref.—250.52(A)(1) and 250.53(A)(1)

PRACTICE EXAM 9

1. Yes, but not advisable. One should isolate any possibility of imposing faults on those circuits. Ref.—680.5(B), 680.22(A)(5), 680.23(A)(3)

2. (a) Yes. Ref.—250.21 and 250.24(A) and (B)

   (b) In each of the parallel raceways. Ref.—250.24(A) and (B)

   (c) 2 AWG copper in each conduit. Ref.—250.24, 250.66

3. 21 inches × 24 inches × 6 inches. Ref.—314.28 (A)(2)

4. Yes, by Exception. Ref.—312.8

5. Yes. Ref.—358.30

6. Either system installation is okay where a three-pole transfer switch is used. Grounding electrode is not required for this installation. Ref.—250.24(D), 250.30

7. No. Ref.—300.5(I) Exception 2, although grouping of the conductors is preferred.

8. This installation is a violation; you cannot mix these conductors, unless the control is a Class 1 or power circuit. Ref.—300.3(C)(1), 725.55

9. Permit; however, it should not be located directly over the switchboard for the best coverage. Ref.—110.26(F)(1)

10. Yes. Ref.—250.50. Steel can be bonded to the water pipe within 5 feet (1.5 m) of where the water pipe enters the building. Ref.—250.50 and 250.52

11. 20-ampere breaker. Ref.—240.4(D) and 210.22

12. Yes. Also, the receptacles required by 210.52(E) still must be installed at direct grade access and be GFCI protected. Ref.—210.8(A)(3) and 210.52(E)

13. Yes, if listed for 2. Ref.—110.3(B) and 312.5(C)

14. No. Ref.—230.42(A), (B), and (C)

15. Yes. Ref.—430.109(F)

16. Article 310. Ref.—Table 310.15(B)(6)

17. No, neutral is not used at the switch. Ref.—200.7(C)(2) and 300.3(B)

18. Yes; but where they enter the building, they must be enclosed in a raceway. Ref.—340.10(4)

19. Yes, area may be unclassified by AHJ where there are four air changes per hour. Ref.—511.3(A) and 511.3(B)(1) Exception

20. No. Neutral must be capable of carrying the maximum unbalanced load. Ref.—250.140 and 220.22

21. Yes. Ref.—348.20(A)

22. Maybe; AHJ is the building inspector. Ref.—Local adopted building code—518.2(C)(FPN)

23. Yes, but it must be connected to the same circuit as supplies the lighting for the area to be served. Ref.—700.12(E)(4)

24. Maybe. This wet location is defined in 517.2; 210.8(B)

25. Yes; No. Ref.—422.12

PRACTICE EXAM 10

1. Yes, but storage of combustibles should be considered. Ref.—Part II, Article 410

2. Hot tubs and spas are defined in Article 680. Ref.—680.2

3. No. Ref.—511.4, 501.15(A)(4) Exception 2; 511.9

4. 150; No. P = EI = 480 × 104 = 49,920 single phase

   49,920/(480 × 1.73) = 60.12 amperes

   60.12 × 250% = 150

   Ref.—Table 450.3(B)(2)

5. Each section 5 feet (1.5 m) or fraction thereof shall be calculated at 180 VA in other than dwellings. Ref.—220.18(H)

6. No. Ref.—Part IX, Article 430 and 430.102(B)

7. Yes, possibly. Ref.—430.72 and Exceptions

8. Yes, except in an industrial facility. Ref.—368.11

9. No. Ref.—340.12(1) and 340.12(10)

10. Yes. Ref.—250.20(B)(2) and 250.24

11. No. Ref.—250.30 covers the requirements for a separately derived system. Nearest building steel is preferred.

12. A tap is defined in 240.2. No, you cannot tap a tap. Ref.—240.2

13. Yes. Ref.—Part III, Article 424, 424.20(B)

14. 2 AWG. Ref.—Chapter 9, Table 8, Table 250.66

15. Yes. However, consideration must be made for corrosive elements. Ref.—300.6, 501.10(A), Article 344—Non-Ferrous Rigid Metal Conduit.

16. No. Ref.—700.12(E)

17. No. Ref.—250.140

18. Yes, but there are installation requirements for disconnecting overcurrent. Ref.—230.82 and 695.3

19. Bond services together. Ref.—250.58

20. No. Ref.—680.71, 410.4(A)–(D)

21. No. Ref.—*NEC*®. Most requirements for GFCI protection appear in 210.8. Other requirements appear in Articles 511, 517, 527, 550, 551, 553, 555, and 680.

22. No. Ref.—410.28(D) and 410.31

23. Yes. Ref.—362.10(5). The 15-minute finish rating is a wall finish rating established by Underwriters Laboratories and can be found in the "UL Fire Resistant Directory."

24. Yes. Ref.—Article 314, 314.16(A) and (B), Table 314.6(B). $4 \times 2\frac{1}{8} = 30.3$ cubic inches. 8 AWG conductor = 3 cubic inches. Therefore, ten 8 AWG conductors would be allowed without considering devices or other appurtenances such as clamps.

25. They must be fastened. Ref.—410.16(C)

PRACTICE EXAM 11

1. Yes. Ref.—340.10(4)

2. No. Ref.—680.23(A)(1)

3. Yes. Ref.—700.12(E) Exception

4. No. Ref.—300.22(C) Exception. Perpendicular, yes; parallel, no.

5. No, they must be equipped with a factory-installed GFCI. Ref.—600.11

6. In each disconnect means. Ref.—250.64(C)

7. No. Ref.—250.64(C) Exception

8. As a wireway in accordance with Article 376 or 378. Ref.—Article 376, Part II and Article 378, Part II.

9. Yes. Ref.—250.148

10. No. It can be insulated or covered. Ref.—680.21(A)(4)

11. Yes. This installation is not in violation of the *Code*. Ref.—240.4(B), 310.16, 220.10, and 408.13.

    240.4(B)—The 400-ampere breaker properly protects 500-kcmil conductors that permit the next higher overcurrent device.

    220.10—Conductors 500 kcmil are large enough to carry the 180-ampere load.

    408.34(B)—It is a power panel, not a lighting and appliance branch-circuit panelboard.

12. Yes, unless not required by NFPA 99. Ref.—517.25 (FPN)

13. Generally, no—Table 220.42, and 220.44. Ref.—220.12, 220.14, and 220.40

14. Yes. No. Ref.—517.2 definition of patient care areas in hospitals. Ref.—517.13(A)

15. Yes, there are no conditions. Ref.—600.7, 250.112

16. No. Ref.—250.30

17. Yes. Ref.—310.4

18. They must be derated to 80 percent. Ref.—Table 310.15(B)(2)

19. Yes. Ref.—210.6(C)

20. Yes. Ref.—Definition "Accessible Readily," definition of "Direct Grade Access"—210.52(E)

21. Yes, maybe. Ref.—210.52(D), 210.11(C)(3)

22. Yes, if they do not exceed allowable load of circuit. They are not continuous loads. Ref.—210.19(C) and 210.23

23. No. Ref.—210.52(G) only applies to dwellings (see definition, Article 100)

24. Yes. Yes. Ref.—210.70(A) Exception 2

25. Overcurrent device. Yes. Ref.—220.40

PRACTICE EXAM 12

1. Only branch-circuits. Ref.—220.51, 424.3(D)

2. No. Table 220.55 Heading Column A applies to all notes as applicable, Columns B and C apply to Note 3. Ref.—Same

3. Yes. Ref.—225.30, 225.32, 225.33, 230.70, and 230.71

4. Location is not specified; where bonded to different electrodes, they must then be bonded together. Ref.—250.58

5. Yes, but may be wired as unclassified in accordance with Chapters 1–4. Ref.—514.1 Scope

6. ½ inch (13 mm). Ref.—410.66(A)

7. System ground. Ref.—810.21(F)

8. No. 90.3. Use Chapters 1–4. Ref.—511.3

9. Yes. Ref.—250.97 Exception

10. Yes. Ref.—680.21(A)(4)

11. No, only as taps (from raceway system to fixture). Ref.—348.20(A)(2)

12. No. Ref.—680.23(C)

13. No. Ref.—680.42, 680.43(C), and 680.43(D)(3)

14. No. Ref.—210.70(A)(2)(b)

15. 65 percent—Column C, Note 3—add 3 + 6 = 9 × 65% = 5.85 kW

16. Yes. Ref.—210.52(F). At least one but only for the laundry area.

17. Yes, all. Ref.—210.8(A)(1)

18. No, except when using Class 1 or power circuits for control. Ref.—725.55

19. Yes. Ref.—800.33

20. Bonding jumper shall be the same size as the grounding electrode conductor. Ref.—250.28 and 250.102

21. Yes, for wet bar sink; no, for laundry sink. Ref.—210.8(A)(7)

22. 25 foot tap = 1 AWG THW, 10 foot tap = 10 AWG THW. Ref.—240.21, Table 310.16, and 240.4(D)

23. No. Ref.—210.70(A). At least one switch

24. 75°C, because there are no circuit breakers rated 90°C. Ref.—110.14(C), 310.15(C)

25. Those areas determined by governing body of facility. Ref.—517.2 Definitions, Patient Care Areas of a Hospital

FINAL EXAMINATION

1. C. Ref.—800.14

2. B. Ref.—General knowledge

3. D. Ref.—General knowledge, *IEEE Dictionary*

4. A. Ref.—Chapter 3

5. A. Ref.—Chapter 3

6. C. Ref.—Chapter 3

7. C. Ref.—Chapter 3
    $P = I(E \times 1.73)$ $P = 83$ kW

8. A. Ref.—General knowledge

9. C. Ref.—Chapter 3
    .5/1 ohm = ⅕
    2.5 = 1

10. A. Ref.—Chapter 3

11. A. Ref.—Chapter 3

12. A. Ref.—210.21(B)

13. C. Ref.—Chapter 3
    $(16 \times 4) [64] \times (10 \times 4) [40] = 2560$

14. A. Ref.—250.102(C)
    Note: 12½ minimum or not smaller than required in Table 250.66, individual bonding

jumpers could be run from each conduit sized 1/0 AWG

15. B. Ref.—Chapter 9, Table 8
    Ohm's law
    Chapter 3
    240 volts × 3% = 7.2 volts drop permitted
    R = Ed/I   R = 7.2/58
    R = .124 ohm for 500 feet
    3 AWG = .254 ohm for 1000 feet

16. A. Ref.—Chapter 9, Tables 1, 5, and 4
    2/0 AWG = .2265 sq. in. × 2 = .4530
    1 AWG = .1590
    .4530 + .1590 = .6120   Table 4 = 1½ inches

17. B. Ref.—Chapter 9, Tables 1, 4, and 5
    2—.1590
    1—.1893
    1—.2265
    1—.7338 square inches   Table 4, 1½ = .82

18. D. Ref.—220.80, Chapter 9 Example

19. B. Ref.—Table 250.66

20. C. Ref.—230.95(A)

21. B with Exception. Ref.—230.9(A)

22. B. Ref.—Duff and Herman, *Alternating Current Fundamentals, 4E,* Thomson Delmar Learning

23. B. Ref.—Table 310.15(B)(2)

24. A. Ref.—Table 310.16

25. A. Ref.—Table 310.16 correction factors

26. A. Ref.—Chapter 9, Note 3

27. B. 342.22 states for conductor fill see Chapter 9, Table 1. Ref.—Chapter 9, Tables 1, 4, and 5
    Solution:
    Table 1: over 2 conduits 40 percent
    Table 5:
    4 AWG   = .0824 sq. in. × 3 = .2472
    1/0 AWG = .2223 sq. in. × 4 = .8892
                        Total = 1.1364 sq. in.
    2" IMC = 1.452   1½ IMC = 0.890
    Therefore 2 (53) IMC would be required

28. B. Ref.—352.22, Chapter 9, Table 1, Note 1 and Annex C, Table C10

29. C. Authority having jurisdiction. Ref.—90.4

30. C. Ref.—230.70(A)

31. C. Ref.—Appendix 1 (text)
32. D. Ref.—Appendix 1 (text)
33. D. Ref.—Basic Math, Chapter 3
34. B. Ref.—110.9
35. B. Ref.—430.52
36. D. Ref.—Article 430, 430.37, 38, and 39, Table 430.37
37. B. Ref.—430.226
38. A. Ref.—Article 430, Part XI
39. D. Ref.—430.89
40. D. Ref.—250.140 (existing installations only)
41. B. Ref.—250.110(1)
42. C. Ref.—250.53(G)
43. C. Ref.—250.114(2) Exception 2
44. D. Ref.—Article 430.7
45. D. Ref.—225.18
46. C. Ref.—230.95(C)
47. A. Ref.—430.24

Largest $5.6 \times 1.25 = 7$
$4.5 \times 1.00 = 4.5$
$4.5 \times 1.00 = \underline{4.5}$
16.0 amperes—minimum ampacity of conductors

48. C. Ref.—690.4(A), Article 705, 230.82(6)
49. C. Ref.—250.24, 250.28
50. D. Ref.—Electrician's Handbook
51. B. Ref.—404.8(A)
52. A. Ref.—430.36
53. D. Ref.—General knowledge; see Annex 2
54. D. Ref.—General knowledge; Annex 2

55. A. Ref.—General knowledge; Chapter 3
56. B. Ref.—Chapter 3; Ohm's law; Annex 2
57. C. Ref.—Article 310
58. A. Ref.—General knowledge; Chapter 3
59. A. Ref.—Definitions, Article 100
60. D. Ref.—Ohm's law; Annex 2
61. D. Ref.—Ohm's law; General knowledge; $P=I^2R$
62. B. Ref.—General knowledge
63. C. Ref.—Ohm's law; Annex 2
64. A. Ref.—Definitions, Article 100
65. D. Ref.—230.9(A) Exception
66. D. Not a *Code* requirement
67. A. Ref.—Chapter 3
68. B. Ref.—General knowledge
69. A. Ref.—*Direct Current Fundamentals,* Thomson Delmar Learning
70. B. Ref.—*Direct Current Fundamentals*, Thomson Delmar Learning
71. A. Ref.—Basic Math, Power Formula, Ohm's law
Clue: Resistance is constant
$I = P/E$ $2400/240 = 10$
$R = E/I$ $240/10 = 24$
$I = E/R$ $120/24 = 5$
$P = EI$ $5 \times 120 = 600$
72. B. Ref.—Basic Math, Ohm's law; Annex 2
73. C. Ref.—Basic Math; Annex 2
74. C. Ref.—Tables 430.250 and 430.252; 150% $\times$ 2/0 amperes
75. D. Ref.—Ohm's law, 65 percent load, 220.56; Annex 2. Calculation is based on 1Ø units; if calculation based on 3Ø units, answer is "B".

Index